Classroom Companion: Economics

The Classroom Companion series in Economics includes undergraduate and graduate textbooks alike. It welcomes fundamental textbooks aimed at introducing students to the core concepts, empirical methods, theories and tools of the field, as well as advanced textbooks written for students at the Master and PhD level seeking a deeper understanding of economic theory, mathematical tools and quantitative methods.

Valérie Mignon

Principles of Econometrics

Theory and Applications

Springer

Valérie Mignon
EconomiX-CNRS
University of Paris Nanterre
Nanterre Cedex, France

ISSN 2662-2882 ISSN 2662-2890 (electronic)
Classroom Companion: Economics
ISBN 978-3-031-52534-6 ISBN 978-3-031-52535-3 (eBook)
https://doi.org/10.1007/978-3-031-52535-3

The translation was done with the help of an artificial intelligence machine translation tool. A subsequent human revision was done primarily in terms of content.
Translation from the French language edition: "Économétrie - Théorie et applications - 2e éd." by Valérie Mignon, © 2022. Published by Economica. All Rights Reserved.

This Springer imprint is published by the registered company Springer Nature Switzerland AG
The registered company address is: Gewerbestrasse 11, 6330 Cham, Switzerland

Paper in this product is recyclable.

Preface

Econometrics is the study and measurement of economic phenomena based on the statistical observation of relevant quantities describing them. Econometrics is a branch of economic science that draws jointly on economic theory, statistics, mathematics, and computer science. In particular, it is used to analyze and verify, i.e., to test, economic phenomena and theories.

Econometrics, as a discipline, was born in 1930 with the creation of the Econometric Society by Ragnar Frisch, Charles Roos, and Irving Fisher. Frisch (1933) defines econometrics as follows: "econometrics is by no means the same as economic statistics. Nor is it identical with what we call general economic theory, although a considerable portion of this theory has a definitely quantitative character. Nor should econometrics be taken as synonymous with the application of mathematics to economics. Experience has shown that each of these three viewpoints, that of statistics, economic theory, and mathematics, is a necessary, but not by itself a sufficient, condition for a real understanding of the quantitative relations in modern economic life. It is the unification of all three that is powerful. And it is this unification that constitutes econometrics."

The development of databases—particularly at a very fine level and at high frequency—combined with the development of computer tools has enabled this unification of economic theory, statistics, and mathematics. Moreover, as Pirotte (2004) reminds us "econometrics provides economists with a fundamental basis for studying the prospects and consequences of economic policies that can be applied. More specifically, it is the only method that provides both quantitative and qualitative information." Thus, through macroeconometric models in particular, econometrics is characterized by a high level of operational content, especially for macroeconomists, economic analysts, and policymakers. Macroeconometric models, the aim of which is to describe economic activity, are used as a simulation tool and thus provide an aid to policy decision-making. Similarly, in the field of finance, econometrics has undergone considerable developments, enabling us to better understand the dynamics of financial markets.

Work with econometric content has developed substantially during the twentieth century, as demonstrated by the large number of journals on econometrics.[1] Examples include: *Biometrika*, *Econometrica*, *Econometric Theory*, *Econometric Reviews*, *Journal of Econometrics*, *Journal of the American Statistical Association*, *Journal of Time Series Analysis*, and *Quantitative Economics*. There are also journals with more applied content such as *Empirical Economics*, *International Journal of Forecasting*, *Journal of Applied Econometrics*, *Journal of Business and Economic Statistics*, and *Journal of Financial Econometrics*. In addition, many general economic journals publish articles with strong econometric content: *American Economic Review*, *Economics Letters*, *European Economic Review*, *International Economic Review*, *International Economics*, *Journal of the European Economic Association*, *Quarterly Journal of Economics*, and *Review of Economic Studies*.

The rise of econometrics can also be illustrated by the fact that recent Nobel Prizes in economics have been awarded to econometricians. James Heckman and Daniel McFadden received the Nobel Prize in Economics in 2000 for their work on theories and methods for the analysis of selective samples and on discrete choice models. Similarly, in 2003, the Nobel Prize in Economics was awarded to Robert Engle and Clive Granger for their work on methods of analyzing economic time series with (i) time-varying volatility (R. Engle) and (ii) common trends (C. Granger), which has contributed to improved forecasts of economic growth, interest rates, and stock prices. The Prize was also awarded to Christopher Sims and Thomas Sargent in 2011 for their empirical work on cause and effect in the macroeconomy, and to Eugene Fama, Lars Peter Hansen, and Robert Shiller in 2013 for their empirical analysis of asset prices.

These different points testify that econometrics is a discipline in its own right and a fundamental branch of economics.

This book aims to provide readers with the basics of econometrics. It is composed of eight chapters. The first, introductory chapter recalls some essential concepts in statistics and econometrics. Chapter 2 deals with the simple regression model. Chapter 3 generalizes the previous chapter to the case of the multiple regression model, in which more than one explanatory variable is included. In Chap. 4, the fundamental themes of heteroskedasticity and autocorrelation of errors are addressed in detail. Chapter 5 brings together a set of problems related to explanatory variables. It deals successively with dependence between explanatory variables and the error term, the problem of multicollinearity, and the question of stability of the estimated models. Chapter 6 introduces dynamics into the models and presents distributed lag models. Chapter 7 extends the previous chapter by presenting time series models, a branch of econometrics that has undergone numerous developments over the last 40 years. Finally, Chap. 8 deals with structural models by studying simultaneous equations models.

[1] Pirotte's (2004) book gives a history of econometrics, from the origins of the discipline to its recent developments. See also Morgan (1990) and Hendry and Morgan (1995).

While providing a detailed introduction to econometrics, this book also focuses on some recent developments in the discipline, particularly in time series econometrics. The choice to focus on contemporary advances means that some topics have been deliberately omitted. This is notably the case for panel data econometrics (Matyas and Sevestre, 2008; Wooldridge, 2010; Baltagi, 2021), spatial econometrics (LeSage and Pace, 2008; Elhorst, 2014), econometrics of qualitative variables (Gouriéroux, 2000; Greene, 2020), and models with unobservable variables (Florens, Marimoutou, and Péguin-Feissolle, 2007), and nonlinear models (see in particular Florens et al., 2007; Greene, 2020).

All the theoretical developments in this book are illustrated by numerous applications to macroeconomics and finance. Each chapter contains several concrete empirical applications, using *Eviews* software. This permanent combination of theoretical and applied aspects will allow readers to quickly put into practice the different concepts presented.

This book is the fruit of various econometrics courses taught by the author at the University of Paris Nanterre in France. It is primarily intended for undergraduates and graduates in economics, management, and in mathematics and computer science applied to the social sciences, as well as for students at business and engineering schools. It will also be useful for professionals who work with econometric techniques. They will find in it practical solutions to the various problems they face.

I would like to thank Agnès Bénassy-Quéré, Hubert Kempf, and Jean Pavlevski for encouraging me to write this textbook, the first edition of which was published in French in 2008. I am particularly indebted to Hubert Kempf for prompting me to write this new edition in English, and to my publisher, Springer. I would also like to thank Emmanuel Dubois for his constant support and for the help he gave me in formatting this book.

To Tania and Emmanuel

Paris, France Valérie Mignon

About This Book

Bringing together theory and practice, this book presents the basics of econometrics in a clear and pedagogical way. It focuses on the acquisition of the methods and skills that are essential for all students wishing to succeed in their studies and for all practitioners wishing to apply econometric techniques. The approach adopted in this textbook is resolutely applied. Through this book, the author aims to meet a pedagogical and operational need to quickly put into practice the various concepts presented (statistics, tests, methods, etc.). This is why, after each theoretical presentation, numerous examples are given, as well as empirical applications carried out on the computer using existing econometric and statistical software.

This textbook is primarily intended for students of bachelor's and master's degrees in Economics, Management, and Mathematics and Computer Sciences, as well as for students of Engineering and Business schools. It will also be useful for professionals who will find practical solutions to the various problems they face.

Contents

About the Author

Valérie Mignon is Professor of Economics at the University of Paris Nanterre (France), Member of the EconomiX-CNRS research center, and Scientific Advisor to the leading French center for research and expertise on the world economy, CEPII (Paris, France). She teaches econometrics at undergraduate and graduate levels. Her econometric research focuses mainly on macroeconomics, finance, international macroeconomics and finance, and energy, fields in which she has published numerous articles and books.

Introductory Developments 1

After defining the concepts of model and variable, this chapter offers some statistical reminders about the mean, variance, standard deviation, covariance, and linear correlation coefficient. A brief introduction to the concept of stationarity is also provided. Finally, this chapter lists the main databases in economics and finance, as well as the most commonly used software packages. Beforehand, we give some introductory examples to illustrate in a simple way what econometrics can do.

1.1 What Is Econometrics? Some Introductory Examples

Econometrics is a discipline with a strong operational content. It enables us to quantify a phenomenon, establish a relationship between several variables, validate or invalidate a theory, evaluate the effects of an economic policy measure, etc.

1.1.1 Answers to Many Questions

Econometrics provides answers to a wide range of questions. Let us take some simple examples.

- Are the terms of trade a determinant of the value of exchange rates? Do other economic variables have more impact?
- Is the purchasing power parity theory empirically verified?
- Do rising oil prices have a significant impact on car sales?
- Is the depreciation of the dollar compatible with rising oil prices?
- Is the euro overvalued? If so, by how much? In other words, what is the equilibrium value of the euro?
- Are international financial markets integrated?

V. Mignon, *Principles of Econometrics*, Classroom Companion: Economics,
https://doi.org/10.1007/978-3-031-52535-3_1

- Is the efficient capital market hypothesis confirmed?
- Is there international convergence in GDP per capita?
- What is the impact of the 35-hour work week on unemployment?
- Does higher inflation reduce unemployment?
- Does their parents' socio-occupational category have an impact on children's level of education?
- What is the impact of air pollution on children's health?
- What are the effects of global warming on economic growth?
- etc.

To answer these questions, the econometrician must build a **model** to relate the **variables** of interest. Consider, for example, the question "What is the impact of an increase of 10 monetary units in income on household consumption?"

1.1.2 The Example of Consumption and Income

To answer this question, two variables need to be taken into account: household consumption and household income (gross disposable income). To relate these two variables, we write an equation of the following type:

$$CONS = \alpha + \beta \times INC \tag{1.1}$$

where $CONS$ denotes consumption and INC income. The impact of a variation in income on consumption is taken into account by the parameter β. To quantify this impact, it is necessary to have a numerical value for the coefficient β. To this end, an **estimation** of the model is performed: estimating a model thus amounts to quantifying it, i.e., quantifying the relationship between two or more variables. In the following, we will detail the methods available for estimating a model. For the moment, let us restrict ourselves to a few illustrations.

Consider two countries: Finland and Italy. For each of the two countries, we want to assess the impact of a 10-unit increase in the gross disposable income of Finnish (resp. Italian) households on their consumption. Figures 1.1 and 1.2 show the evolution of real consumption ($CONS$) and income (INC) of households for each of the two countries.[1] The data are annual and cover the period from 1995 to 2020.[2] Regardless of which figure we look at, we see that the series move in the same direction: consumption and income show an overall upward trend in the case of Finland, and the two series move in tandem, alternating between bullish and bearish phases, in the case of Italy. If there is a relationship between the two variables,

[1] The series are expressed in real terms, i.e., they are deflated by the consumer price index of each country.

[2] The data are extracted from the national statistical institutes of the two countries: Statistics Finland and the Italian National Institute of Statistics (Istat).

Fig. 1.1 Evolution of consumption and gross disposable income of Finnish households (euros), 1995–2020

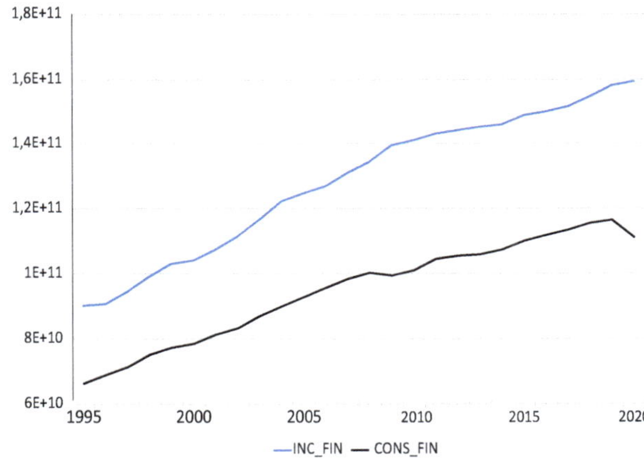

Fig. 1.2 Evolution of consumption and gross disposable income of Italian households (euros), 1995–2020

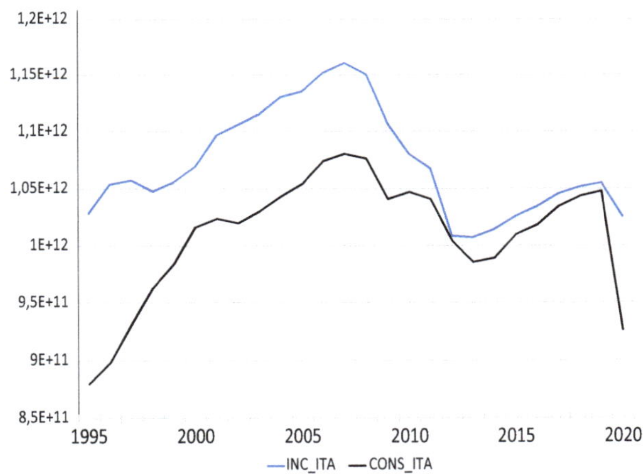

it should therefore be positive. In other words, we expect the value obtained for the coefficient β to be positive. More specifically, if we estimate model (1.1), we obtain the following values for the coefficient β associated with income: 0.690 for Finland and 0.721 for Italy. These values are positive, which means that an increase in income is accompanied by an increase in consumption in both countries, all other things being equal. We can also quantify this increase:

- A €10 increase in income in Finland translates into a €6.90 increase in consumption of Finnish households, all other things being equal.
- A €10 increase in income in Italy generates, all other things being equal, an increase in consumption of Italian households of around €7.21.

Although different, these two values are quite close, which means that household consumption behavior, in relation to the change in income, is similar in Finland and

Italy, even though the economic characteristics of the two countries differ. In the rest of this book, we will see that it is possible to refine these comments by studying whether or not the values obtained are **significantly** different. This will be done using statistical tests.

1.1.3 The Answers to the Other Questions Asked

To conduct their analysis, econometricians have to find the **data** they need. In the case of the example previously studied, the following series are needed: household consumption, household gross disposable income, and the consumer price indexes for Finland and Italy, i.e., a total of six series. For this purpose, econometricians need access to **databases**. Nowadays, there are many such databases, some of which are freely accessible. A non-exhaustive list of the main economic and financial databases is given at the end of this chapter. Once the data have been collected, it is possible to proceed with the study in question.

Let us now consider the various questions posed in Sect. 1.1.1 and give some possible answers.

- Are the terms of trade a determinant of the value of exchange rates? Do other economic variables have more impact?
  The following data are required for the country under consideration: export prices, import prices, and the exchange rate, the ratio between export prices and import prices being used to measure the terms of trade. To assess whether the terms of trade are a determinant of the exchange rate, it is necessary to estimate a model that relates the exchange rate and the terms of trade and to test whether the coefficient associated with the variable "terms of trade" is significantly different from zero. To determine whether other economic variables have more impact, we need to add them to the previous model and study their statistical significance. Other potential determinants include the country's net foreign asset position, productivity, interest rate differential, etc.
- Is the purchasing power parity theory empirically confirmed?
  According to the purchasing power parity (PPP) theory, each country's currency provides the same purchasing power in all countries. In other words, if the products traded are physically identical (without transport costs), the nominal exchange rate (indirect quote) is determined by the relative price of the good, i.e., $Q_t = P_t / P_t^*$, which can be written in logarithmic form: $q_t = p_t - p_t^*$ where the lowercase variables are the logarithms of the uppercase variables, Q_t is the nominal exchange rate, P_t is the domestic consumer price index, and P_t^* is the foreign consumer price index. In order to grasp the empirical validity of PPP, we can estimate a relationship of the type $q_t = \alpha + \beta_1 p_t - \beta_2 p_t^*$ and check that $\alpha = 0$, $\beta_1 = \beta_2 = 1$. This is done by statistically testing that the coefficients take certain specific values.

- Do rising oil prices have a significant impact on car sales?
  This question can be answered by estimating an equation linking car sales to oil prices. The value obtained for the coefficient assigned to oil prices will quantify the effect of their increase on car sales. If a significant impact is detected, it is expected to be negative, as higher oil prices generate an additional cost.
- Is the depreciation of the dollar compatible with rising oil prices?
  This question about the link between oil prices and the dollar exchange rate is essential because oil prices are denominated in dollars. Traditionally, it is assumed that there is a positive relationship between the two variables, in the sense that a rise in oil prices is generally accompanied by an appreciation of the US currency. To understand the link between the two variables, it is necessary to estimate a relationship explaining the dollar exchange rate by oil prices. The coefficient assigned to the oil price variable should therefore be positive, and its value makes it possible to quantify the impact of oil prices on the dollar.
- Is the euro overvalued? If so, by how much? In other words, what is the equilibrium value of the euro?
  To answer these questions, we need to define a "standard" corresponding to the equilibrium value of the euro. Among the theories for determining equilibrium exchange rates is the BEER (behavioral equilibrium exchange rate) framework. By this approach, the exchange rate is linked in the long term to a set of economic fundamentals, such as the net foreign asset position, the relative price level or any other measure of productivity, the terms of trade, and the interest rate differential. Estimating an equation that explains the euro exchange rate by these different fundamentals allows us to define the equilibrium value of the European currency. The question of overvaluation is then addressed by comparing the observed value of the euro with its estimated equilibrium value. In Chap. 7, we will see that estimating an equilibrium relationship, or long-term relationship, is based on cointegration theory.
- Are international financial markets integrated?
  There are, of course, many ways of approaching this fundamental question. One possible approach is to adopt the work of Feldstein and Horioka (1980): if financial markets are perfectly integrated, then capital is perfectly mobile, which implies that capital should move to wherever the rate of return is highest. Consequently, for a given country, the investment rate should be totally uncorrelated with its savings rate. To understand this hypothesis, we need to estimate a relationship linking the investment rate to the savings rate and to consider the value of the coefficient assigned to the savings rate. The farther from 1, the weaker the correlation and the more this suggests a high degree of financial integration.
- Is the efficient capital market hypothesis confirmed?
  In line with the weak form of informational efficiency, prices observed on a market follow a random walk. In other words, price changes, or so-called returns, are unpredictable in the sense that it is impossible to predict future returns from past returns. A simple way to test this hypothesis is to estimate a relationship of the type $R_t = \alpha + \beta R_{t-1}$ and test whether the coefficient β assigned to past

returns R_{t-1} is zero or not. If it is zero, the efficient capital market hypothesis is not called into question, since past values of returns do not provide any information to explain the current change in returns.

- Is there international convergence in GDP per capita?
  Analyzing the convergence of GDP per capita is fundamental to studying inequalities between nations. In particular, this question raises the issue of poor countries catching up with rich ones. If we are interested in conditional convergence, the Solow model can be used. In this model, the growth rate of a country's per capita income depends on the level at which this income is situated in relation to the long-run equilibrium path of the economy. It is then possible to estimate a relationship to explain the GDP growth rate between the current date and the initial date by the level of GDP at the initial date. If the coefficient assigned to the level of GDP is zero, this indicates an absence of convergence.
- What is the impact of the 35-hour work week on unemployment?
  There are several ways to approach this question. One is to estimate a relationship to explain the unemployment rate by working hours, by varying those working hours. If the impact of the 35-hour work week on the unemployment rate is neutral, the coefficient assigned to the duration variable should be similar, whether the duration is 35 or 39 hours.
- Can higher inflation reduce unemployment?
  This question is linked to a relationship that is widely studied in macroeconomics, namely, the Phillips curve, according to which there is a negative relationship between the unemployment rate and the inflation rate. This relationship will be studied in Chap. 2 in order to determine whether inflation has a beneficial effect on unemployment.
- Does their parents' socio-occupational category have an impact on children's level of education?
  Such a question can again be addressed by estimating a relationship between children's level of education and their parents' socio-occupational category (SOC). If the coefficient assigned to SOC differs with the SOC, this indicates an impact of SOC considered on children's level of education.
- Does air pollution have an impact on children's health?
  Answering this question first requires some way of measuring air pollution and children's health. Once these two measures have been established, the analysis is carried out in a standard way, by estimating a relationship linking children's health to air pollution.
- What are the effects of global warming on economic growth?
  As before, once the means of measuring global warming (e.g., greenhouse gas emissions) has been found, a relationship between economic growth and this variable must be estimated.

Having presented these examples and introductory points, let us formalize the various concepts, such as the notions of model and variable in more detail.

1.2 Model and Variable

An essential part of econometrics is the construction and estimation of models. A **model** relates various variables, which are often economic quantities. It is a formalized representation of a phenomenon or theory in the form of equations. We speak of **modeling**, its aims being to understand, explain, and possibly predict the phenomenon under study.

First of all, it is necessary to define the concept of model, as well as the types of variables that can be involved in it.

1.2.1 The Concept of Model

A model is a simplified representation of reality which consists in representing a phenomenon in the form of one or more equations. It makes it possible to specify relationships between variables and to explain the way in which certain variables are determined by others. Consider, for example, the Keynesian consumption function. In accordance with Keynes' (1936) "fundamental psychological law," "men are disposed, as a rule and on average, to increase their consumption as their income increases, but not as much as the increase in their income." According to this law, consumption is an increasing function of income. By noting C consumption and Y income, we have:

$$C = f(Y) \tag{1.2}$$

where f is such that $f' > 0$. However, three types of functions, or models, are compatible with the fundamental psychological law:

- A linear proportional model: $C = cY$, with $0 < c < 1$. The parameter c designates the average propensity to consume $\left(c = \frac{C}{Y}\right)$, but also the marginal propensity to consume, since $\frac{dC}{dY} = c$. In line with this formulation, the variation of the marginal propensity to consume as a function of income is zero: $\frac{d^2C}{dY^2} = 0$;
- A linear affine model: $C = cY + C_0$, with $0 < c < 1$ and $C_0 > 0$. The average propensity to consume is now given by $c + \frac{C_0}{Y}$, while the marginal propensity remains equal to c. Furthermore, as before, we have: $\frac{d^2C}{dY^2} = 0$;
- A concave function: $C = f(Y)$ with $f'' < 0$. Under these conditions, the marginal propensity to consume $\left(f'\right)$ is lower than the average propensity $\left(\frac{C}{Y}\right)$. Because the function is concave, we have $\frac{d^2C}{dY^2} < 0$, reflecting the fact that the variation in the marginal propensity to consume as a function of income is negative.

As an approximation,[3] the affine linear model is frequently used as a representation of the Keynesian consumption function. The model:

$$C = cY + C_0 \tag{1.3}$$

thus represents the consumption behavior of agents from a Keynesian perspective. c and C_0 are **parameters** (or **coefficients**) that must be estimated. In the next chapter, we will see that the **ordinary least squares (OLS)** method is used to estimate these coefficients; its purpose is to attribute values to the coefficients, i.e., to quantify the relationship between consumption and income. As an example, suppose that the application of this method yields the following estimates: 0.86 for the estimated value of c and 200 000 for the estimated value of C_0. We then have:

$$\hat{C} = 0.86Y + 200{,}000 \tag{1.4}$$

where $\hat{C}$ designates the estimated consumption.[4] By virtue of Eq. (1.4), it appears that the estimated value of c is positive: the relationship between C and Y is indeed increasing. Furthermore, the value 0.86 of the marginal propensity to consume allows us to write that, all other things being equal, an increase of one monetary unit in income Y is accompanied by an average increase of 0.86 monetary units in consumption C.

Remark 1.1 The model (1.3) has only one equation describing the relationship between consumption and income. This is a **behavioral equation** in the sense that behavior, i.e., household consumption decisions, depends on changes in income. The models may also contain **technological relationships**: these arise, for example, from constraints imposed by existing technology, or from constraints due to limited budgetary resources. In addition to these two types of relationships—behavioral and technological relationships—models frequently include **identities**, i.e., technological **accounting relationships** between variables. For example, the relationship $Y = C + I + G$, where Y denotes output, C consumption expenditure, I investment expenditure, and G government spending, frequently used in economic models, is an identity. No parameter needs to be estimated.

---

[3] Strictly speaking, a reading of the *General Theory* suggests that the concave function seems closest to Keynes' words; the affine form, however, is the most frequently chosen for practical reasons.

[4] The circumflex (or hat) notation is a simple convention indicating that this is an estimate (and not an observed value). This convention will be adopted throughout the book.

1.2.2 Different Types of Data

Having specified the model and in order to estimate it, it is necessary to have **data** representative of the economic phenomena being analyzed. In the case of the Keynesian consumption function, we need the consumption and income data for the households studied. The main types of data are:

- **Time series** are variables observed at regular time intervals. For example, the quarterly series of consumption of French households over the period 1970–2022 constitutes a time series in the sense that an observation of French household consumption is available for each quarter between 1970 and 2022. The regularity of observations is called the **frequency**. In our example, the frequency of the series is quarterly. A time series can also be observed at annual, monthly, weekly, daily, intra-daily, etc. frequency.
- **Cross-sectional data** are variables observed at the same moment in time and which concern a specific group of individuals (in the statistical sense of the term).[5] An example would be a data set composed of the consumption of French households in 2022, the consumption of German households in 2022, the consumption of Spanish households in 2022, etc.
- **Panel data** are variables that concern a specific group of individuals and are measured at regular time intervals. An example would be a data set composed of the consumption of French households over the period 1970–2022, the consumption of German households over the period 1970–2022, the consumption of Spanish households over the period 1970–2022, etc. Panel data thus have a double dimension: individual and temporal.

1.2.3 Explained Variable/Explanatory Variable

In the model representing the Keynesian consumption function, two variables are involved: consumption and income. In accordance with relationship (1.3), income appears to be the determinant of consumption. In other words, income explains consumption. We then say that income is an **explanatory variable** and consumption is an **explained variable**.

More generally, the variable we are trying to explain is called the explained variable or **endogenous variable** or **dependent variable**. The explanatory variable or **exogenous variable** or **independent variable** is the variable that explains the endogenous variable. The values of the explained variable thus depend on the values of the explanatory variable.

If the model consists of a single equation, there is only one dependent variable. On the other hand, there may be several explanatory variables. For example, household consumption can be explained not only by income, but also by the

[5] Remember that an individual, or a statistical unit, is an element of the population studied.

unemployment rate. We can write the following model:

$$C = cY + aU + C_0 \tag{1.5}$$

where U is the unemployment rate and a is a parameter. In this model, the dependent variable is consumption C, the explanatory variables are income Y and the unemployment rate U.

Remark 1.2 In the model $C = cY + C_0$, time is not explicitly involved. Suppose that the consumption and income data are time series. If we assume that income at date t explains consumption at the same date, then we have:

$$C_t = cY_t + C_0 \tag{1.6}$$

where t denotes time. Such a model relates variables located at the same moment in time. However, it is possible to introduce **dynamics** into the models. Let us consider, for example, the following model:

$$C_t = cY_t + \alpha C_{t-1} + C_0 \tag{1.7}$$

Past consumption (i.e., consumption at date $t-1$) acts as an explanatory variable for current consumption (i.e., consumption at date t). The explanatory variable C_{t-1} is also called the **lagged endogenous variable**. The coefficient α represents the **degree of inertia** of consumption. Assuming that $\alpha < 1$, the closer α is to 1, the greater the degree of consumption inertia. In other words, a value of α close to 1 means that past consumption has a strong influence on current consumption. We also speak of **persistence**.

1.2.4 Error Term

In the model (1.3), it has been assumed that consumption is explained solely by income. If such a relationship is true, it is straightforward to obtain the values of the parameters c and C_0: it suffices to have two observations and join them by a straight line, the other observations lying on this same line. However, such a relationship is not representative of economic reality. The fact that income alone is used as an explanatory variable in the model may indeed seem very restrictive, as it is highly likely that other variables contribute to explaining consumption. We therefore add a term ε which represents all other explanatory variables not included in the model. The model is written:

$$C = cY + C_0 + \varepsilon \tag{1.8}$$

The term ε is a random variable called the **error** or **disturbance**. It is the error in the specification of the model, in that it collects all the variables, other than income, that have been ignored in explaining consumption. The error term thus provides a measure of the difference between the observed values of consumption and those that would be observed if the model were correctly specified. The error term includes not only the model specification error, but it can also represent a measurement error due to problems in measuring the variables under consideration.

1.3 Statistics Reminders

The purpose of this section is to recall the definition of some basic statistical concepts that will be used in the remainder of the book: mean, variance, standard deviation, covariance, and linear correlation coefficient.

1.3.1 Mean

The (arithmetic) **mean** of a variable is equal to the sum of the values taken by this variable, divided by the number of observations. Consider a variable X with T observations: $X_1, X_2, \ldots, X_T$. The (empirical) mean of this series, noted $\bar{X}$, is given by:

$$\bar{X} = \frac{1}{T}\left(X_1 + X_2 + \ldots + X_T\right) = \frac{1}{T}\sum_{t=1}^{T} X_t \tag{1.9}$$

Example 1.1 The six employees of a small company received the following wages X (in euros): 1 200, 1 200, 1 300, 1 500, 1 500, and 2 500. The mean wage $\bar{X}$ is therefore: $\bar{X} = \frac{1}{6}(1{,}200 + 1{,}200 + 1{,}300 + 1{,}500 + 1{,}500 + 2{,}500) = 1{,}533.33$ euros. The mean could also have been calculated by weighting the wages by the number of employees, i.e.: $\bar{X} = \frac{1}{6}(1{,}200 \times 2 + 1{,}300 \times 1 + 1{,}500 \times 2 + 2{,}500 \times 1) = 1{,}533.33$ euros. This is a **weighted arithmetic mean**.

1.3.2 Variance, Standard Deviation, and Covariance

The **variance** $V(X)$ of a variable X is equal to the average of the squares of the deviations from the mean:

$$V(X) = \frac{1}{T}\left[\left(X_1 - \bar{X}\right)^2 + \left(X_2 - \bar{X}\right)^2 + \ldots + \left(X_T - \bar{X}\right)^2\right] = \frac{1}{T}\sum_{t=1}^{T}\left(X_t - \bar{X}\right)^2 \tag{1.10}$$

The **standard deviation**, noted σ_X, is the square root of the variance, i.e.:

$$\sigma_X = \sqrt{\frac{1}{T}\sum_{t=1}^{T}\left(X_t - \bar{X}\right)^2} \tag{1.11}$$

In practice, we often use the following formula, obtained by expanding (1.10):

$$V(X) = \frac{1}{T}\sum_{t=1}^{T} X_t^2 - \bar{X}^2 \tag{1.12}$$

The use of this formula simplifies the calculations in that it is no longer necessary to calculate deviations from the mean.

The relationships (1.10), (1.11), and (1.12) are valid when studying a **population**.[6] In practice, the study of a population is rare, and we are often limited to studying a sub-part of the population, i.e., a **sample**. In this case, a slightly different measure of variance is used, called the **empirical variance**, which is given by:[7]

$$s_X^2 = \frac{1}{T-1}\sum_{t=1}^{T}\left(X_t - \bar{X}\right)^2 \tag{1.13}$$

or:

$$s_X^2 = \frac{1}{T-1}\sum_{t=1}^{T} X_t^2 - \frac{T}{T-1}\bar{X}^2 \tag{1.14}$$

The **empirical standard deviation** is then:

$$s_X = \sqrt{\frac{1}{T-1}\sum_{t=1}^{T}\left(X_t - \bar{X}\right)^2} \tag{1.15}$$

Consider two variables X and Y each comprising T observations. The **covariance** between these two variables, noted $Cov(X, Y)$, is given by:

$$Cov(X, Y) = \frac{1}{T}\sum_{t=1}^{T}\left(X_t - \bar{X}\right)\left(Y_t - \bar{Y}\right) = \frac{1}{T}\sum_{t=1}^{T} X_t Y_t - \bar{X}\bar{Y} \tag{1.16}$$

[6] A population is a set of elements, called statistical units or individuals, that we wish to study.

[7] The division by $(T-1)$ instead of T comes from the loss of one degree of freedom since the empirical mean (and not the true population mean) is used in calculating the variance.

1.3.3 Linear Correlation Coefficient

The **correlation coefficient** is an indicator of the link between two variables.[8] Thus, when two variables move together, i.e., vary in the same direction, they are said to be correlated.

Consider two variables X and Y. The linear correlation coefficient between these two variables, noted r_{XY}, is given by:

$$r_{XY} = \frac{Cov(X, Y)}{\sigma_X \sigma_Y} \tag{1.17}$$

or:

$$r_{XY} = \frac{\sum_{t=1}^{T} (X_t - \bar{X})(Y_t - \bar{Y})}{\sqrt{\sum_{t=1}^{T} (X_t - \bar{X})^2} \sqrt{\sum_{t=1}^{T} (Y_t - \bar{Y})^2}} \tag{1.18}$$

or alternatively:

$$r_{XY} = \frac{T \sum_{t=1}^{T} X_t Y_t - \sum_{t=1}^{T} X_t \sum_{t=1}^{T} Y_t}{\sqrt{T \sum_{t=1}^{T} X_t^2 - \left(\sum_{t=1}^{T} X_t\right)^2} \sqrt{T \sum_{t=1}^{T} Y_t^2 - \left(\sum_{t=1}^{T} Y_t\right)^2}} \tag{1.19}$$

The linear correlation coefficient is such that:

$$-1 \leq r_{XY} \leq 1 \tag{1.20}$$

Thus, the linear correlation coefficient can be positive, negative, or zero. If it is positive, it means that the variables X and Y move in the same direction: both variables increase (or decrease) simultaneously. If it is negative, the two variables move in opposite directions: if one variable increases (respectively decreases), the other variable decreases (respectively increases). Finally, if it is zero, the covariance between X and Y equals zero, and the variables are not correlated: there is no linear relationship between X and Y. More precisely, if the linear correlation coefficient is close to 1, the variables are strongly positively correlated, and if it is close to -1, the variables are strongly negatively correlated.

Figures 1.3, 1.4, and 1.5 schematically illustrate the cases of positive, negative, and zero linear correlation between two variables X and Y.

[8] If more than two variables are studied, the concept of **multiple correlation** must be used (see below).

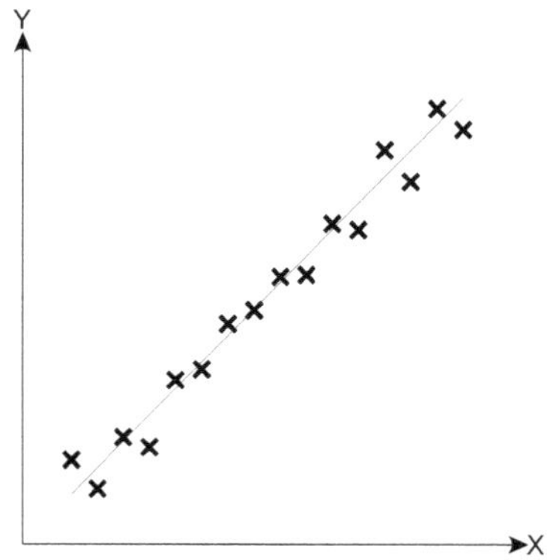

Fig. 1.3 Positive linear correlation

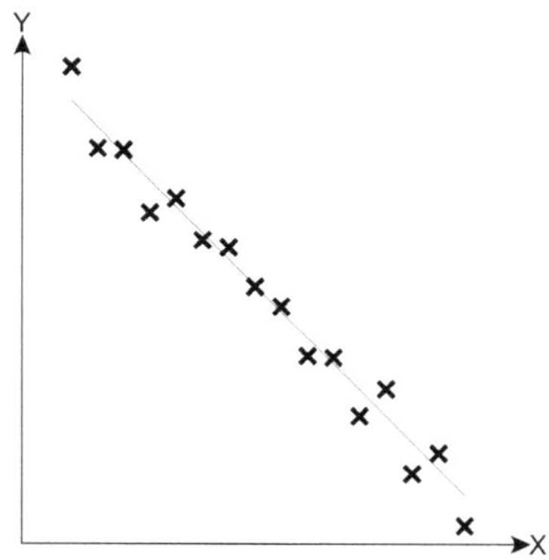

Fig. 1.4 Negative linear correlation

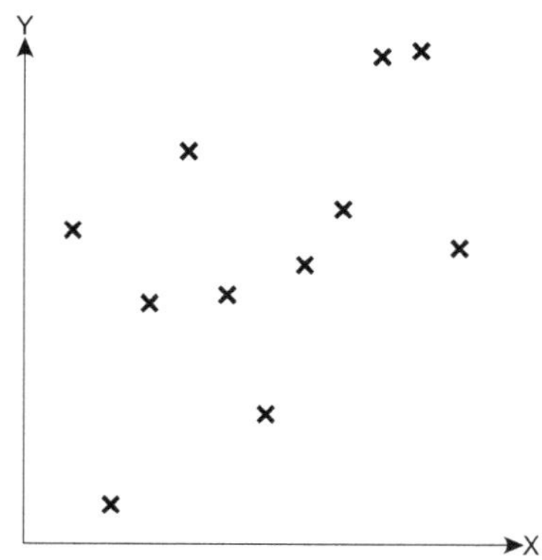

Fig. 1.5 No linear correlation

Remark 1.3 So far, we have considered a **linear** correlation between two variables X and Y: the values of the pair (X, Y) appear to lie on a straight line (see Figs. 1.3 and 1.4). When these values are no longer on a straight line, but on a curve of any shape, we speak of **nonlinear correlation**. Positive and negative nonlinear correlations are illustrated in Figs. 1.6 and 1.7.

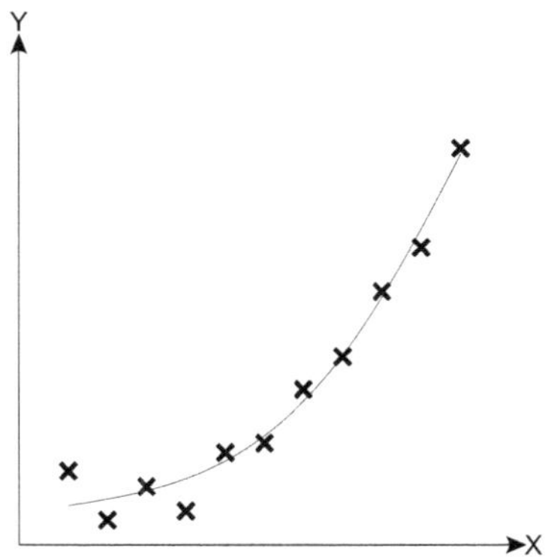

Fig. 1.6 Positive nonlinear correlation

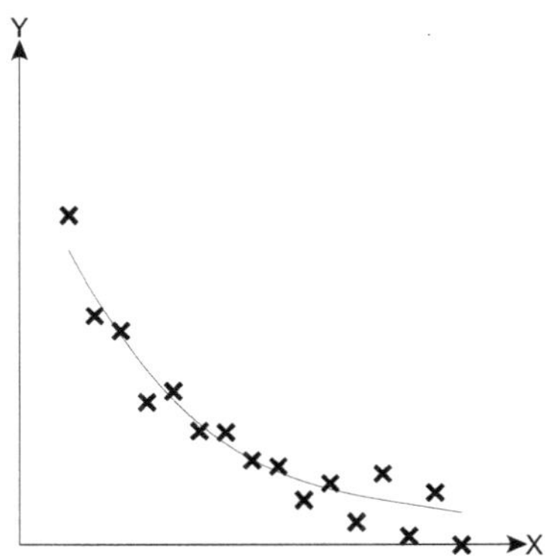

Fig. 1.7 Negative nonlinear correlation

1.3.4 Empirical Application

Consider the following two annual series (see Table 1.1): the household consumption series (noted C) and the household gross disposable income series (noted Y) for France over the period 1990–2019. These two series are expressed in real terms, i.e., they have been deflated by the French consumer price index. The number of observations is 30.

From the values in Table 1.1, it is possible to calculate the following quantities, which are necessary to determine the statistics presented above:

- $\sum_{t=1}^{30} C_t = 30{,}455{,}596.93$
- $\sum_{t=1}^{30} Y_t = 34{,}519{,}740.64$
- $\sum_{t=1}^{30} C_t^2 = 3.14 \times 10^{13}$

Table 1.1 Consumption and gross disposable income of households in France (in € million). Annual data, 1990–2019

	C	Y		C	Y
1990	870,338.41	830,572.81	2005	1,187,709.02	1,049,755.63
1991	868,923.49	832,883.96	2006	1,228,476.20	1,081,354.36
1992	913,134.25	844,679.04	2007	1,260,465.07	1,103,446.70
1993	943,367.50	840,328.00	2008	1,286,487.36	1,128,089.30
1994	950,773.58	847,173.42	2009	1,278,688.70	1,102,747.33
1995	971,315.23	852,111.23	2010	1,292,234.68	1,115,634.36
1996	987,452.68	866,004.67	2011	1,284,970.35	1,114,405.93
1997	982,724.94	867,822.35	2012	1,281,460.14	1,109,228.05
1998	1,018,159.61	902,064.20	2013	1,267,030.06	1,116,070.14
1999	1,039,090.56	916,606.04	2014	1,282,764.87	1,125,426.60
2000	1,083,578.97	957,094.57	2015	1,295,592.76	1,142,198.12
2001	1,126,231.00	985,857.04	2016	1,311,829.11	1,156,441.13
2002	1,146,598.73	991,312.79	2017	1,328,847.32	1,171,560.14
2003	1,148,028.86	1,002,269.15	2018	1,345,881.90	1,183,006.67
2004	1,171,763.22	1,021,751.72	2019	1,365,822.06	1,197,701.47

Data sources: Insee for the consumption and consumer price index series, European Commission for the gross disposable income series

- $\sum_{t=1}^{30} Y_t^2 = 4.04 \times 10^{13}$
- $\sum_{t=1}^{30} C_t Y_t = 3.56 \times 10^{13}$

From these preliminary calculations, we deduce:

- The mean of consumption and income series:

$$\bar{C} = \frac{1}{30} 30{,}455{,}596.93 = 1{,}015{,}186.56 \tag{1.21}$$

$$\bar{Y} = \frac{1}{30} 34{,}519{,}740.64 = 1{,}150{,}658.02 \tag{1.22}$$

- The standard deviation of the consumption and income series:

$$s_C = \sqrt{\frac{1}{29} 3.14 \times 10^{13} - \frac{30}{29} (1{,}015{,}186.56)^2} = 125{,}970.16 \tag{1.23}$$

and:

$$\sigma_C = 123{,}852.87 \tag{1.24}$$

$$s_Y = \sqrt{\frac{1}{29}4.04 \times 10^{13} - \frac{30}{29}(1{,}150{,}658.02)^2} = 157{,}952.45 \tag{1.25}$$

and:

$$\sigma_Y = 155{,}297,60 \tag{1.26}$$

– The covariance between consumption and income series:

$$\begin{aligned} Cov\,(C,Y) &= \frac{1}{30}3.56 \times 10^{13} - 1{,}015{,}186.56 \times 1{,}150{,}658.02 \\ &= 19{,}084{,}753{,}775.26 \end{aligned} \tag{1.27}$$

By calculating the covariance, we can determine the linear correlation coefficient between the consumption and income series:

$$r_{CY} = \frac{Cov(C,Y)}{\sigma_C \sigma_Y} = \frac{19{,}084{,}753{,}775.26}{123{,}852.87 \times 155{,}297.60} = 0.9922 \tag{1.28}$$

We can see that the linear correlation coefficient is positive and very close to 1. This indicates a strong positive correlation between consumption and income: the two series move in the same direction.

This result can be illustrated graphically. Figure 1.8 clearly shows that the series move together; they share a common trend. Figure 1.9 shows the values of the pair (C, Y). These values are well represented by a straight line, illustrating the fact that the linear correlation coefficient is very close to 1.

1.4 A Brief Introduction to the Concept of Stationarity

When working on time series, one must be careful to ensure that they are stationary over time. The methods described in this book, particularly the ordinary least squares method, are valid only if the time series are stationary. Only a graphical intuition of the concept of stationarity will be given here; for more details, readers can refer to Chap. 7. We distinguish between stationarity in the mean and stationarity in the variance.

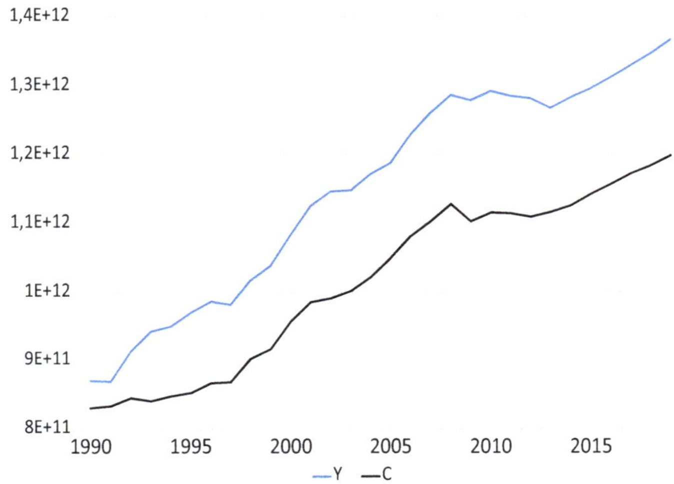

Fig. 1.8 Consumption (C) and gross disposable income (Y) series of French households (euros)

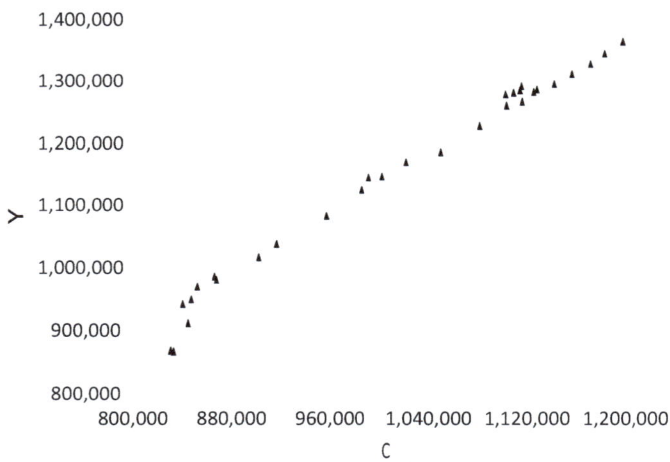

Fig. 1.9 Representation of the values of the pair (consumption, income)

1.4.1 Stationarity in the Mean

A time series is **stationary in the mean** if its mean remains stable over time. As an illustration, we have reproduced in a very schematic way a nonstationary series in Fig. 1.10. We can see that the mean, represented by the dotted line, increases over time.

In Fig. 1.11, the mean of the series is now represented by a straight line parallel to the x-axis: the mean is stable over time, suggesting that the series is stationary in

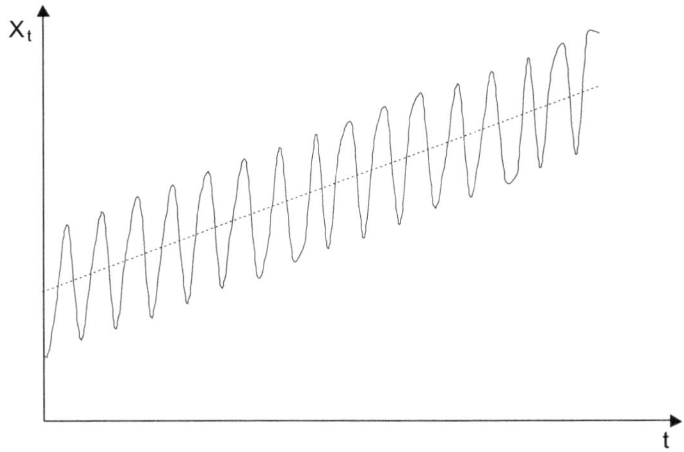

Fig. 1.10 Nonstationary series in the mean

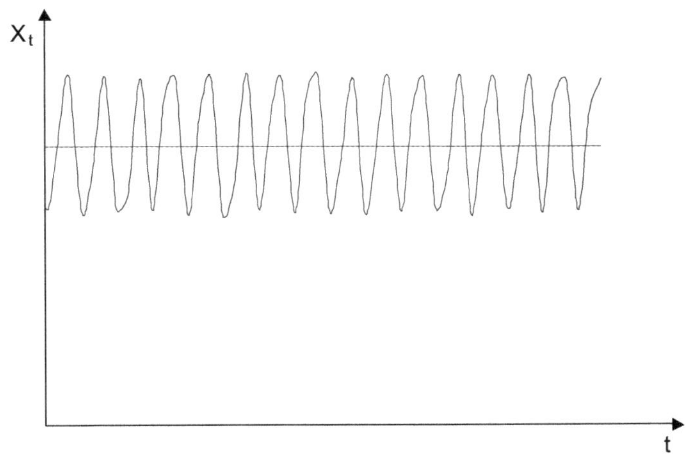

Fig. 1.11 Stationary series in the mean

the mean. Of course, this intuition must be verified statistically by applying specific tests, called **unit root tests** (see Chap. 7).

In order to apply the usual econometric methods, the series studied must be mean stationary. Otherwise, it is necessary to stationarize the series, i.e., to make it stationary. The technique commonly used in practice consists in differentiating the nonstationary series X_t, i.e., in applying the **first difference operator** Δ:

$$\Delta X_t = X_t - X_{t-1}$$

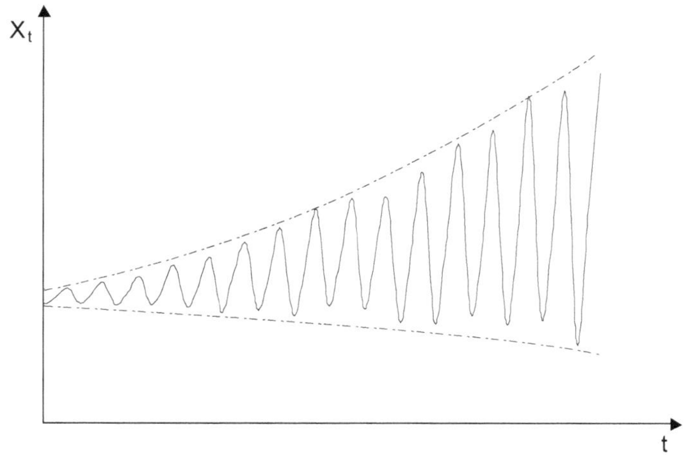

Fig. 1.12 Nonstationary series in the variance

Thus, very often, to make a series stationary in the mean, it is sufficient to differentiate it. Here again, the stationarity of the differentiated series must be verified by applying unit root tests.

1.4.2 Stationarity in the Variance

A stationary time series in the variance is such that its variance is constant over time.

It is also possible to graphically apprehend the concept of stationarity in the variance. The series shown in Fig. 1.12 is nonstationary in the variance: graphically, we can see a "funnel-like phenomenon," indicating that the variance of the series tends to increase over time. In order to reduce the variability of a series, the logarithmic transformation is frequently used.[9] The logarithm allows the series to be framed between two lines, i.e., to eliminate the funneling phenomenon, as shown schematically in Fig. 1.13.

Remark 1.4 In practice, when we want to make a series stationary in both the mean and the variance, we must first make it stationary in the variance and, then, in the mean. The result is a series in **logarithmic difference**. This logarithmic difference

[9] The logarithmic transformation is a special case of the **Box-Cox transformation** used to reduce the variability of a time series (see Box and Cox 1964, and Chap. 2) below.

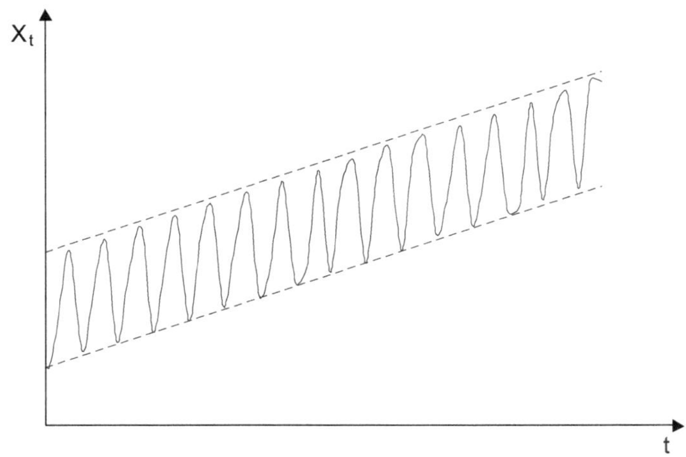

Fig. 1.13 Stationary series in the variance

also has an economic interpretation:

$$Y_t = \Delta \log X_t = \log X_t - \log X_{t-1} = \log \frac{X_t}{X_{t-1}} = \log \left(1 + \frac{X_t - X_{t-1}}{X_{t-1}}\right)$$
$$\cong \frac{X_t - X_{t-1}}{X_{t-1}} \tag{1.29}$$

because $\log(1 + x) \cong x$ for x small compared to 1; log denotes the Napierian logarithm. The logarithmic difference can be interpreted as a growth rate. If X_t is a stock price, Y_t can be interpreted as stock returns.

1.4.3 Empirical Application: A Study of the Nikkei Index

To illustrate the concept of stationarity, let us consider the Japanese stock market index series: the Nikkei 225 index. This series, extracted from the Macrobond database, has a quarterly frequency and covers the period from the third quarter of 1949 to the second quarter of 2021 (1949.3–2021.2). The Nikkei index series is reproduced in Fig. 1.14, whereas Fig. 1.15 represents the dynamics of this same series in logarithms. These graphs highlight an upward trend in the first half of the sample, followed by a general downward trend, and then an increasing trend from the early 2010s. The mean therefore changes over time, reflecting that the Japanese stock market index series seems nonstationary in the mean.

Fig. 1.14 Nikkei 225 index, 1949.3–2021.2

Fig. 1.15 Nikkei 225 index in logarithms, 1949.3–2021.2

Faced with the apparent non-stationarity (in the mean) of the Nikkei index series, we differentiate it by applying the first difference operator. We then obtain the series of returns R_t of the Nikkei index:

$$R_t = \Delta \log X_t = \log X_t - \log X_{t-1} = \log \frac{X_t}{X_{t-1}} \cong \frac{X_t - X_{t-1}}{X_{t-1}} \tag{1.30}$$

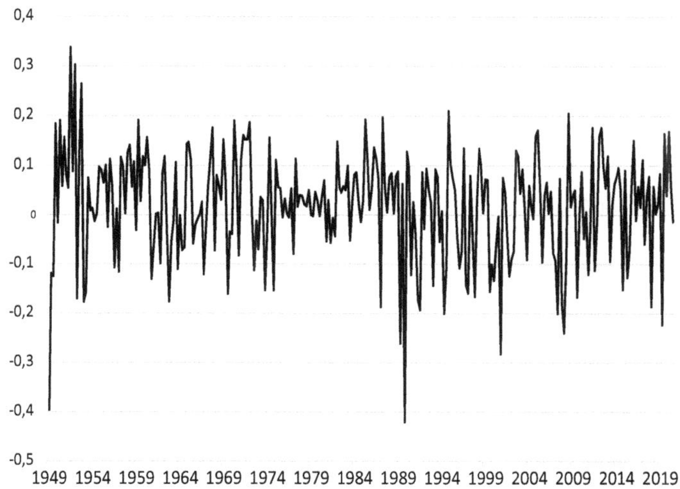

Fig. 1.16 Nikkei 225 returns, 1949.4–2021.2

where X_t denotes the Nikkei 225 stock index. The series of returns is displayed in Fig. 1.16. As shown, the upward trend in the mean has been suppressed by the differentiation operation, suggesting that the returns series is a priori mean stationary.

1.5 Databases and Software

As we have already mentioned, there are many databases in the field of economics and finance, which have expanded considerably in recent decades. The aim here is not to give an exhaustive list, but to provide some reference points concerning a number of frequently used databases. Similarly, we will mention some of the econometric software that practitioners often use.

1.5.1 Databases

We provide below some indications concerning various databases frequently used in economics and finance, remembering that this list—arranged alphabetically—is by no means exhaustive:

- *Bank for International Settlements* (open access): financial and monetary data
- *Banque de France* (free access): economic, monetary, banking, and financial data for France and the eurozone

- *British Petroleum* (open access): energy data (oil, gas, electricity, biofuels, coal, nuclear, etc.)
- *CEPII* (open access): databases in international macroeconomics and international trade
- *Datastream/Eikon*: economic and financial database with many series for all countries
- *DB.nomics* (open access): many economic data sets provided by national and international institutions for most countries
- *ECONDATA* (free access): server on databases available online
- *Economagic* (free access): numerous macroeconomic and financial series, on the United States, the eurozone, and Japan
- *Euronext* (free access): data and statistics on stock markets
- *European Central Bank* (*ECB Statistical Data Warehouse*, open access): economic and financial data for Europe
- *Eurostat* (free access): socio-economic indicators for European countries, aggregated by theme, country, region, or sector
- *Eurozone Statistics* (ESCB, free access): eurozone and national central bank statistics
- *FAO* (*Food and Agriculture Organization of the United Nations*, FAOSTAT, open access): food and agricultural data for most countries
- *Insee* (free access): statistics and data series for the French economy, quarterly national accounts
- *International Monetary Fund* (IMF, partly open access): numerous databases, including *International Financial Statistics* (IFS) and *World Economic Outlook* (WEO) covering most countries
- *Macrobond*: economic and financial database with a wide range of series for all countries
- *National Bureau of Economic Research* (NBER, open access): various macroeconomic, sectoral, and international series
- *OECD* (open access): statistics and data at national and sectoral levels for OECD countries, China, India, Indonesia, Russia, and South Africa
- *Penn World Table* (free access): annual national accounts series for many countries
- *UN* (open access): macroeconomic and demographic series and statistics
- *UNCTAD* (open access): data on international trade, foreign direct investments, commodity prices, population, macroeconomic indicators, etc.
- *WebEc World Wide Web Resources in Economics* (free access): server on economics and econometrics resources
- *Worldbank, World Development Indicators* (WDI, free access): annual macroeconomic and financial series for most countries, numerous economic development indicators
- *World Inequality Database* (WID, open access): database on global inequalities

Many other databases are available for macroeconomic, socio-economic, microeconomic, and financial data, and it is, of course, impossible to list them all here.

1.5.2 Econometric Software

Most of the applications presented in this book have been processed with *Eviews* software, this choice being here guided by pedagogical considerations. Of course, there are many other econometric and statistical software packages, some of which are freely available. We mention a few of them below, in alphabetical order, emphasizing once again that these lists—one of which concerns commercial software, the other open-source software—are by no means intended to be exhaustive.

Let us start by mentioning some software packages that require a paid license:

- *EViews*: econometric software, more particularly adapted for time series analysis
- *GAUSS*: programming language widely used in statistics and econometrics
- *LIMDEP* and *NLOGIT*: econometric software adapted for panel data, discrete choice, and multinomial choice models
- *Matlab*: programming language for data analysis, modeling, and algorithmic programming
- *RATS*: econometric software, more particularly adapted for time series analysis
- *S*: statistical programming language; an open-access version of which is *R* (see below)
- *SAS*: statistical and econometric software, allowing the processing of very large databases
- *SPAD*: software for data analysis, statistics, data mining, and textual data analysis
- *SPSS*: statistical software for advanced analysis
- *Stata*: general statistical and econometric software, widely used, especially in panel data econometrics

Open-source software includes:

- *Gretl* (*Gnu Regression, Econometrics and Time-Series Library*): general econometric software.
- *Grocer*: library of econometric programs, developed from *Scilab* and *Matlab* software and languages.
- *JMulTi*: econometric software, specialized in the analysis of univariate and multivariate time series, including in the nonlinear domain.
- *Ox*: programming language used in econometrics and matrix calculation.
- *Python*: a general-purpose programming language, widely used in econometrics and in the field of *big data* thanks to its complementary modules like NumPy, Pandas, StatsModels, etc. Also worth mentioning is the Jupyter application, mainly based on the Python language, which is part of the reproducible research field.
- *R*: this is the open-access version of the *S* language. *R* is widely used and has become a reference language in statistics and econometrics, with the development of many packages in all fields of econometrics.

– *RunMyCode*: a user-friendly platform allowing authors to make their data and codes (programs) freely available to everyone to promote reproducible research.

Conclusion

This introductory chapter has recalled some basic concepts in statistics and econometrics. In particular, it has highlighted the importance of the correlation coefficient in determining whether two variables move together. The next chapter extends this with a detailed presentation of the basic econometric model: the simple regression model. This model links the behavior of two variables, in the sense that one of them explains the other. The notion of correlation is thus deepened, as we study not only whether two variables move together, but also whether one of them has explanatory power over the other.

The Gist of the Chapter

Let X and Y be two variables with T observations.

Mean	$\bar{X} = \frac{1}{T}\sum_{t=1}^{T} X_t$
Variance	$V(X) = \frac{1}{T}\sum_{t=1}^{T} (X_t - \bar{X})^2$
Standard deviation	$\sigma_X = \sqrt{V(X)}$
Empirical variance	$s_X^2 = \frac{1}{T-1}\sum_{t=1}^{T} (X_t - \bar{X})^2$
Empirical standard deviation	$s_X = \sqrt{s_X^2}$
Covariance	$Cov(X, Y) = \frac{1}{T}\sum_{t=1}^{T} (X_t - \bar{X})(Y_t - \bar{Y})$
Correlation coefficient	$r_{XY} = \frac{Cov(X,Y)}{\sigma_X \sigma_Y}, \ -1 \leq r_{XY} \leq 1$

Further Reading

For further information on econometric methodology, see Hendry (1995) or Spanos (1999). For a simple presentation of the different types of data, refer to Intriligator (1978), and for a critical review of the content and accuracy of economic data, see Morgenstern (1963).

As for statistics, there are many books available. Among the works in English, readers can refer to Newbold (1984) for a simple and applied presentation of statistics, or to Hoel (1974) for an introduction. Mood et al. (1974) also provide a fairly comprehensive introduction to statistical methods.

The Simple Regression Model

2

Regression analysis consists in studying the dependence of a variable (the **explained variable**) on one or more other variables (the **explanatory variables**).

Let us look at some examples. When a company, or a brand owner, advertises one of its products, does it increase its sales? In other words, is there a relationship between product sales and advertising expenditure? Does a family's consumer spending depend on its size? To what extent does an increase in household income affect consumption? Is there a link between mortality or morbidity rates and the number of cigarettes consumed? Are children's school results dependent on parental income? All these questions, and others of the kind, can be answered using **regression analysis**.

When only one explanatory variable is considered, we speak of **simple regression**. When there are several explanatory variables, we speak of **multiple regression**. The simple regression model is thus a linear model comprising a single equation linking an explained variable to an explanatory variable. It is therefore a bivariate model.

The simple regression model is a **random model** in the sense that an error term is included in the equation linking the dependent variable to the explanatory variable. It should be recalled that this error term allows us to take into account discrepancies between the explanation given by the model and reality.

2.1 General

2.1.1 The Linearity Assumption

Consider two variables X and Y. We distinguish between linearity in the variables and linearity in the parameters.

V. Mignon, *Principles of Econometrics*, Classroom Companion: Economics,
https://doi.org/10.1007/978-3-031-52535-3_2

Linearity in the Variables

Let f be a function such that:

$$Y = f(X) \tag{2.1}$$

where Y is the dependent variable and X the explanatory variable. The function f is said to be **linear** in X if the power of X is equal to unity and if X is not multiplied or divided by another variable. In other words, Y is linearly related to X if the derivative of Y with respect to X—i.e., the slope of the regression line—is independent of X.

As an example, the model:

$$Y = 3X \tag{2.2}$$

is linear since $\frac{dY}{dX} = 3$: the derivative of Y with respect to X is independent of X.

More generally, the model:

$$Y = \alpha + \beta X \tag{2.3}$$

is a linear model with respect to X and Y.

Now consider the following model:

$$\log Y = \alpha + \beta \log X \tag{2.4}$$

This model is not linear with respect to X and Y, but it is linear with respect to $\log X$ and $\log Y$. Similarly the model:

$$\log Y = \alpha + \beta X \tag{2.5}$$

is linear with respect to X and $\log Y$. The model:

$$Y = \exp\left(\alpha + \beta \frac{1}{X}\right) \tag{2.6}$$

can also be written using the logarithmic transformation:

$$\log Y = \alpha + \beta \frac{1}{X} \tag{2.7}$$

which is a linear model in $1/X$ and $\log Y$.

Remark 2.1 Some models can be linearized. The model:

$$Y = \beta X^2 \tag{2.8}$$

is not linear in X because X is assigned a power of 2. This model can, however, be linearized by applying the logarithmic transformation:

$$\log Y = \log\left(\beta X^2\right) = \log\beta + 2\log X \tag{2.9}$$

The model (2.9) thus becomes a linear model in $\log X$ and $\log Y$.

Linearity in the Parameters

A function is said to be linear in the parameters if they are assigned a power equal to unity and are not multiplied or divided by one or more other parameters. Thus, the model:

$$Y = \alpha + \beta X \tag{2.10}$$

is linear in the parameters α and β. Similarly, the model:

$$Y = \alpha + \beta X^2 \tag{2.11}$$

is also linear in the parameters. In contrast, the models:

$$Y = \alpha + \beta^2 X \tag{2.12}$$

and:

$$Y = \alpha + \frac{\beta}{\alpha} X \tag{2.13}$$

are not linear in the parameters.

Linear Model

We wrote in the introduction to this chapter that the simple regression model is a linear model. The linearity discussed here is the linearity in the parameters. The methods described in this chapter therefore apply to models that are linear in the parameters. Of course, the model under study can also be linear in the variables, but this is not necessary in the sense that it is sufficient that the model can be linearized. In other words, the model can be linear in X or in any transformation of X.

2.1.2 Specification of the Simple Regression Model and Properties of the Error Term

The simple regression model studied in this chapter is written as:

$$Y = \alpha + \beta X + \varepsilon \tag{2.14}$$

where Y is the dependent variable, X is the explanatory variable, and ε is the **error term** (or **disturbance**). The **parameters** (or **coefficients**) of the model are α and β. It is assumed that the variable X is observed without error, i.e., that X is a certain variable. Therefore, the variable X is independent of the error term ε. The variable Y is a **random variable**, its random nature coming from the presence of the error term in the model.

Suppose that the variables X and Y each include T observations: we note $X_t, t = 1, \ldots, T$, and $Y_t, t = 1, \ldots, T$. The simple regression model is then written:

$$Y_t = \alpha + \beta X_t + \varepsilon_t \tag{2.15}$$

t may designate:

- Time: in which case, we speak of a time series model
- An individual: in which case, we speak of a cross-sectional model with the number of observations T representing the number of individuals

The error term cannot be predicted for every observation, but a number of assumptions can be made, which are described below.

The Nullity of the Mean Error

First, the error term can take on negative and positive values. There is no reason for positive (respectively negative) values to be higher or lower than negative (respectively positive) values. In other words, there is no bias in favor of positive values, nor in favor of negative values. We deduce that the mathematical expectation E of the error is zero, i.e.:

$$E(\varepsilon_t) = 0 \quad \forall t \tag{2.16}$$

This assumption means that, on average, the model is correctly specified and therefore that, on average, the error is zero.

The Absence of Autocorrelation in Errors

Second, it is assumed that the error term is not autocorrelated: the value in t does not depend on the value in t' for $t \neq t'$. In other words, if we consider a time series model, this means that the error made at one date t is not correlated with the error made at another date. For example, if the error made at t is positive, the probability of observing a positive error at $t + 1$ is neither increased nor decreased. This hypothesis of **uncorrelated errors** is written as follows:

$$E(\varepsilon_t \varepsilon_{t'}) = 0 \quad \forall t \neq t' \tag{2.17}$$

The Homoskedasticity of Errors

Third, it is assumed that the variance of the error term is constant regardless of the sample. If we consider a time series model, this means that the variance of the error term is constant over time. In the case of a cross-sectional model, this refers to the fact that the variance does not differ between individuals. The constant variance assumption is the **homoskedasticity** hypothesis. A series whose variance is constant is said to be **homoskedastic**.[1] Mathematically, this hypothesis is written as follows:

$$E\left(\varepsilon_t^2\right) = \sigma_\varepsilon^2 \quad \forall t \tag{2.18}$$

where σ_ε^2 represents the variance of the error term.

Remark 2.2 The assumptions of no autocorrelation and homoskedasticity of errors can be gathered under the following expression:

$$E\left(\varepsilon_t \varepsilon_{t'}\right) = \begin{cases} 0 & \forall t \neq t' \\ \sigma_\varepsilon^2 & \forall t = t' \end{cases} \tag{2.19}$$

The errors that simultaneously satisfy the assumptions of homoskedasticity and no autocorrelation are called **spherical errors**.

In addition, a series ε_t verifying the relationships (2.16) and (2.19) is called **white noise**. More generally, the following definition can be used.

Definition 2.1 A stationary process ε_t is **white noise** if:

$$E\left(\varepsilon_t\right) = 0 \quad \forall t \tag{2.20}$$

$$E\left(\varepsilon_t \varepsilon_{t'}\right) = \begin{cases} 0 & \forall t \neq t' \\ \sigma_\varepsilon^2 & \forall t = t' \end{cases} \tag{2.21}$$

White noise is thus a zero mean, constant variance, and non-autocorrelated process. We note:

$$\varepsilon_t \sim WN\left(0, \sigma_\varepsilon^2\right) \tag{2.22}$$

The Normality of Errors

Under the **central limit theorem**,[2] it is assumed that the error term follows a **normal distribution** with zero mean (or expectation) and constant variance (see

[1] The hypothesis of homoskedasticity is opposed to that of **heteroskedasticity**. A series whose variance evolves over time (for a time series model) or differs between individuals (for a cross-sectional model) is called a **heteroskedastic series**.

[2] *Central limit theorem*: let $X_1, X_2, \ldots, X_n$, be n independent random variables with the same probability density function of mean m and variance σ^2. When n tends to infinity, then the sample mean $\bar{X} = \frac{1}{n} \sum_{i=1}^{n} X_i$ tends towards a normal distribution with mean m and variance σ^2/n.

Appendix 2.2 for a detailed presentation of the normal distribution). We thus add the assumption of normality of the distribution of the error term to the assumptions of nullity of the expectation (Eq. (2.16)) and of homoskedasticity (Eq. (2.18)), which can be written as follows:

$$\varepsilon_t \sim N\left(0, \sigma_\varepsilon^2\right) \tag{2.23}$$

where N denotes the normal distribution, and the sign "$\sim$" means "follow the law."

Remark 2.3 The assumption that the errors follow a normal distribution with zero expectation and constant variance and that they are not autocorrelated can also be formulated by writing that the errors are **normally and independently distributed** (Nid), which is noted:

$$\varepsilon_t \sim Nid\left(0, \sigma_\varepsilon^2\right) \tag{2.24}$$

If the errors follow the same distribution other than the normal distribution, we speak of **identically and independently distributed** (iid) errors, which is noted:

$$\varepsilon_t \sim iid\left(0, \sigma_\varepsilon^2\right) \tag{2.25}$$

Remark 2.4 The assumption of normality of the errors is not necessary to establish the results of the regression model. However, it does allow us to derive statistical results and construct test statistics (see below).

2.1.3 Summary: Specification of the Simple Regression Model

The complete specification of the simple regression model studied in this chapter is written as:

$$Y_t = \alpha + \beta X_t + \varepsilon_t \tag{2.26}$$

with:

$$E\left(\varepsilon_t\right) = 0 \quad \forall t \tag{2.27}$$

$$E\left(\varepsilon_t \varepsilon_{t'}\right) = \begin{cases} 0 & \forall t \neq t' \\ \sigma_\varepsilon^2 & \forall t = t' \end{cases} \tag{2.28}$$

and:

$$\varepsilon_t \sim N\left(0, \sigma_\varepsilon^2\right) \tag{2.29}$$

We can also write the complete specification of the simple regression model by combining the relations (2.27), (2.28), and (2.29):

$$Y_t = \alpha + \beta X_t + \varepsilon_t \tag{2.30}$$

with:

$$\varepsilon_t \sim Nid\left(0, \sigma_\varepsilon^2\right) \tag{2.31}$$

2.2 The Ordinary Least Squares (OLS) Method

2.2.1 Objective and Reminder of Hypotheses

The parameters α and β of the simple regression model between X and Y are unknown. If we wish to quantify this relationship between X and Y, we need to **estimate** these parameters. This is our objective.

More precisely, from the observed values of the series X_t and Y_t, the aim is to find the quantified relationship between these two variables, i.e.:

$$\hat{Y}_t = \hat{\alpha} + \hat{\beta} X_t \tag{2.32}$$

where $\hat{\alpha}$ and $\hat{\beta}$ are the **estimators** of the parameters α and β. $\hat{Y}_t$ is the estimated (or adjusted of fitted) value of Y_t. The most frequently used method for estimating the parameters α and β is the **ordinary least squares (OLS) method**.

The implementation of the OLS method requires a certain number of assumptions set out previously and recalled below:

- The variable X_t is observed without error and is generated by a mechanism unrelated to the error term ε_t. In other words, the correlation between X_t and ε_t is zero,[3] i.e.: $Cov\left(X_t, \varepsilon_t\right) = 0\ \forall t$.
- The expectation of the error term is zero: $E\left(\varepsilon_t\right) = 0\ \forall t$.
- The errors are homoskedastic and not autocorrelated, i.e., $E\left(\varepsilon_t \varepsilon_{t'}\right) = \begin{cases} 0 & \forall t \neq t' \\ \sigma_\varepsilon^2 & \forall t = t' \end{cases}$.

[3] Assuming that the variable X_t is nonrandom simplifies the analysis in the sense that it allows us to use mathematical statistical results by considering X_t as a known variable for the probability distribution of the variable Y_t. However, such an assumption is sometimes difficult to maintain in practice, and the fundamental assumption is, in fact, the absence of correlation between the variable X_t and the error term.

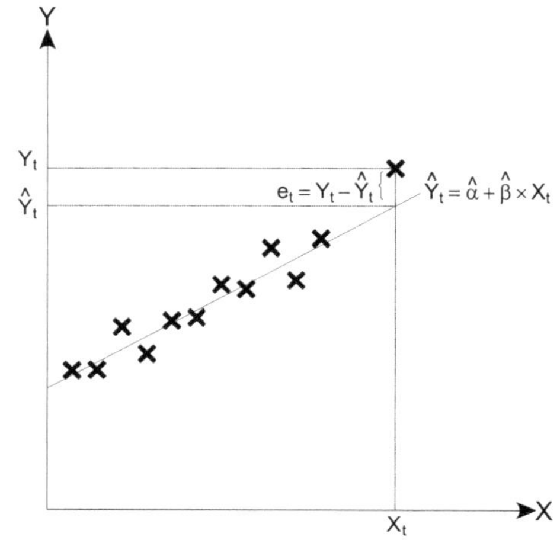

Fig. 2.1 The OLS principle

2.2.2 The OLS Principle

Figure 2.1 plots the values of the pair (X_t, Y_t) for $t = 1, \ldots, T$. We obtain a **scatter plot** that we try to fit with a line. Any line drawn through this scatter plot may be considered as an estimate of the linear relationship under consideration:

$$Y_t = \alpha + \beta X_t + \varepsilon_t \tag{2.33}$$

The equation of such a line, called the **regression line** or **OLS line**, is:

$$\hat{Y}_t = \hat{\alpha} + \hat{\beta} X_t \tag{2.34}$$

where $\hat{\alpha}$ and $\hat{\beta}$ are the estimators of the parameters α and β. The estimated value $\hat{Y}_t$ of Y_t is the ordinate of a point on the line whose abscissa is X_t. As shown in Fig. 2.1, some points of the pair (X_t, Y_t) lie above the line (2.34), and others lie below it. There are therefore deviations, noted e_t, from this line:

$$e_t = Y_t - \hat{Y}_t = Y_t - \hat{\alpha} - \hat{\beta} X_t \tag{2.35}$$

for $t = 1, \ldots, T$. These deviations are called **residuals**.

Intuitively, it seems logical to think that the better a line fits the scatter plot, the smaller the deviations e_t. The OLS method thus consists in finding the estimators $\hat{\alpha}$ and $\hat{\beta}$ such that the sum of the squares of the differences between the values of Y_t and those of $\hat{Y}_t$ is minimal. In other words, the method consists in minimizing the squared distance between each observation and the line (2.34), which is equivalent to **minimizing the sum of squared residuals**. The OLS principle can then be stated:

$$OLS \Longleftrightarrow Min \sum_{t=1}^{T} e_t^2 \tag{2.36}$$

The objective is to find $\hat{\alpha}$ and $\hat{\beta}$ such that the sum of squared residuals is minimal.

2.2.3 The OLS Estimators

Searching for Estimators

The OLS estimators $\hat{\alpha}$ and $\hat{\beta}$ of the parameters α and β are given by:

$$\hat{\alpha} = \bar{Y} - \hat{\beta}\bar{X} \tag{2.37}$$

and:

$$\hat{\beta} = \frac{Cov(X_t, Y_t)}{V(X_t)} \tag{2.38}$$

Let us demonstrate these formulas. Using Eq. (2.35), we can write the sum of squared residuals as:

$$\sum_{t=1}^{T} e_t^2 = \sum_{t=1}^{T} \left(Y_t - \hat{\alpha} - \hat{\beta}X_t\right)^2 \tag{2.39}$$

To obtain the estimators $\hat{\alpha}$ and $\hat{\beta}$, we have to minimize this expression with respect to the parameters $\hat{\alpha}$ and $\hat{\beta}$. We are therefore looking for the values $\hat{\alpha}$ and $\hat{\beta}$ such that:

$$\frac{\partial\left(\sum_{t=1}^{T} e_t^2\right)}{\partial\hat{\alpha}} = \frac{\partial\left(\sum_{t=1}^{T} e_t^2\right)}{\partial\hat{\beta}} = 0 \tag{2.40}$$

First, let us calculate the derivative of the sum of squared residuals with respect to $\hat{\alpha}$:

$$\frac{\partial\left(\sum_{t=1}^{T} e_t^2\right)}{\partial\hat{\alpha}} = \frac{\partial\left(\sum_{t=1}^{T} \left(Y_t - \hat{\alpha} - \hat{\beta}X_t\right)^2\right)}{\partial\hat{\alpha}} = 0 \tag{2.41}$$

That is:

$$-2\sum_{t=1}^{T}\left(Y_t-\hat{\alpha}-\hat{\beta}X_t\right)=0 \tag{2.42}$$

Hence:

$$\sum_{t=1}^{T}\left(Y_t-\hat{\alpha}-\hat{\beta}X_t\right)=0 \tag{2.43}$$

Noting that $\sum\limits_{t=1}^{T}\hat{\alpha}=T\hat{\alpha}$, we deduce:

$$\sum_{t=1}^{T}Y_t=T\hat{\alpha}+\hat{\beta}\sum_{t=1}^{T}X_t \tag{2.44}$$

Now let us determine the derivative of the sum of squared residuals with respect to $\hat{\beta}$:

$$\frac{\partial\left(\sum\limits_{t=1}^{T}e_t^2\right)}{\partial\hat{\beta}}=\frac{\partial\left(\sum\limits_{t=1}^{T}\left(Y_t-\hat{\alpha}-\hat{\beta}X_t\right)^2\right)}{\partial\hat{\beta}}=0 \tag{2.45}$$

That is:

$$-2\sum_{t=1}^{T}\left(Y_t-\hat{\alpha}-\hat{\beta}X_t\right)X_t=0 \tag{2.46}$$

Hence:

$$\sum_{t=1}^{T}\left(Y_t-\hat{\alpha}-\hat{\beta}X_t\right)X_t=0 \tag{2.47}$$

Expanding this expression, we obtain:

$$\sum_{t=1}^{T}X_tY_t=\hat{\alpha}\sum_{t=1}^{T}X_t+\hat{\beta}\sum_{t=1}^{T}X_t^2 \tag{2.48}$$

Equations (2.44) and (2.48), called the **normal equations**, form a system of two equations with two unknowns $\left(\hat{\alpha}\text{ and }\hat{\beta}\right)$ that we have to solve. By dividing

Eq. (2.44) by T, we get:

$$\frac{1}{T}\sum_{t=1}^{T} Y_t = \hat{\alpha} + \hat{\beta}\frac{1}{T}\sum_{t=1}^{T} X_t \tag{2.49}$$

Hence:

$$\bar{Y} = \hat{\alpha} + \hat{\beta}\bar{X} \Longleftrightarrow \hat{\alpha} = \bar{Y} - \hat{\beta}\bar{X} \tag{2.50}$$

Equation (2.50) gives us the OLS estimator $\hat{\alpha}$ of α and states that the regression line passes through the mean point $\left(\bar{X}, \bar{Y}\right)$.

Let us now determine the expression of the OLS estimator $\hat{\beta}$ of β. For this purpose, we replace $\hat{\alpha}$ by its value given in (2.50) in Eq. (2.48):

$$\sum_{t=1}^{T} X_t Y_t = \left(\bar{Y} - \hat{\beta}\bar{X}\right)\sum_{t=1}^{T} X_t + \hat{\beta}\sum_{t=1}^{T} X_t^2 \tag{2.51}$$

That is:

$$\sum_{t=1}^{T} X_t Y_t = \hat{\beta}\left(\sum_{t=1}^{T} X_t^2 - \bar{X}\sum_{t=1}^{T} X_t\right) + \bar{Y}\sum_{t=1}^{T} X_t \tag{2.52}$$

We deduce:

$$\hat{\beta}\left(\sum_{t=1}^{T} X_t^2 - \bar{X}\sum_{t=1}^{T} X_t\right) = \sum_{t=1}^{T} X_t Y_t - \bar{Y}\sum_{t=1}^{T} X_t \tag{2.53}$$

Hence:

$$\hat{\beta} = \frac{\sum\limits_{t=1}^{T} X_t Y_t - \frac{1}{T}\sum\limits_{t=1}^{T} X_t \sum\limits_{t=1}^{T} Y_t}{\sum\limits_{t=1}^{T} X_t^2 - \frac{1}{T}\left(\sum\limits_{t=1}^{T} X_t\right)^2} \tag{2.54}$$

We have:

$$V\left(X_t\right) = \frac{1}{T}\sum_{t=1}^{T} X_t^2 - \bar{X}^2 = \frac{1}{T}\sum_{t=1}^{T} X_t^2 - \left(\frac{1}{T}\sum_{t=1}^{T} X_t\right)^2 \tag{2.55}$$

Hence:

$$TV\left(X_t\right)=\sum_{t=1}^{T} X_t^2-\frac{1}{T}\left(\sum_{t=1}^{T} X_t\right)^2 \tag{2.56}$$

We deduce that the denominator of (2.54) is equal to $TV\left(X_t\right)$.
It is also known that the covariance between X_t and Y_t is given by:

$$Cov(X_t, Y_t)=\frac{1}{T}\sum_{t=1}^{T} X_t Y_t-\bar{X}\bar{Y} \tag{2.57}$$

That is:

$$Cov(X_t, Y_t)=\frac{1}{T}\sum_{t=1}^{T} X_t Y_t-\frac{1}{T}\sum_{t=1}^{T} X_t \frac{1}{T}\sum_{t=1}^{T} Y_t \tag{2.58}$$

Hence:

$$TCov(X_t, Y_t)=\sum_{t=1}^{T} X_t Y_t-\frac{1}{T}\sum_{t=1}^{T} X_t \sum_{t=1}^{T} Y_t \tag{2.59}$$

We deduce that the numerator of (2.54) is $TCov(X_t, Y_t)$. Therefore, we have:

$$\hat{\beta}=\frac{TCov(X_t, Y_t)}{TV\left(X_t\right)} \tag{2.60}$$

Finally, the OLS estimator $\hat{\beta}$ of β is given by:

$$\hat{\beta}=\frac{Cov(X_t, Y_t)}{V\left(X_t\right)} \tag{2.61}$$

Remark 2.5 (Case of Centered Variables) When the variables are centered, i.e., when observations are centered on their mean:

$$x_t=X_t-\bar{X} \text{ and } y_t=Y_t-\bar{Y} \tag{2.62}$$

the OLS estimators $\hat{\alpha}$ and $\hat{\beta}$ are, respectively, given by:

$$\hat{\alpha}=\bar{Y}-\hat{\beta}\bar{X} \tag{2.63}$$

and:

$$\hat{\beta} = \frac{\sum_{t=1}^{T} x_t y_t}{\sum_{t=1}^{T} x_t^2} \tag{2.64}$$

Remark 2.6 Here we have focused on estimating the regression model using the OLS method. Another estimation method is the **maximum likelihood** procedure. This method is presented in the appendix to this chapter. It leads to the same estimators of the coefficients α and β as the OLS method. However, the maximum likelihood estimator of the error variance is biased (see Appendix 2.3).

Example: The Phillips Curve and the Natural Unemployment Rate

The Phillips curve is one of the most widely studied relationships in macroeconomics. According to the modified version[4] of the Phillips curve, there is a negative relationship between the inflation rate and the unemployment rate. Taking into account inflation expectations, this relationship can be written in the following form:

$$\pi_t - E\left[\pi_t \left| I_{t-1}\right.\right] = \gamma\left(u_t - u^*\right) + \varepsilon_t \tag{2.65}$$

where π_t is the inflation rate (measured as the growth rate of the consumer price index) at date t, $E\left[\pi_t \left| I_{t-1}\right.\right]$ is the expectation (made at date $t-1$) for the inflation rate π_t given the set of information I available at date $(t-1)$, u_t is the unemployment rate at date t, and u^* is the natural rate of unemployment. In order to make this model operational, we need to make an assumption about the formation of expectations. Let us assume that the expected inflation rate is equal to the inflation rate of the previous period, i.e.:

$$E\left[\pi_t \left| I_{t-1}\right.\right] = \pi_{t-1} \tag{2.66}$$

The model to be estimated can therefore be written:

$$\pi_t - \pi_{t-1} = \alpha + \beta u_t + \varepsilon_t \tag{2.67}$$

where $\beta = \gamma$ and $\alpha = -\gamma u^*$. This equation shows that the variation in the inflation rate between t and $t-1$ is a function of the unemployment rate at date t. It is also

[4] The original version related the rate of change of nominal wages to the unemployment rate. Let us recall that this was originally a relationship estimated by Phillips (1958) for the British economy for the period 1861–1957.

Table 2.1 US inflation and unemployment rates, 1957–2020

t	π_t	$\pi_t - \pi_{t-1}$	u_t
1957	2.8986	−0.0865	5.2
1958	1.7606	−1.1380	6.2
1959	1.7301	−0.0305	5.3
1960	1.3605	−0.3696	6.6
...	...	...	...
2017	2.1091	0.0345	4.1
2018	1.9102	−0.1989	3.9
2019	2.2851	0.3750	3.6
2020	1.3620	−0.9231	6.7

Data sources: *US Bureau of Labor Statistics (BLS)* for the unemployment rate (noted u_t) and IMF, *International Financial Statistics*, for the inflation rate (noted π_t)

possible to calculate the natural rate of unemployment:

$$u^* = \frac{\hat{\alpha}}{\hat{\beta}} \tag{2.68}$$

Equation (2.67) is a simple regression model since it explains the variation in the inflation rate by a single explanatory variable, the unemployment rate. To illustrate this, let us consider annual data for the inflation rate and the unemployment rate in the United States over the period 1956–2020. Of course, calculating the change in the inflation rate at t requires the value of the inflation rate at $(t-1)$ to be known. Given that this series only begins in 1957, the estimation of Eq. (2.67) will therefore cover the period 1957–2020. Table 2.1 shows the first and last values of each series.

Before proceeding with the estimation, let us graphically represent the series in order to get a first idea of the potential relationship between the two variables. Figure 2.2 reproduces the dynamics of the unemployment rate (denoted $UNEMP$) and the variation in the inflation rate (denoted $DINF$) over the period 1957–2020. Generally, this graph shows that there seems to be a negative relationship between the two variables, in the sense that periods of rising unemployment are frequently associated with periods of falling inflation and vice versa. We would therefore expect to find a negative relationship between the two variables.

To extend this intuition, we can graphically represent the scatter plot, i.e., the values of the pair (unemployment rate, change in the inflation rate). Figure 2.3 shows that the scatter plot appears to be concentrated around a line with a generally decreasing trend, confirming the negative nature of the relationship between the two variables. Let us now proceed to the OLS estimation of the relationship between the two variables to confirm these intuitions.

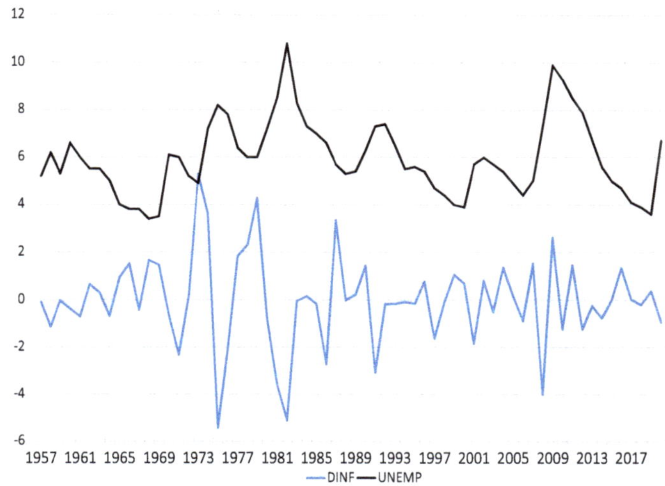

Fig. 2.2 Unemployment rate ($UNEMP$) and change in the inflation rate ($DINF$), United States, 1957–2020

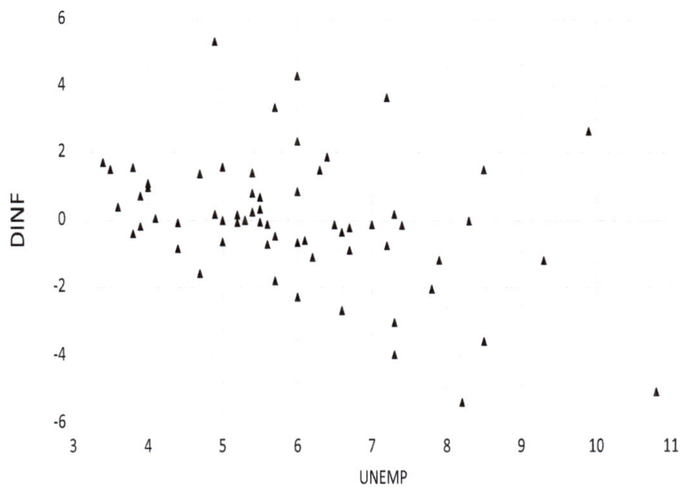

Fig. 2.3 Values of the pair (UNEMP, DINF)

Estimating Eq. (2.67) performed over the period 1957–2020 leads to the following result:

$$\widehat{\pi_t - \pi_{t-1}} = 2.70 - 0.46u_t \tag{2.69}$$

This model shows us that the coefficient assigned to the unemployment rate is negative: there is indeed a decreasing relationship between the unemployment rate

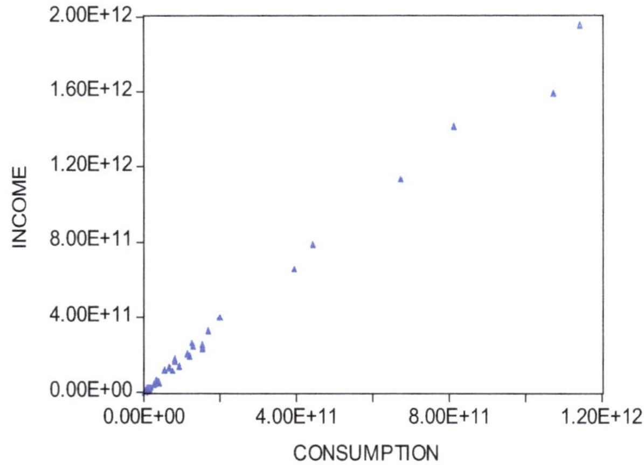

Fig. 2.4 Scatter plot, household consumption and income series

and the change in the inflation rate. The estimated value, -0.46, also allows us to write that if the unemployment rate falls by 1 point, the change in the inflation rate increases by 0.46 points on average. The ratio 2.70/0.46 gives us the estimated value of the natural unemployment rate, i.e., 5.87. Over the period under consideration, the natural unemployment rate is therefore equal to 5.87%. Note in particular that, while between 2014 and 2019 the observed unemployment rate was lower than its natural level, this was no longer the case in 2020—a result that may well be explained by the effects of the Covid-19 pandemic.

A Cross-Sectional Example: The Consumption-Income Relationship

To illustrate that the OLS method also applies to cross-sectional data, consider household consumption and gross disposable income data for various countries for the year 2004. The data are expressed in real terms[5] and converted to dollars for consistency. Figure 2.4 shows the scatter plot for the 43 countries considered.[6] It is clear that the points are distributed around a straight line, suggesting the existence of a linear relationship between the two variables for all countries. Furthermore, the relationship is increasing, showing that when income increases, consumption tends to follow a similar upward trend.

[5] The series were deflated by the consumer price index of each country.

[6] The data are from the World Bank. The 43 countries considered are Albania, Armenia, Austria, Azerbaijan, Belarus, Belgium, Bulgaria, Canada, Croatia, Czech Republic, Denmark, Estonia, Finland, France, Georgia, Germany, Greece, Hungary, Iceland, Ireland, Italy, Kazakhstan, Kyrgyzstan, Latvia, Lithuania, Luxembourg, Macedonia, Moldova, Netherlands, Norway, Poland, Portugal, Romania, Russia, Serbia and Montenegro, Slovakia, Slovenia, Spain, Sweden, Switzerland, Turkey, Ukraine, and United Kingdom.

These intuitions can be confirmed by estimating the regression of consumption on income for households in the 43 countries studied. The OLS estimation leads to the following relationship:

$$\widehat{CONSUMPTION}_{2004} = 3.98.10^9 + 0.61 INCOME_{2004} \tag{2.70}$$

This estimation shows that the relationship between consumption and income is indeed increasing, since the value of the coefficient assigned to income is positive. This coefficient represents the marginal propensity to consume: an increase of 10 monetary units in gross disposable income in 2004 leads, all other things being equal, to an increase of 6.1 monetary units in consumption the same year.

Summary and Properties

Let us summarize the main results obtained so far. According to the previous developments, the OLS estimators $\hat{\alpha}$ and $\hat{\beta}$ of the parameters α and β are given by:

$$\hat{\alpha} = \bar{Y} - \hat{\beta}\bar{X} \tag{2.71}$$

$$\hat{\beta} = \frac{Cov(X_t, Y_t)}{V(X_t)} \tag{2.72}$$

The expression:

$$\hat{Y}_t = \hat{\alpha} + \hat{\beta} X_t \tag{2.73}$$

is the regression line or OLS line. $\hat{\beta}$ is the slope of the regression line. The variable $\hat{Y}_t$ is the estimated variable (or adjusted or fitted variable). The difference between the observed value and the estimated value of the dependent variable is called the residual:

$$e_t = Y_t - \hat{Y}_t = Y_t - \hat{\alpha} - \hat{\beta} X_t \tag{2.74}$$

for $t = 1, \ldots, T$, and is a measure of the error ε_t.

We have also highlighted some properties of the linear regression, which we summarize below.

Property 2.1 The regression line passes through the mean point $(\bar{X}, \bar{Y})$.

This property, as we have seen, is derived from the relationship $\bar{Y} = \hat{\alpha} + \hat{\beta}\bar{X}$. Furthermore, knowing that the regression line is given by:

$$\hat{Y}_t = \hat{\alpha} + \hat{\beta} X_t \tag{2.75}$$

we deduce:

$$\overline{\hat{Y}} = \hat{\alpha} + \hat{\beta}\bar{X} = \bar{Y} \tag{2.76}$$

which can be formulated by the following property.

Property 2.2 The observed Y_t and estimated $\hat{Y}_t$ variables have the same mean: $\overline{\hat{Y}} = \bar{Y}$.

Knowing that the residuals are given by the difference between the observed and estimated variables, i.e., $e_t = Y_t - \hat{Y}_t$, we have:

$$\bar{e} = \bar{Y} - \overline{\hat{Y}} \tag{2.77}$$

By virtue of Property 2.2, we deduce that $\bar{e} = 0$, which is expressed by the following property.

Property 2.3 On average, the residuals are zero:

$$\bar{e} = 0 \tag{2.78}$$

i.e., the sum of residuals is zero:

$$\sum_{t=1}^{T} e_t = 0 \tag{2.79}$$

This property means that, on average, the model is correctly estimated.

Property 2.4 The covariance between the residuals and the explanatory variable X_t is zero, as is the covariance between the residuals and the estimated variable $\hat{Y}_t$:

$$Cov\left(X_t, e_t\right) = 0 \text{ and } Cov\left(\hat{Y}_t, e_t\right) = 0 \tag{2.80}$$

Let us prove this property. We have (see Box 2.1):

$$Cov\left(X_t, e_t\right) = Cov\left(X_t, Y_t - \hat{Y}_t\right) = Cov\left(X_t, Y_t\right) - Cov\left(X_t, \hat{Y}_t\right) \tag{2.81}$$

Moreover:

$$\begin{aligned} Cov\left(X_t, \hat{Y}_t\right) &= Cov\left(X_t, \hat{\alpha} + \hat{\beta}X_t\right) = Cov\left(X_t, \hat{\beta}X_t\right) = \hat{\beta}Cov\left(X_t, X_t\right) \\ &= \hat{\beta}V(X_t) \end{aligned} \tag{2.82}$$

According to the expression of $\hat{\beta}$ (Eq. (2.72)), we have:

$$Cov(X_t, Y_t) = \hat{\beta} V(X_t) \tag{2.83}$$

Hence:

$$Cov\left(X_t, \hat{Y}_t\right) = Cov(X_t, Y_t) \tag{2.84}$$

Equation (2.81) therefore gives us the following result:

$$Cov(X_t, e_t) = 0 \tag{2.85}$$

stipulating the absence of correlation between the explanatory variable and the residuals.

Let us now show that $Cov\left(\hat{Y}_t, e_t\right) = 0$. We have:

$$Cov\left(\hat{Y}_t, e_t\right) = Cov\left(\hat{\alpha} + \hat{\beta} X_t, e_t\right) = Cov\left(\hat{\beta} X_t, e_t\right) = \hat{\beta} Cov(X_t, e_t) \tag{2.86}$$

Using Eq. (2.85), we deduce:

$$Cov\left(\hat{Y}_t, e_t\right) = 0 \tag{2.87}$$

which means that the estimated variable and the residuals are not correlated.

Box 2.1 Properties of the variance and the covariance

Consider two variables X and Y and two constants a and b:

$V(X + Y) = V(X) + V(Y) + 2Cov(X, Y)$
$V(X - Y) = V(X) + V(Y) - 2Cov(X, Y)$
$V(aX) = a^2 V(X)$
$V(a + X) = V(X)$
$V(aX + bY) = a^2 V(X) + b^2 V(Y) + 2abCov(X, Y)$
$V(aX - bY) = a^2 V(X) + b^2 V(Y) - 2abCov(X, Y)$
$Cov(X, X) = V(X)$
$Cov(aX, bY) = abCov(X, Y)$
$Cov(a + X, b + Y) = Cov(X, Y)$

Property 2.5 A change of origin does not modify the parameter $\hat{\beta}$.

To demonstrate this property, let us perform the following change of origin:

$$W_t = X_t + a \text{ and } Z_t = Y_t + b \tag{2.88}$$

where a and b are constants. The regression model $Y_t = \alpha + \beta X_t + \varepsilon_t$ is then written as:

$$Z_t - b = \alpha + \beta (W_t - a) + \varepsilon_t \tag{2.89}$$

Hence:

$$Z_t = \alpha + b - \beta a + \beta W_t + \varepsilon_t \tag{2.90}$$

Let us note $\alpha' = \alpha + b - \beta a$. We have:

$$Z_t = \alpha' + \beta W_t + \varepsilon_t \tag{2.91}$$

It appears that the intercept is modified, but not the parameter β. We can also note that:

$$\hat{\beta} = \frac{Cov(W_t, Z_t)}{V(W_t)} = \frac{Cov(X_t + a, Y_t + b)}{V(X_t + a)} = \frac{Cov(X_t, Y_t)}{V(X_t)} \tag{2.92}$$

Property 2.6 A change of scale generally modifies the parameter $\hat{\beta}$.

Consider the following two variables:

$$W_t = aX_t \text{ and } Z_t = bY_t \tag{2.93}$$

where a and b are constants. The regression model $Y_t = \alpha + \beta X_t + \varepsilon_t$ is then written:

$$\frac{Z_t}{b} = \alpha + \beta \frac{W_t}{a} + \varepsilon_t \tag{2.94}$$

Hence:

$$Z_t = b\alpha + b\beta \frac{W_t}{a} + b\varepsilon_t \tag{2.95}$$

Or again, by noting $\alpha' = b\alpha$ and $\beta' = \frac{b\beta}{a}$:

$$Z_t = \alpha' + \beta' W_t + b\varepsilon_t \tag{2.96}$$

The estimator $\hat{\beta}'$ of β' is thus given by:

$$\hat{\beta}' = \frac{Cov(W_t, Z_t)}{V(W_t)} = \frac{Cov(aX_t, bY_t)}{V(aX_t)} = \frac{abCov(X_t, Y_t)}{a^2 V(X_t)} = \frac{b}{a}\hat{\beta} \tag{2.97}$$

As shown, $\hat{\beta}'$ differs from $\hat{\beta}$ if $a \neq b$.

2.2.4 Properties of OLS Estimators

The OLS estimators $\hat{\alpha}$ and $\hat{\beta}$ of the parameters α and β are:

- **Linear estimators**; in other words, they are functions of the dependent variable Y_t.
- **Unbiased estimators**; this means that $E(\hat{\alpha}) = \alpha$ and $E(\hat{\beta}) = \beta$: the bias of each of the estimators ($Bias(\hat{\alpha}) = E(\hat{\alpha}) - \alpha$ and $Bias(\hat{\beta}) = E(\hat{\beta}) - \beta$) is zero.
- **Minimum variance estimators**. The estimators $\hat{\alpha}$ and $\hat{\beta}$ are the unbiased estimators with the lowest variance among all the possible linear unbiased estimators.

The OLS estimators $\hat{\alpha}$ and $\hat{\beta}$ are therefore **BLUE** (the *best linear unbiased estimators*). Let us now demonstrate each of these properties.

Linear Estimators

Consider the centered variables $x_t = X_t - \bar{X}$, $y_t = Y_t - \bar{Y}$, and let w_t be defined as:

$$w_t = \frac{x_t}{\sum_{t=1}^{T} x_t^2} \tag{2.98}$$

It can then be shown (see Appendix 2.1.1) that:

$$\hat{\beta} = \frac{\sum_{t=1}^{T} x_t Y_t}{\sum_{t=1}^{T} x_t^2} = \sum_{t=1}^{T} w_t Y_t \tag{2.99}$$

and:

$$\hat{\alpha} = \sum_{t=1}^{T} \left(\frac{1}{T} - \bar{X} w_t\right) Y_t \tag{2.100}$$

The expression (2.99) reflects the fact that $\hat{\beta}$ is a **linear** estimator of β: $\hat{\beta}$ indeed appears as a linear function of the dependent variable Y_t. It is the same for $\hat{\alpha}$ which is expressed as a linear function of Y_t according to Eq. (2.100): $\hat{\alpha}$ is thus a linear estimator of α.

Let us summarize this first result concerning the properties of the OLS estimators as follows.

Property 2.7 The OLS estimators $\hat{\alpha}$ and $\hat{\beta}$ are linear estimators of the parameters α and β.

Unbiased Estimators

Starting from the linearity property of estimators, it is possible to show that $E\left(\hat{\beta}\right) = \beta$ and $E\left(\hat{\alpha}\right) = \alpha$, leading to the following property (the proof is given in Appendix 2.1.2).

Property 2.8 The OLS estimators $\hat{\alpha}$ and $\hat{\beta}$ are unbiased estimators of the parameters α and β:

$$E\left(\hat{\alpha}\right) = \alpha \tag{2.101}$$

$$E\left(\hat{\beta}\right) = \beta \tag{2.102}$$

Consistent and Minimum Variance Estimators

Starting from the formulas of the variances of the OLS estimators (see demonstration of the formulas in Appendix 2.1.3):

$$V(\hat{\beta}) = \frac{\sigma_\varepsilon^2}{\sum_{t=1}^{T} x_t^2} = \frac{\sigma_\varepsilon^2}{TV\left(X_t\right)} \tag{2.103}$$

and:

$$V\left(\hat{\alpha}\right) = \sigma_\varepsilon^2 \left(\frac{1}{T} + \frac{\bar{X}^2}{\sum_{t=1}^{T} x_t^2} \right) = \sigma_\varepsilon^2 \frac{\sum_{t=1}^{T} x_t^2 + T\bar{X}^2}{T\sum_{t=1}^{T} x_t^2} = \sigma_\varepsilon^2 \frac{\sum_{t=1}^{T} X_t^2}{T^2 V(X_t)} \tag{2.104}$$

we notice that if $T \to \infty$, then $V(\hat{\beta}) \to 0$ and $V(\hat{\alpha}) \to 0$ (see Appendix 2.1.3), which can be summarized as follows.

Property 2.9 The OLS estimators $\hat{\alpha}$ and $\hat{\beta}$ are consistent estimators of the parameters α and β:

$$\lim_{T\to\infty} V(\hat{\alpha}) = 0 \text{ and } \lim_{T\to\infty} V(\hat{\beta}) = 0 \tag{2.105}$$

It can also be shown that the OLS estimators $\hat{\alpha}$ and $\hat{\beta}$ are estimators of minimum variance among the class of linear unbiased estimators (see demonstration in Appendix 2.1.3).

Property 2.10 In the class of linear unbiased estimators, the OLS estimators $\hat{\alpha}$ and $\hat{\beta}$ are of minimum variance.

Putting together all the properties of the OLS estimators presented in this section, we can finally state the following fundamental property.

Property 2.11 The OLS estimators $\hat{\alpha}$ and $\hat{\beta}$ are the best linear unbiased estimators of the parameters α and β: they are **BLUE**.

It is because of this property that the OLS method is very frequently used.

2.2.5 OLS Estimator of the Variance of the Error Term

Finding the Estimator of the Error Variance

We now seek to determine an estimator $\hat{\sigma}_\varepsilon^2$ of the error variance σ_ε^2. Starting from the definition of the residuals:

$$e_t = Y_t - \hat{Y}_t = \alpha + \beta X_t + \varepsilon_t - \hat{\alpha} - \hat{\beta} X_t \tag{2.106}$$

that is:

$$e_t = \varepsilon_t - (\hat{\alpha} - \alpha) - \left(\hat{\beta} - \beta\right) X_t \tag{2.107}$$

we can show that such an estimator is written (see Appendix 2.1.4):

$$\hat{\sigma}_\varepsilon^2 = \frac{1}{T-2} \sum_{t=1}^{T} e_t^2 \tag{2.108}$$

This is an unbiased estimator of σ_ε^2.

Estimation of the Variances of the OLS Estimators

Determining the estimator $\hat{\sigma}_\varepsilon^2$ of the variance of the error term (Eq. (2.108)) allows us to give the estimates of the variances of the OLS estimators $\hat{\alpha}$ and $\hat{\beta}$. Using

Eq. (2.103), the estimator of the variance of $\hat{\beta}$ is written:

$$\widehat{V(\hat{\beta})} = \frac{\hat{\sigma}_{\varepsilon}^{2}}{\sum_{t=1}^{T} x_t^2} = \frac{\hat{\sigma}_{\varepsilon}^{2}}{T V(X_t)} \tag{2.109}$$

Similarly, from Eq. (2.104), we have the estimator of the variance of $\hat{\alpha}$:

$$\widehat{V(\hat{\alpha})} = \hat{\sigma}_{\varepsilon}^{2} \frac{\sum_{t=1}^{T} X_t^2}{T \sum_{t=1}^{T} x_t^2} = \hat{\sigma}_{\varepsilon}^{2} \frac{\sum_{t=1}^{T} X_t^2}{T^2 V(X_t)} \tag{2.110}$$

Calculating these expressions allows us to assess the precision of the estimators.

2.2.6 Empirical Application

To illustrate the OLS method, let us consider the following two series:

- The series of returns of the US Dow Jones Industrial Average index, denoted RDJ
- The series of returns of the Euro Stoxx 50, i.e., the European stock market index, denoted $REURO$

These two series, taken from the Macrobond database, have a quarterly frequency over the period from the second quarter of 1987 to the second quarter of 2021, i.e., a total of 137 observations.

Figure 2.5 shows that the returns series move in much the same way, which is not surprising given the international integration of financial markets. Figure 2.6 further shows that the scatter plot can be reasonably adjusted by a regression line of the type:

$$\widehat{REURO}_t = \hat{\alpha} + \hat{\beta} RDJ_t \tag{2.111}$$

We assume here that the dependent variable corresponds to the returns of the European index, the explanatory variable being the returns of the US index. This choice can be justified by the fact that it is frequently admitted that the US stock market has an influence on all the other international stock markets.

Our purpose is to obtain the estimated values $\hat{\alpha}$ and $\hat{\beta}$ by applying the OLS method:

$$\hat{\alpha} = \overline{REURO} - \hat{\beta}\overline{RDJ} \tag{2.112}$$

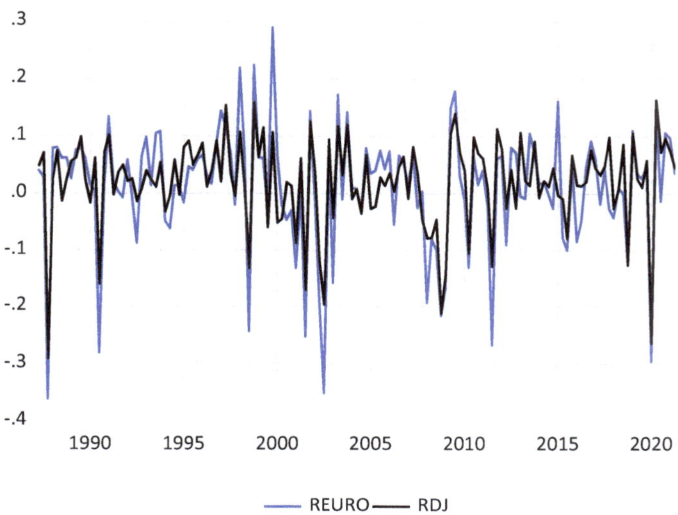

Fig. 2.5 Dow Jones and Euro Stoxx 50 returns, 1987.2–2021.2

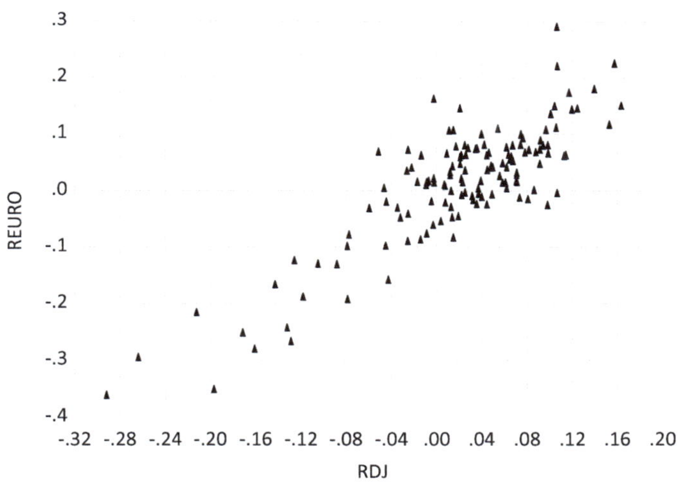

Fig. 2.6 Representation of the values of the pair (RDJ,REURO)

Table 2.2 OLS estimation of the relationship between $REURO$ and RDJ

	RDJ	$REURO$	$(RDJ)^2$	$RDJ \times REURO$
1987.2	0.0482	0.0404	0.0023	0.0019
1987.3	0.0709	0.0277	0.0050	0.0020
1987.4	−0.2920	−0.3619	0.0853	0.1057
1988.1	0.0251	0.0792	0.0006	0.0020
1988.2	0.0744	0.0807	0.0055	0.0060
...	...	...	...	...
2020.2	0.1636	0.1488	0.0268	0.0243
2020.3	0.0735	−0.0126	0.0054	−0.0009
2020.4	0.0968	0.1065	0.0094	0.0103
2021.1	0.0747	0.0982	0.0056	0.0073
2021.2	0.0451	0.0364	0.0020	0.0016
Sum	2.7061	1.5421	0.9080	1.0183

and:

$$\hat{\beta} = \frac{Cov(RDJ, REURO)}{V(RDJ)} \tag{2.113}$$

Table 2.2 presents the calculations required to determine the estimators $\hat{\alpha}$ and $\hat{\beta}$. We thus have:

$$\overline{RDJ} = \frac{1}{137} 2.7061 = 0.0196 \tag{2.114}$$

$$\overline{REURO} = \frac{1}{137} 1.5421 = 0.0113 \tag{2.115}$$

$$V(RDJ) = \frac{1}{137} 0.9080 - (0.0196)^2 = 0.0062 \tag{2.116}$$

$$Cov\,(RDJ, REURO) = \frac{1}{137} 1.0183 - 0.0196 \times 0.0113 = 0.0072 \tag{2.117}$$

From these calculations, we derive the values of the estimators $\hat{\alpha}$ and $\hat{\beta}$:

$$\hat{\beta} = \frac{0.0072}{0.0062} = 1.1559 \tag{2.118}$$

and:

$$\hat{\alpha} = 0.0113 - 1.1559 \times 0.0196 = -0.0116 \tag{2.119}$$

The equation of the regression line is therefore given by:

$$\widehat{REURO}_t = -0.0116 + 1.1559 RDJ_t \tag{2.120}$$

By virtue of (2.120), we find that there is a positive relationship between the US and European stock returns insofar as $\hat{\beta} > 0$. More precisely, we note that a 1-point increase in the Dow Jones returns translates, all other things being equal, into a 1.1559 points increase in the returns of the Euro Stoxx index.

2.3 Tests on the Regression Parameters

So far, the assumption that the error term follows a normal distribution has not been made, since it was not necessary to establish the main results of the regression analysis. This assumption can now be introduced to determine the distribution followed by the estimators $\hat{\alpha}$ and $\hat{\beta}$, as well as by the estimator $\hat{\sigma}_\varepsilon^2$ of the variance of the error term.

2.3.1 Determining the Distributions Followed by the OLS Estimators

Since $\hat{\alpha}$ and $\hat{\beta}$ are linear functions of the error term ε, they are also normally distributed. The expectation and variance of these two normal distributions still have to be specified.

We know that $\hat{\alpha}$ and $\hat{\beta}$ are unbiased estimators of α and β, that is: $E\left(\hat{\alpha}\right) = \alpha$ and $E\left(\hat{\beta}\right) = \beta$. Moreover, we have shown that the variances of the two estimators are given by (Eqs. (2.104) and (2.103)):

$$V(\hat{\alpha}) = \sigma_\varepsilon^2 \left(\frac{1}{T} + \frac{\bar{X}^2}{\sum\limits_{t=1}^{T} x_t^2} \right) = \sigma_\varepsilon^2 \frac{\sum\limits_{t=1}^{T} x_t^2 + T\bar{X}^2}{T \sum\limits_{t=1}^{T} x_t^2} = \sigma_\varepsilon^2 \frac{\sum\limits_{t=1}^{T} X_t^2}{T^2 V(X_t)} \tag{2.121}$$

and:

$$V(\hat{\beta}) = \frac{\sigma_\varepsilon^2}{\sum\limits_{t=1}^{T} x_t^2} \tag{2.122}$$

We deduce the distributions followed by the two estimators $\hat{\alpha}$ and $\hat{\beta}$:

$$\hat{\alpha} \sim N\left(\alpha, \sigma_\varepsilon^2\left(\frac{1}{T} + \frac{\bar{X}^2}{\sum\limits_{t=1}^{T} x_t^2}\right)\right) \tag{2.123}$$

and:

$$\hat{\beta} \sim N\left(\beta, \frac{\sigma_\varepsilon^2}{\sum\limits_{t=1}^{T} x_t^2}\right) \tag{2.124}$$

These expressions are a function of σ_ε^2 which is unknown. In order to make them operational, it is necessary to replace σ_ε^2 by its estimator $\hat{\sigma}_\varepsilon^2$ given by (Eq. (2.108)):

$$\hat{\sigma}_\varepsilon^2 = \frac{1}{T-2}\sum_{t=1}^{T} e_t^2 \tag{2.125}$$

However, such an operation requires knowledge of the distribution followed by $\hat{\sigma}_\varepsilon^2$ to deduce the distributions followed by the estimators $\hat{\alpha}$ and $\hat{\beta}$. Since the error term ε_t is normally distributed, we have (see Box 2.2):

$$(T-2)\frac{\hat{\sigma}_\varepsilon^2}{\sigma_\varepsilon^2} \sim \chi_{T-2}^2 \tag{2.126}$$

where χ_x^2 designates the Chi-squared distribution with x degrees of freedom. It follows that:

$$\frac{\sum\limits_{t=1}^{T} e_t^2}{\sigma_\varepsilon^2} \sim \chi_{T-2}^2 \tag{2.127}$$

Box 2.2 Relationships between the normal, Chi-squared, and Student's *t* distributions

Consider a random variable z following a **standard normal distribution**, that is: $z \sim N(0, 1)$. Let $z_1, z_2, \ldots, z_T$ T be independent random draws of this variable, which can be likened to T observations of the variable z. The sum of

(continued)

Box 2.2 (continued)
the squares of the z_i, $i = 1, \ldots, T$, follows a **Chi-squared distribution** with T degrees of freedom, i.e.:

$$\left(z_1^2 + z_2^2 + \ldots + z_T^2\right) \sim \chi_T^2 \tag{2.128}$$

When the number of degrees of freedom T tends to infinity, the Chi-squared distribution tends to a normal distribution. Let us now consider two independent random variables z and v. Assume that z has a standard normal distribution and v a Chi-squared distribution with r degrees of freedom: $z \sim N(0, 1)$ and $v \sim \chi_r^2$. Under these conditions, the quantity:

$$t = \frac{z\sqrt{r}}{\sqrt{v}} \tag{2.129}$$

follows a **Student's *t* distribution** with r degrees of freedom, i.e.:

$$t = \frac{z\sqrt{r}}{\sqrt{v}} \sim t(r) \tag{2.130}$$

Consider two random variables w and v each following a Chi-squared distribution with s and r degrees of freedom, respectively, and suppose that these two distributions are independent, i.e.:

$$w \sim \chi_s^2 \text{ and } v \sim \chi_r^2 \tag{2.131}$$

The statistics:

$$F = \frac{w/s}{v/r} \tag{2.132}$$

follows a **Fisher distribution** with (s, r) degrees of freedom, i.e.:

$$F \sim F(s, r) \tag{2.133}$$

According to Eqs. (2.123) and (2.124), we can write:

$$\frac{\hat{\alpha} - \alpha}{\sigma_\varepsilon \sqrt{\frac{1}{T} + \frac{\bar{X}^2}{\sum_{t=1}^{T} x_t^2}}} \sim N(0, 1) \tag{2.134}$$

and:

$$\frac{\hat{\beta}-\beta}{\sigma_\varepsilon / \sqrt{\sum_{t=1}^{T} x_t^2}} \sim N(0,1) \tag{2.135}$$

Let us examine what happens to these expressions when we replace σ_ε by its estimator $\hat{\sigma}_\varepsilon$. Using the results given in Box 2.2, let us posit:

$$t = \left(\frac{(\hat{\alpha}-\alpha)}{\sigma_\varepsilon \sqrt{\frac{1}{T} + \frac{\bar{X}^2}{\sum_{t=1}^{T} x_t^2}}} \right) / \frac{\left(\sqrt{\sum_{t=1}^{T} e_t^2}\right) / \sigma_\varepsilon}{\sqrt{T-2}} \tag{2.136}$$

Hence:

$$\frac{\hat{\alpha}-\alpha}{\hat{\sigma}_\varepsilon \sqrt{\frac{1}{T} + \frac{\bar{X}^2}{\sum_{t=1}^{T} x_t^2}}} \sim t(T-2) \tag{2.137}$$

Let us apply the same reasoning to $\hat{\beta}$. Thus, by positing:

$$t = \left(\frac{\hat{\beta}-\beta}{\sigma_\varepsilon / \sqrt{\sum_{t=1}^{T} x_t^2}} \right) / \frac{\left(\sqrt{\sum_{t=1}^{T} e_t^2}\right) / \sigma_\varepsilon}{\sqrt{T-2}} \tag{2.138}$$

we deduce that:

$$t = \left(\frac{\hat{\beta}-\beta}{\sigma_\varepsilon / \sqrt{\sum_{t=1}^{T} x_t^2}} \right) / \frac{\left(\sqrt{\sum_{t=1}^{T} e_t^2}\right) / \sigma_\varepsilon}{\sqrt{T-2}} \tag{2.139}$$

Equations (2.137) and (2.139) highlight the fact that replacing σ_ε^2 by its estimator $\hat{\sigma}_\varepsilon^2$ amounts to replacing a normal distribution by a Student's t distribution. When the sample size T is sufficiently large, the Student's t distribution tends to a standard normal distribution. In practice, when the number of observations exceeds 30

($T > 30$), we consider that the Student's t distribution in Eqs. (2.137) and (2.139) can be replaced by a standard normal distribution. From expressions (2.137) and (2.139), it is possible to derive statistical tests on the regression coefficients.

2.3.2 Tests on the Regression Coefficients

We present the tests on the two parameters α and β, even if the tests on β are in practice more frequently used.

Test on α

By virtue of (2.137), it is possible to construct a $100(1-p)\%$ **confidence interval** for α, that is:

$$\hat{\alpha} \pm t_{p/2}\hat{\sigma}_{\varepsilon}\sqrt{\frac{1}{T}+\frac{\bar{X}^2}{\sum_{t=1}^{T} x_t^2}} \quad (2.140)$$

where $t_{p/2}$ is the value obtained from the Student's t distribution for the $100\,(p/2)\%$ significance level. This value is called the **critical value** of the Student's t law at the $100(p/2)\%$ significance level. We often use $p = 0.05$, which corresponds to a 95% confidence interval.

Remark 2.7 The significance level corresponds to the probability of rejecting the null hypothesis when it is true. It is also called the **size of the test**.

Remark 2.8 The confidence interval (2.140) can also be written as:

$$Prob\left[\hat{\alpha} - t_{p/2}\hat{\sigma}_{\varepsilon}\sqrt{\frac{1}{T}+\frac{\bar{X}^2}{\sum_{t=1}^{T} x_t^2}} < \alpha < \hat{\alpha} + t_{p/2}\hat{\sigma}_{\varepsilon}\sqrt{\frac{1}{T}+\frac{\bar{X}^2}{\sum_{t=1}^{T} x_t^2}}\right] = 100(1-p)\%$$

It is then possible to test the null hypothesis that the coefficient α is equal to a given value α_0:

$$H_0 : \alpha = \alpha_0 \quad (2.141)$$

against the alternative hypothesis:

$$H_1 : \alpha \neq \alpha_0 \quad (2.142)$$

If the null hypothesis is true, then:

$$\frac{\hat{\alpha}-\alpha_0}{\hat{\sigma}_\varepsilon\sqrt{\frac{1}{T}+\frac{\bar{X}^2}{\sum\limits_{t=1}^{T}x_t^2}}} \sim t\,(T-2) \tag{2.143}$$

The decision rule is:

- If $\left|\frac{\hat{\alpha}-\alpha_0}{\hat{\sigma}_\varepsilon\sqrt{\frac{1}{T}+\frac{\bar{X}^2}{\sum\limits_{t=1}^{T}x_t^2}}}\right| \leq t_{p/2}$: the null hypothesis is not rejected at the $100p\%$ significance level; therefore, $\alpha=\alpha_0$.
- If $\left|\frac{\hat{\alpha}-\alpha_0}{\hat{\sigma}_\varepsilon\sqrt{\frac{1}{T}+\frac{\bar{X}^2}{\sum\limits_{t=1}^{T}x_t^2}}}\right| > t_{p/2}$: the null hypothesis is rejected at the $100p\%$ significance level; therefore, $\alpha \neq \alpha_0$.

Test on β

By virtue of (2.139), we can construct a $100(1-p)\%$ confidence interval for β, that is:

$$\hat{\beta} \pm t_{p/2}\hat{\sigma}_\varepsilon / \sqrt{\sum_{t=1}^{T}x_t^2} \tag{2.144}$$

As for α, it is possible to test the null hypothesis that the coefficient β is equal to a given value β_0:

$$H_0 : \beta = \beta_0 \tag{2.145}$$

against the alternative hypothesis:

$$H_1 : \beta \neq \beta_0 \tag{2.146}$$

If the null hypothesis is true, then:

$$\frac{\hat{\beta}-\beta_0}{\hat{\sigma}_\varepsilon / \sqrt{\sum\limits_{t=1}^{T}x_t^2}} \sim t\,(T-2) \tag{2.147}$$

The decision rule is given by:

– If $\left| \frac{\hat{\alpha}-\alpha_0}{\hat{\sigma}_\varepsilon \sqrt{\frac{1}{T}+\frac{\bar{X}^2}{\sum_{t=1}^{T} x_t^2}}} \right| \leq t_{p/2}$: the null hypothesis is not rejected at the $100p\%$ significance level; therefore, $\beta = \beta_0$.

– If $\left| \frac{\hat{\alpha}-\alpha_0}{\hat{\sigma}_\varepsilon \sqrt{\frac{1}{T}+\frac{\bar{X}^2}{\sum_{t=1}^{T} x_t^2}}} \right| > t_{p/2}$: the null hypothesis is rejected at the $100p\%$ significance level; therefore, $\beta \neq \beta_0$.

The commonest practice is to test the null hypothesis:

$$H_0 : \beta = 0 \tag{2.148}$$

against the alternative hypothesis:

$$H_0 : \beta \neq 0 \tag{2.149}$$

This is a **test of coefficient significance**, also called the ***t*-test**. Thus, under the null hypothesis, the coefficient associated with the variable X_t is not significant: X_t plays no role in determining the dependent variable Y_t. The test is performed by replacing β_0 by 0 in (2.147). The test statistic is then given by:

$$\frac{\hat{\beta}}{\hat{\sigma}_\varepsilon / \sqrt{\sum_{t=1}^{T} x_t^2}} \tag{2.150}$$

This expression corresponds to the ratio of the estimated coefficient $\hat{\beta}$ on its estimated standard deviation $\widehat{\sigma_{\hat{\beta}}}$, which is noted $t_{\hat{\beta}}$. The quantity:

$$t_{\hat{\beta}} = \frac{\hat{\beta}}{\widehat{\sigma_{\hat{\beta}}}} \tag{2.151}$$

is the **calculated *t*-statistic** of the coefficient $\hat{\beta}$.

The decision rule of the significance test of the coefficient β is:

- If $\left|t_{\hat{\beta}}\right| \leq t_{p/2}$: the null hypothesis is not rejected at the $100p\%$ significance level; therefore, $\beta = 0$: the coefficient associated with the variable X_t is not significant and X_t does not contribute to explaining Y_t.
- If $\left|t_{\hat{\beta}}\right| > t_{p/2}$: the null hypothesis is rejected at the $100p\%$ significance level; therefore, $\beta \neq 0$: the coefficient associated with the variable X_t is significant, meaning that X_t contributes to explaining the dependent variable Y_t.

As said, it is very common to use $p = 0.05$. For a sufficiently large number of observations, the value of the Student's t distribution at the 5% significance level is 1.96. Consequently:

- If $\left|t_{\hat{\beta}}\right| \leq 1.96$: the null hypothesis $\beta = 0$ is not rejected at the 5% significance level.
- If $\left|t_{\hat{\beta}}\right| > 1.96$: the null hypothesis $\beta = 0$ is rejected at the 5% significance level.

This t-test is widely used in practice. It can of course be applied in a similar way to the coefficient α.

Test on σ_ε^2

It is also possible to construct a test on the variance of the error term from the equation:

$$(T-2)\,\frac{\hat{\sigma}_\varepsilon^2}{\sigma_\varepsilon^2} \sim \chi^2_{T-2} \tag{2.152}$$

The confidence interval is given by:

$$Prob\left[\chi^2_{p/2} < (T-2)\,\frac{\hat{\sigma}_\varepsilon^2}{\sigma_\varepsilon^2} < \chi^2_{1-p/2}\right] = 100(1-p)\% \tag{2.153}$$

or:

$$Prob\left[\frac{(T-2)\,\hat{\sigma}_\varepsilon^2}{\chi^2_{1-p/2}} < \sigma_\varepsilon^2 < \frac{(T-2)\,\hat{\sigma}_\varepsilon^2}{\chi^2_{p/2}}\right] = 100(1-p)\% \tag{2.154}$$

It is then possible to carry out a test of the type:

$$H_0 : \sigma_\varepsilon^2 = \sigma_0^2 \tag{2.155}$$

2.3.3 Empirical Application

Let us go back to the previous example linking the following two series:

- The series of returns of the Dow Jones Industrial Average index, RDJ
- The series of returns of the Euro Stoxx 50 index, $REURO$

We obtained the following estimated relationship:

$$\widehat{REURO}_t = -0.0116 + 1.1559 RDJ_t \tag{2.156}$$

We can now ask whether or not the constant and the coefficient of the slope of the regression line are significantly different from zero. To this end, let us calculate the t-statistics of these two coefficients:

$$t_{\hat{\alpha}} = \frac{\hat{\alpha}}{\widehat{\sigma_{\hat{\alpha}}}} \text{ and } t_{\hat{\beta}} = \frac{\hat{\beta}}{\widehat{\sigma_{\hat{\beta}}}} \tag{2.157}$$

First, we need to determine the standard deviations of the estimated coefficients. We have seen that:

$$\widehat{V(\hat{\alpha})} = \hat{\sigma}_\varepsilon^2 \frac{\sum_{t=1}^{T} RDJ_t^2}{T^2 V(RDJ_t)} \tag{2.158}$$

and:

$$\widehat{V(\hat{\beta})} = \frac{\hat{\sigma}_\varepsilon^2}{TV(RDJ_t)} \tag{2.159}$$

It is therefore necessary to determine $\hat{\sigma}_\varepsilon^2$:

$$\hat{\sigma}_\varepsilon^2 = \frac{1}{T-2} \sum_{t=1}^{T} e_t^2 \tag{2.160}$$

Calculating $\hat{\sigma}_\varepsilon^2$ first involves determining the residuals $e_t, t = 1, \ldots, T$:

$$e_t = REURO_t - \widehat{REURO}_t \tag{2.161}$$

Table 2.3 presents the calculations needed to obtain the residuals and the sum of squared residuals.

Table 2.3 Calculation of the residuals

	$REURO$	$\widehat{REURO}_t$	e_t	e_t^2
1987.2	0.0404	0.0442	−0.0037	1.3978E-05
1987.3	0.0277	0.0704	−0.0427	1.8251E-03
1987.4	−0.3619	−0.3491	−0.0123	1.6271E-04
1988.1	0.0792	0.0174	0.0618	3.8201E-03
1988.2	0.0060	0.0745	0.0062	3.8997E-05
...	...	...	...	...
2020.2	0.1488	0.1775	−0.0287	8.2524E-04
2020.3	−0.0126	0.0734	−0.0860	7.3924E-03
2020.4	0.1065	0.1004	0.0062	3.8212E-05
2021.1	0.0982	0.0748	0.0234	5.4685E-04
2021.2	0.0364	0.0405	−0.0042	1.7524E-05
Sum	1.5421	1.5421	0.0000	0.4322

The estimated values $\widehat{REURO}_t$ of $REURO_t$ are determined as follows:

- $\widehat{REURO}_{1987.2} = -0.0116 + 1.1559 \times 0.0482 = 0.0442$
- $\widehat{REURO}_{1987.3} = -0.0116 + 1.1559 \times 0.0709 = 0.0704$
- ...
- $\widehat{REURO}_{2021.2} = -0.0116 + 1.1559 \times 0.0451 = 0.0405$

It can be seen from Table 2.3 that the sum of the values of $REURO_t$ is equal to the sum of the values of $\widehat{REURO}_t$, illustrating that the observed series and the estimated series have the same mean.

We derive the values of the residuals:

- $e_{1987.2} = 0.0404 - 0.0442 = -0.0037$
- $e_{1987.3} = 0.0277 - 0.0704 = -0.0427$
- ...
- $e_{2021.2} = 0.0364 - 0.0405 = -0.0042$

We find that $\sum_{t=1}^{137} e_t^2 = 0.4322$. Hence:

$$\hat{\sigma}_\varepsilon^2 = \frac{1}{137 - 2} 0.4322 = 0.0032 \tag{2.162}$$

Moreover, we had previously calculated the variance of RDJ, i.e.: $V(RDJ) = 0.0062$ and $\sum (RDJ)^2 = 0.9080$. According to (2.158), we therefore have:

$$\widehat{V(\hat{\alpha})} = 0.0032 \frac{0.9080}{137^2 \times 0.0062} = 2.4828.10^{-5} \tag{2.163}$$

Hence:

$$\hat{\sigma}_{\hat{\alpha}} = 0.0050 \tag{2.164}$$

So finally:

$$t_{\hat{\alpha}} = \frac{-0.0116}{0.0050} = -2.3232 \tag{2.165}$$

Similarly, we determine $\widehat{V(\hat{\beta})}$:

$$\widehat{V(\hat{\beta})} = \frac{0.0032}{137 \times 0.0062} = 0.0037 \tag{2.166}$$

and:

$$t_{\hat{\beta}} = \frac{1.1559}{\sqrt{0.0037}} = 18.8861 \tag{2.167}$$

Having determined the t-statistics of the coefficients $\hat{\alpha}$ and $\hat{\beta}$, given by Eqs. (2.165) and (2.167), we can perform the significance tests:

$$H_0 : \alpha = 0 \text{ against } H_1 : \alpha \neq 0 \tag{2.168}$$

and:

$$H_0 : \beta = 0 \text{ against } H_1 : \beta \neq 0 \tag{2.169}$$

The number of observations is $T = 137$. Recall that, under the null hypothesis, the $t_{\hat{\alpha}}$ and $t_{\hat{\beta}}$ statistics follow Student's t distributions with $(T - 2)$ degrees of freedom. Reading the Student's t table, for a number of degrees of freedom equal to 135 and for a 5% significance level, gives us the critical value: $t_{0.025}(135) = 1.96$. It can be seen that:

- $|t_{\hat{\alpha}}| = 2.3232 > 1.96$: we reject the null hypothesis that $\alpha = 0$. The constant term is therefore significantly different from zero.
- $\left|t_{\hat{\beta}}\right| = 18.8861 > 1.96$: we reject the null hypothesis that $\beta = 0$. The slope coefficient of the regression is therefore significantly different from zero, indicating that the variable RDJ contributes to explaining the variable $REURO$.

It is possible to construct confidence intervals for α and β:

- The 95% confidence interval for α is given by $\hat{\alpha} \pm t_{0.025} \times \widehat{\sigma_{\hat{\alpha}}}$, or $-0.0116 \pm 1.96 \times 0.0050$, which corresponds to the interval $[-0.0214; -0.0018]$. We can see that 0 does not belong to this interval, thus confirming the rejection of the null hypothesis for the coefficient α.

- The 95% confidence interval for β is given by $\hat{\beta} \pm t_{0.025} \times \widehat{\sigma_{\hat{\beta}}}$, or $1.1559 \pm 1.96 \times 0.0612$, which corresponds to the interval $[1.0359; 1.2759]$. We can see that 0 does not belong to this interval, thus confirming the rejection of the null hypothesis for the coefficient β.

2.4 Analysis of Variance and Coefficient of Determination

Once the regression parameters have been estimated and tested for statistical significance, the goodness of fit remains to be assessed. In other words, it is necessary to study whether the observed scatter plot is concentrated or, on the contrary, dispersed around the regression line. For this purpose, the analysis of the variance (analysis of variance [ANOVA]) of the regression is performed and the coefficient of determination is calculated.

2.4.1 Analysis of Variance (ANOVA)

From the definition of residuals:

$$e_t = Y_t - \hat{Y}_t = Y_t - \hat{\alpha} - \hat{\beta} X_t \tag{2.170}$$

we have:

$$Y_t = \hat{Y}_t + e_t \tag{2.171}$$

This relationship can be written in terms of variance:

$$V(Y_t) = V\left(\hat{Y}_t + e_t\right) = V\left(\hat{Y}_t\right) + V(e_t) + 2cov\left(\hat{Y}_t, e_t\right) \tag{2.172}$$

By virtue of Property 2.4, we know that: $cov\left(\hat{Y}_t, e_t\right) = 0$. So, we have:

$$V(Y_t) = V\left(\hat{Y}_t\right) + V(e_t) \tag{2.173}$$

This equation can also be expressed in terms of sums of squares by replacing the variances by their definitions:

$$\sum_{t=1}^{T} \left(Y_t - \bar{Y}\right)^2 = \sum_{t=1}^{T} \left(\hat{Y}_t - \overline{\hat{Y}}\right)^2 + \sum_{t=1}^{T} (e_t - \bar{e})^2 \tag{2.174}$$

which can also be written, noting that $\bar{e} = 0$ and $\overline{\hat{Y}} = \bar{Y}$ (see Property 2.2):

$$\sum_{t=1}^{T} (Y_t - \bar{Y})^2 = \sum_{t=1}^{T} \left(\hat{Y}_t - \bar{Y}\right)^2 + \sum_{t=1}^{T} e_t^2 \tag{2.175}$$

Equation (2.173) or (2.175) is called the **analysis-of-variance (ANOVA) equation**. In accordance with Eq. (2.173), we see that the **total variance** $V(Y_t)$ can be expressed as the sum of two terms:

- The explained variance, which corresponds to the variance of the estimated variable $\left(V\left(\hat{Y}_t\right)\right)$: this is the **variance explained** by the model, i.e., by the explanatory variable X_t.
- The variance of the residuals, called **residual variance** $(V(e_t))$. This is the variance that is not explained by the model.

In a similar way, Eq. (2.175) involves three terms:

- The sum of the squares of the deviations of the explained variable from its mean, known as the **total sum of squares**, noted TSS
- The **explained sum of squares**, noted ESS
- The **residual sum of squares** (also called sum of squared residuals), noted RSS

Equation (2.175) can thus be schematically written as follows:

$$TSS = ESS + RSS \tag{2.176}$$

Example 2.1 Let us take the example of the relationship between the returns of the Dow Jones Industrial Average index (RDJ) and the returns of the Euro Stoxx 50 index ($REURO$). We have already calculated the residual variance, i.e., $V(e_t) = 0.0032$. Furthermore, we have $V(REURO) = 0.0115$ and $V\left(\widehat{REURO}\right) = 0.0083$. We can therefore write the ANOVA equation:

$$0.0115 = 0.0083 + 0.0032 \tag{2.177}$$

We deduce that the part of the variation of $REURO$ explained by the model is given by:

$$\frac{V\left(\widehat{REURO}\right)}{V(REURO)} = \frac{0.0083}{0.0115} \simeq 0.7254 \tag{2.178}$$

Thus, 72.54% of the variation in $REURO$ is explained by the model.

2.4.2 Coefficient of Determination

The ANOVA equation enables us to judge the quality of a regression. The closer the explained variance is to the total variance, i.e., the lower the residual variance, the better the regression. In order to quantify this, we calculate the ratio between the explained variance and the total variance, which is called the **coefficient of determination** denoted as R^2 (R-squared):

$$R^2 = \frac{V\left(\hat{Y}_t\right)}{V\left(Y_t\right)} = \frac{\sum_{t=1}^{T}\left(\hat{Y}_t - \bar{Y}\right)^2}{\sum_{t=1}^{T}\left(Y_t - \bar{Y}\right)^2} = 1 - \frac{\sum_{t=1}^{T} e_t^2}{\sum_{t=1}^{T}\left(Y_t - \bar{Y}\right)^2} \tag{2.179}$$

or:

$$R^2 = \frac{ESS}{TSS} = 1 - \frac{RSS}{TSS} \tag{2.180}$$

The coefficient of determination thus measures the proportion of the variance of Y_t explained by the model. By definition, we have:

$$0 \leq R^2 \leq 1 \tag{2.181}$$

The closer the coefficient of determination is to 1, the better the model. A coefficient of determination equal to 1 indicates a perfect fit: $\hat{Y}_t = Y_t \ \forall t$. A coefficient of determination of zero indicates that there is no relationship between the dependent variable and the explanatory variable: $\hat{\beta} = 0$. In the latter case, the best estimate of Y_t is equal to its mean value, i.e., $\hat{Y}_t = \hat{\alpha} = \bar{Y}$. Figures 2.7, 2.8, 2.9, and 2.10 illustrate schematically the case of a coefficient of determination starting from zero and tending towards 1.

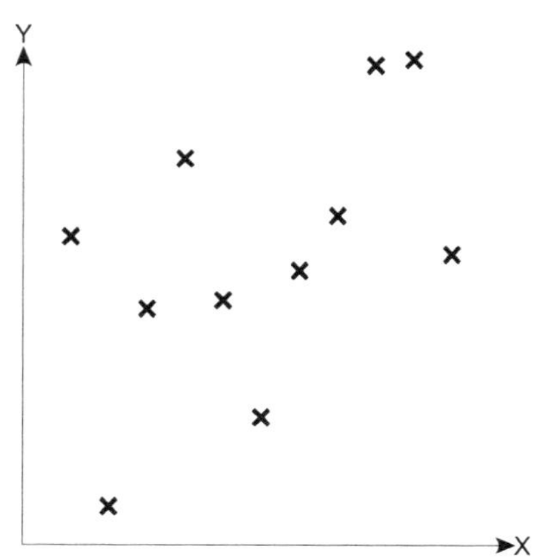

Fig. 2.7 Coefficient of determination close to zero

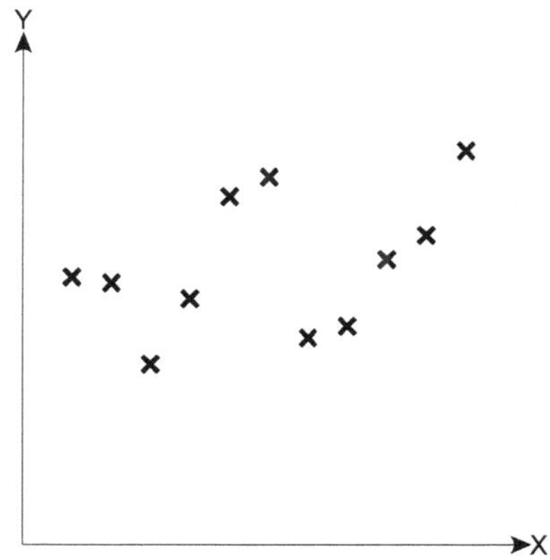

Fig. 2.8 Coefficient of determination moving away from zero

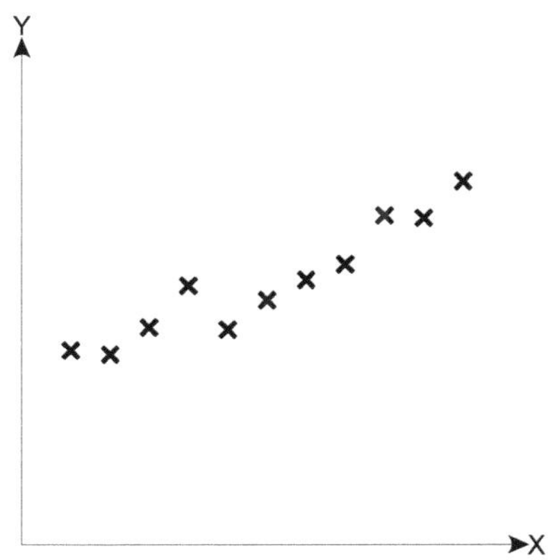

Fig. 2.9 Increasing coefficient of determination

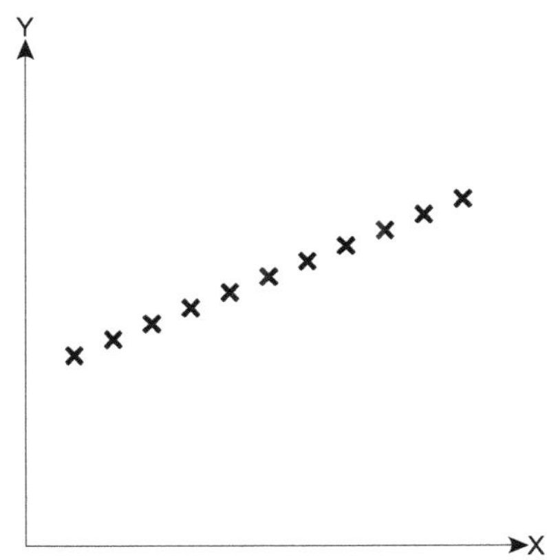

Fig. 2.10 Coefficient of determination close to 1

Remark 2.9 Since $V\left(\hat{Y}_t\right) = V\left(\hat{\alpha} + \hat{\beta}X_t\right) = \hat{\beta}^2 V\left(X_t\right)$, the coefficient of determination can be written:

$$R^2 = \frac{\hat{\beta}^2 V\left(X_t\right)}{V\left(Y_t\right)} \tag{2.182}$$

Furthermore, since $\hat{\beta} = \frac{Cov(X_t, Y_t)}{V(X_t)}$, we can also give the following expression for the coefficient of determination:

$$R^2 = \frac{\left[Cov\left(X_t, Y_t\right)\right]^2}{V\left(X_t\right) V\left(Y_t\right)} \tag{2.183}$$

Example 2.2 Let us go back to our example relating to the regression of $REURO$ on RDJ and a constant, i.e.:

$$\widehat{REURO}_t = -0.0116 + 1.1559 RDJ_t \tag{2.184}$$

Let us determine the coefficient of determination of this regression. We have already calculated:

$$R^2 = \frac{V\left(\widehat{REURO}\right)}{V(REURO)} \simeq 0.7254 \tag{2.185}$$

We can also use Eq. (2.182):

$$R^2 = \frac{(1.1559)^2 \times 0.0062}{0.0115} \simeq 0.7254 \tag{2.186}$$

or Eq. (2.183):

$$R^2 = \frac{[0.0072]^2}{0.0062 \times 0.0115} \simeq 0.7254 \tag{2.187}$$

it can be deduced that the selected model explains about 72.5% of the variation of $REURO$.

Remark 2.10 The coefficient of determination can be used to compare the quality of models having the same dependent variable. On the other hand, it cannot be used to compare models with different dependent variables. For example, the coefficient of determination can be used to compare the models:

$$Y_t = \alpha + \beta X_t + \varepsilon_t \text{ and } Y_t = a + bZ_t + u_t \tag{2.188}$$

where Z_t is an explanatory variable (other than X_t) and u_t an error term, but it cannot be used to compare:

$$Y_t = \alpha + \beta X_t + \varepsilon_t \text{ and } \log Y_t = a + bX_t + u_t \tag{2.189}$$

Thus, if we take the models in Eq. (2.188) and if the coefficient of determination of the model $Y_t = a + bZ_t + u_t$ is higher than that of the model $Y_t = \alpha + \beta X_t + \varepsilon_t$, the model $Y_t = a + bZ_t + u_t$ is preferred to the model $Y_t = \alpha + \beta X_t + \varepsilon_t$. On the other hand, if the coefficient of determination associated with the model $\log Y_t = a + bX_t + u_t$ is greater than that of the model $Y_t = \alpha + \beta X_t + \varepsilon_t$, we cannot conclude that the model $Y_t = a + bX_t + u_t$ is better, because the dependent variable is not the same in the two models.

2.4.3 Analysis of Variance and Significance Test of the Coefficient β

The significance test of the coefficient β, that is, the test of the null hypothesis $H_0 : \beta = 0$, can be approached in the ANOVA framework. Recall that we have (Eq. (2.135)):

$$\frac{\hat{\beta} - \beta}{\sigma_\varepsilon / \sqrt{\sum_{t=1}^{T} x_t^2}} \sim N(0, 1) \tag{2.190}$$

Furthermore, by virtue of the property that the sum of the squares of the terms of a normally distributed series follows a Chi-squared distribution, we can write by squaring the previous expression (see Box 2.2):

$$\frac{\left(\hat{\beta} - \beta\right)^2}{\sigma_\varepsilon^2 / \sum_{t=1}^{T} x_t^2} \sim \chi_1^2 \tag{2.191}$$

We also know from Eq. (2.127) that:

$$\frac{\sum_{t=1}^{T} e_t^2}{\sigma_\varepsilon^2} \sim \chi_{T-2}^2 \tag{2.192}$$

By relating Eqs. (2.191) and (2.192), we obtain:

$$F = \frac{\left(\hat{\beta} - \beta\right)^2 \sum\limits_{t=1}^{T} x_t^2}{\sum\limits_{t=1}^{T} e_t^2 / (T-2)} \sim F(1, T-2) \tag{2.193}$$

where $F(1, T-2)$ denotes a Fisher distribution with $(1, T-2)$ degrees of freedom. This result arises because the ratio of two independent Chi-squared distributions, divided by their number of degrees of freedom, follows a Fisher distribution (see Box 2.2).

We can then proceed to the significance test on the coefficient β. Under the null hypothesis, $H_0 : \beta = 0$, we can write:

$$F = \frac{\hat{\beta}^2 \sum\limits_{t=1}^{T} x_t^2}{\sum\limits_{t=1}^{T} e_t^2 / (T-2)} \sim F(1, T-2) \tag{2.194}$$

Let us consider the analysis-of-variance Eq. (2.175):

$$\sum_{t=1}^{T} \left(Y_t - \bar{Y}\right)^2 = \sum_{t=1}^{T} \left(\hat{Y}_t - \bar{Y}\right)^2 + \sum_{t=1}^{T} e_t^2 \tag{2.195}$$

which can also be written using the centered variables:

$$\sum_{t=1}^{T} y_t^2 = \sum_{t=1}^{T} \hat{y}_t^2 + \sum_{t=1}^{T} e_t^2 = \hat{\beta}^2 \sum_{t=1}^{T} x_t^2 + \sum_{t=1}^{T} e_t^2 \tag{2.196}$$

Thus, we have $\hat{\beta}^2 \sum\limits_{t=1}^{T} x_t^2 = ESS$ and $\sum\limits_{t=1}^{T} e_t^2 = RSS$, and Eq. (2.194) becomes:

$$F = \frac{ESS}{RSS/(T-2)} \sim F(1, T-2) \tag{2.197}$$

This statistic can be used to perform a test of significance of the coefficient β:

- If $F \leq F(1, T-2)$, the null hypothesis is not rejected, i.e., $\beta = 0$: the coefficient associated with the variable X_t is significant, indicating that X_t does not contribute to the explanation of the dependent variable.

– If $F > F(1, T-2)$, the null hypothesis is rejected. We deduce that β is significantly different from 0, which implies that the variable X_t contributes to explaining Y_t.

Remark 2.11 By virtue of the definition of the coefficient of determination, it is also possible to write Eq. (2.197) as follows:

$$F = \frac{R^2}{\left(1 - R^2\right)/(T-2)} \sim F(1, T-2) \tag{2.198}$$

This test can then be used as a test of significance of the coefficient of determination, i.e., as a test of the null hypothesis $H_0 : R^2 = 0$. Of course, since the simple regression model has only one explanatory variable—the variable X_t—testing the significance of the coefficient of determination amounts to testing the significance of the coefficient β assigned to X_t.

2.4.4 Empirical Application

Let us go back to our example linking the returns of the European stock index ($REURO$) and the returns of the US stock index (RDJ). The purpose is to apply the tests of significance of β and of the R-squared based on Fisher statistics.

Table 2.4 presents the calculations required to determine the explained sum of squares (ESS) and the sum of squared residuals (RSS), the latter having already been calculated.

Table 2.4 Fisher test

	$\widehat{REURO}_t$	$\left(\widehat{REURO}_t - \overline{REURO}\right)^2$	e_t	e_t^2
1987.2	0.0404	0.0011	−0.0037	1.3978E-05
1987.3	0.0277	0.0035	−0.0427	1.8251E-03
1987.4	−0.3619	0.1299	−0.0128	1.6271E-04
1988.1	0.0792	0.0000	0.0618	3.8201E-03
1988.2	0.0807	0.0040	0.0062	3.8997E-05
...	...	...	...	...
2020.2	0.1488	0.0276	−0.0287	8.2524E-04
2020.3	−0.0126	0.0039	−0.0860	7.3924E-03
2020.4	0.1065	0.0079	0.0062	3.8212E-05
2021.1	0.0982	0.0040	0.0234	5.4685E-04
2021.2	0.0364	0.0009	−0.0042	1.7524E-05
Sum	1.5421	1.1418	0.0000	0.4322

The explained sum of squares is equal to $ESS = 1.1418$ and the sum of squared residuals is given by $RSS = 0.4322$. The application of the formula (2.197) leads to the following result:

$$F = \frac{1.1418}{0.4322/135} \simeq 356.68 \tag{2.199}$$

At the 5% significance level, the value of the Fisher distribution $F\,(1135)$ read from the table is 3.842. Thus, we have $F \simeq 356.68 > 3.842$, which means that we reject the null hypothesis that $\beta = 0$. The variable RDJ contributes significantly to explaining $REURO$, which of course confirms the results previously obtained.

It is also possible to calculate the F statistic from expression (2.198). We have previously shown that $R^2 \simeq 0.7254$. We thus have:

$$F = \frac{0.7254}{(1 - 0.7254)\,/135} \simeq 356.68 \tag{2.200}$$

We obviously obtain the same value as with Eq. (2.197). Comparing, as before, this value to the critical value at the 5% significance level, i.e., $F\,(1135) = 3.842$, we have $356.68 > 3.842$. We therefore reject the null hypothesis of nonsignificance of the coefficient of determination. The coefficient of determination is significant, which is equivalent to concluding that the variable RDJ matters in the explanation of $REURO$, since our model contains only one explanatory variable.

2.5 Prediction

Once the model has been estimated by the OLS method, it is possible to predict the dependent variable. Suppose that the following model has been estimated for $t = 1, \ldots, T$:

$$Y_t = \alpha + \beta X_t + \varepsilon_t \tag{2.201}$$

that is:

$$\hat{Y}_t = \hat{\alpha} + \hat{\beta} X_t \tag{2.202}$$

for $t = 1, \ldots, T$.

We seek to determine the forecast of the dependent variable for a horizon h, i.e., $\hat{Y}_{T+h}$. Assuming that the relationship generating the explained variable remains identical and the value of the explanatory variable is known in $T + h$, we have:

$$\hat{Y}_{T+h} = \hat{\alpha} + \hat{\beta} X_{T+h} \tag{2.203}$$

It is possible to define the **forecast error**, noted e_{T+h}, by:

$$e_{T+h} = Y_{T+h} - \hat{Y}_{T+h} = \alpha + \beta X_{T+h} + \varepsilon_{T+h} - \hat{\alpha} - \hat{\beta} X_{T+h} \tag{2.204}$$

which can also be expressed as:

$$e_{T+h} = \varepsilon_{T+h} - (\hat{\alpha} - \alpha) - \left(\hat{\beta} - \beta\right) X_{T+h} \tag{2.205}$$

In order to show that the forecast given by Eq. (2.203) is unbiased, let us calculate the expectation of the expression (2.205):

$$E\left(e_{T+h}\right) = E\left(\varepsilon_{T+h} - (\hat{\alpha} - \alpha) - \left(\hat{\beta} - \beta\right) X_{T+h}\right) \tag{2.206}$$

Since $\hat{\alpha}$ and $\hat{\beta}$ are unbiased estimators of α and β and given that $E\left(\varepsilon_{T+h}\right) = 0$, we have:

$$E\left(e_{T+h}\right) = 0 \tag{2.207}$$

The forecast given by Eq. (2.203) is therefore unbiased. The **prediction interval** is given by:

$$\left(\hat{\alpha} + \hat{\beta} X_{T+h}\right) \pm \sigma_{e_{T+h}} \tag{2.208}$$

where $\sigma_{e_{T+h}}$ designates the standard deviation of the forecast error. After calculating this standard deviation (see Appendix 2.1.5), we can can write the $100(1-p)\%$ prediction interval[7] for Y_{T+h}:

$$\left(\hat{\alpha} + \hat{\beta} X_{T+h}\right) \pm t_{p/2} \hat{\sigma}_{\varepsilon} \sqrt{1 + \frac{1}{T} + \frac{\left(X_{T+h} - \bar{X}\right)^2}{\sum\limits_{t=1}^{T} x_t^2}} \tag{2.209}$$

It is then possible to give a certain degree of confidence to the forecast if the value of the dependent variable, for the considered horizon, lies within the prediction interval. The length of this interval is not constant: the more the value of X_{T+h} deviates from the mean $\bar{X}$ of the sample under consideration, the wider the interval.

Remark 2.12 The purpose may be not to predict the precise value of Y_{T+h}, but its average value instead. We then consider:

$$E\left(Y_{T+h}\right) = \alpha + \beta X_{T+h} \tag{2.210}$$

[7] The demonstration is given in Appendix 2.1.5.

The forecast error is written:

$$e_{T+h} = E\left(Y_{T+h}\right) - \hat{Y}_{T+h} = -\left[\left(\hat{\alpha} - \alpha\right) + \left(\hat{\beta} - \beta\right) X_{T+h}\right] \tag{2.211}$$

and its variance is given by:

$$V\left(e_{T+h}\right) = \sigma_{\varepsilon}^{2}\left(\frac{1}{T} + \frac{\left(X_{T+h} - \bar{X}\right)^{2}}{\sum_{t=1}^{T} x_{t}^{2}}\right) = \sigma_{\varepsilon}^{2}\left(\frac{1}{T} + \frac{\left(X_{T+h} - \bar{X}\right)^{2}}{\sum_{t=1}^{T}\left(X_{t} - \bar{X}\right)^{2}}\right) \tag{2.212}$$

The $100(1-p)\%$ prediction interval for $E\left(Y_{T+h}\right)$ is therefore given by:

$$\left(\alpha + \beta X_{T+h}\right) \pm t_{p/2}\hat{\sigma}_{\varepsilon}\sqrt{\frac{1}{T} + \frac{\left(X_{T+h} - \bar{X}\right)^{2}}{\sum_{t=1}^{T} x_{t}^{2}}} \tag{2.213}$$

Example 2.3 Consider our example relating the returns of the European stock index ($REURO$) and the returns of the US stock market index (RDJ) over the period from the second quarter of 1987 to the second quarter of 2021. For this period, we estimated the following relationship:

$$\widehat{REURO}_{t} = -0.0116 + 1.1559 RDJ_{t} \tag{2.214}$$

Assume that the returns on the US stock index increase by 2% in the third quarter of 2021 compared to the previous quarter. Given that $RDJ_{2021.2} = 0.0451$, we deduce: $RDJ_{2021.3} = 0.0451 \times 1.02 = 0.0460$. Therefore, we can write:

$$\widehat{REURO}_{2021.3} = -0.0116 + 1.1559 \times 0.0460 = 0.0416 \tag{2.215}$$

Let us now determine the 95% prediction interval:

$$\left(\hat{\alpha} + \hat{\beta} RDJ_{2021.3}\right) \pm t_{0.025}\hat{\sigma}_{\varepsilon}\sqrt{1 + \frac{1}{137} + \frac{\left(RDJ_{2021.3} - \overline{RDJ}\right)^{2}}{\sum_{t=1}^{137}\left(RDJ_{t} - \overline{RDJ}\right)^{2}}} \tag{2.216}$$

We know that $\overline{RDJ} = 0.0196$ and that $\sum_{t=1}^{137} \left(RDJ_t - \overline{RDJ}\right)^2 = 0.8546$. Moreover, we have already calculated $\hat{\sigma}_\varepsilon = \sqrt{0.0032}$. Knowing that $t_{0.025}(135) = 1.96$, we have:

$$(-0.0116 + 1.1559 \times 0.0460) \pm 1.96 \times \sqrt{0.0032}$$
$$\times \sqrt{1 + \frac{1}{137} + \frac{(0.0460 - 0.0196)^2}{0.8546}} \tag{2.217}$$

which corresponds to the interval $[-0.0698; 0.1529]$. If the value taken by $REURO$ in the third quarter of 2021 does not lie within this interval, the forecast is incorrect. This may be the case, for example, if the estimated model, valid until the second quarter of 2021, is no longer valid for the third quarter of the same year. In other words, such a situation may arise if the structure of the model has changed.

2.6 Some Extensions of the Simple Regression Model

So far, we have focused on the model:

$$Y_t = \alpha + \beta X_t + \varepsilon_t \tag{2.218}$$

which is linear with respect to the parameters α and β, but also with respect to the variables Y_t and X_t. We now propose to briefly study models frequently used in economics, which can be nonlinear with respect to the variables Y_t and X_t, but linear with respect to the parameters, or can become so after certain appropriate transformations of the variables. As an example, the model:

$$\log Y_t = \alpha + \beta X_t + \varepsilon_t \tag{2.219}$$

can also be written:

$$Z_t = \alpha + \beta X_t + \varepsilon_t \tag{2.220}$$

with $Z_t = \log Y_t$.

The model (2.219) is nonlinear with respect to the variables Y_t and X_t, but it is linear with respect to the variables Z_t and X_t. The model (2.220) can then be studied using the methodology presented in this chapter. The transformation of the variable Y_t into the variable Z_t has allowed us to obtain, from a nonlinear model with respect to the variables, a linear model with respect to the transformed variables. In our example, only one of the two variables has been transformed, but there are also cases where both variables must undergo transformations in order to obtain a linear model.

2.6.1 Log-Linear Model

The **log-linear model**, also known as **log-log model** or **double-log model**, is given by:

$$\log Y_t = \log \alpha + \beta \log X_t + \varepsilon_t \tag{2.221}$$

By noting $\alpha_0 = \log \alpha$, we get:

$$\log Y_t = \alpha_0 + \beta \log X_t + \varepsilon_t \tag{2.222}$$

This model is linear in the parameters α_0 and β. Furthermore, let us posit:

$$Y_t^* = \log Y_t \ \text{ and } \ X_t^* = \log X_t \tag{2.223}$$

The model (2.221) can therefore be written:

$$Y_t^* = \alpha_0 + \beta X_t^* + \varepsilon_t \tag{2.224}$$

which is a linear model in the variables Y_t^* and X_t^* and in the parameters α_0 and β. It is then possible to apply to this model the methodology presented in this chapter in order to estimate the parameters α_0 and β by OLS.

One of the interests of the log-log model is that the coefficient β measures the **elasticity** of Y_t with respect to X_t, i.e., the percentage change in Y_t for a given percentage of variation in X_t. It is thus a constant elasticity model.

For example, if Y_t denotes the quantity of a given good and X_t the unit price of this good, the coefficient β represents the price elasticity of demand. Similarly, if Y_t designates household consumption and X_t the income of these same households, the coefficient β measures the elasticity of consumption with respect to income: estimating this coefficient allows us to determine how much consumption varies in response to a certain change in income.

Example 2.4 Let us take the example of the consumption and gross disposable income series of French households already studied in Chap. 1 and consider the following model:

$$\log C_t = \log \alpha + \beta \log Y_t + \varepsilon_t \tag{2.225}$$

where C_t denotes consumption and Y_t income. The data are annual and the study period runs from 1990 to 2019. In order to estimate this model, we simply take the logarithm of the raw consumption and income data and apply the OLS method to the transformed model. The estimation leads to the following results:

$$\widehat{\log C_t} = \underset{(4.67)}{1.5552} + \underset{(36.87)}{0.8796} \log Y_t \tag{2.226}$$

where the numbers in parentheses correspond to the t-statistics of the estimated coefficients. These results show that if income increases by 1% on average, consumption increases by 0.88%.

Remark 2.13 The log-log model can be understood from the **Box-Cox transformation** (see Box and Cox, 1964). For a variable Y_t, this transformation is given by:

$$Y_t^{(\lambda)} = \begin{cases} \frac{Y_t^{\lambda}-1}{\lambda} \text{ if } \lambda \neq 0 \\ \log Y_t \text{ if } \lambda = 0 \end{cases} \tag{2.227}$$

where $Y_t^{(\lambda)}$ is the transformed variable. The Box-Cox transformation thus depends on a single parameter, noted λ.

Let $Y_t^{(\lambda_Y)}$ be the transformation of the variable Y_t and let $X_t^{(\lambda_X)}$ be the transformation of the variable X_t:

$$Y_t^{(\lambda_Y)} = \begin{cases} \frac{Y_t^{\lambda_Y}-1}{\lambda_Y} \text{ if } \lambda_Y \neq 0 \\ \log Y_t \text{ if } \lambda_Y = 0 \end{cases} \tag{2.228}$$

$$X_t^{(\lambda_X)} = \begin{cases} \frac{X_t^{\lambda_X}-1}{\lambda_X} \text{ if } \lambda_X \neq 0 \\ \log X_t \text{ if } \lambda_X = 0 \end{cases} \tag{2.229}$$

The log-log model corresponds to the case where $\lambda_Y = \lambda_X = 0$.

2.6.2 Semi-Log Model

The **semi-log model** is given by:

$$\log Y_t = \alpha + \beta X_t + \varepsilon_t \tag{2.230}$$

This is a linear model with respect to the parameters α and β and with respect to the variables $\log Y_t$ and X_t. The special feature of this model lies in the fact that only the dependent variable is in logarithms. After transforming the endogenous variable into a logarithm, it is possible to apply to this model the methodology presented in this chapter to estimate the parameters α and β by OLS.

In the semi-log model, the coefficient β measures the rate of change of Y_t relative to the variation of X_t; this rate of change being constant. In other words, the coefficient β is equal to the ratio between the relative variation of Y_t and the absolute variation of X_t. β is the **semielasticity** of Y_t with respect to X_t.

If the explanatory variable is time, the model is written:

$$\log Y_t = \alpha + \beta t + \varepsilon_t \tag{2.231}$$

or, leaving aside the error term:

$$Y_t = \exp(\alpha + \beta t) \tag{2.232}$$

This model describes the evolution of the variable Y_t, having a constant growth rate if $\beta > 0$, or constant decrease if $\beta < 0$. Let us explain this. The model (2.232) describes an evolution in continuous time and can be written:

$$Y_t = \exp(\alpha + \beta t) = Y_0 \exp(\beta t) \tag{2.233}$$

where $Y_0 = \exp(\alpha)$ is the value of Y_t at date $t = 0$. The coefficient β is thus equal to:

$$\beta = \frac{1}{Y_t} \frac{dY_t}{dt} \tag{2.234}$$

Consequently, the coefficient β represents the instantaneous growth rate of Y at date t.

If we now consider discrete time, assuming that t denotes, for example, months, quarters, or years, we can write:

$$Y_t = Y_0 (1 + g)^t \tag{2.235}$$

where g is the growth rate of Y. Transforming this expression into logarithmic terms gives:

$$\log Y_t = \log Y_0 + t \log(1 + g) \tag{2.236}$$

By positing $\log Y_0 = \alpha$ and $\log(1 + g) = \beta$ and adding the error term, we find model (2.231). The relationship:

$$\log(1 + g) = \beta \tag{2.237}$$

allows us to obtain an estimate of the coefficient β from an estimate of the growth rate g. The coefficient β is interpreted as the continuous growth rate that would give, at the end of a period, the same result as a single increase at the rate g.

Example 2.5 Let us take the example of the French household consumption series (C_t) over the period 1990–2019 at annual frequency and consider the following model:

$$\log C_t = \alpha + \beta t + \varepsilon_t \tag{2.238}$$

where t denotes time, i.e., $t = 0, 1, 2, \ldots, 29$. The OLS estimation of this model leads to the following results:

$$\widehat{\log C_t} = \underset{(1292.53)}{13.6194} + \underset{(22.49)}{0.0140t} \tag{2.239}$$

From this estimation, we deduce that $\log\left(1 + \hat{g}\right) = 0.0140$ where $\hat{g}$ is the estimated growth rate. Hence, $\hat{g} = 0.0141$. Over the period 1990–2019, French household consumption increased annually at a rate of 1.41%.

Remark 2.14 The semi-log model can be understood from the Box-Cox transformation, noting that $\lambda_Y = 0$ and $\lambda_X = 1$.

2.6.3 Reciprocal Model

The **reciprocal model** is written:

$$Y_t = \alpha + \beta\left(\frac{1}{X_t}\right) + \varepsilon_t \tag{2.240}$$

This model is linear with respect to the parameters α and β. Such a model can be estimated by OLS following the methodology described in this chapter and after transforming the variable X_t into its inverse.

According to this model, when the variable X_t tends to infinity, the term $\beta\left(\frac{1}{X_t}\right)$ tends to zero and α is therefore the asymptotic limit of Y_t when X_t tends to infinity. In addition, the slope of the model (2.240) is given by:

$$\frac{dY_t}{dX_t} = -\beta\left(\frac{1}{X_t^2}\right) \tag{2.241}$$

Therefore, if $\beta > 0$, the slope is always negative, and if $\beta < 0$, the slope is always positive.

This type of model, represented in Fig. 2.11 for $\beta > 0$, can be illustrated by the Phillips curve. This curve originally related the growth rate of nominal wages to the unemployment rate. It was subsequently transformed into a relationship between the inflation rate and the unemployment rate. This Phillips curve can be estimated by regressing the inflation rate on the inverse of the unemployment rate, with the inflation rate tending asymptotically towards the estimated value of α.

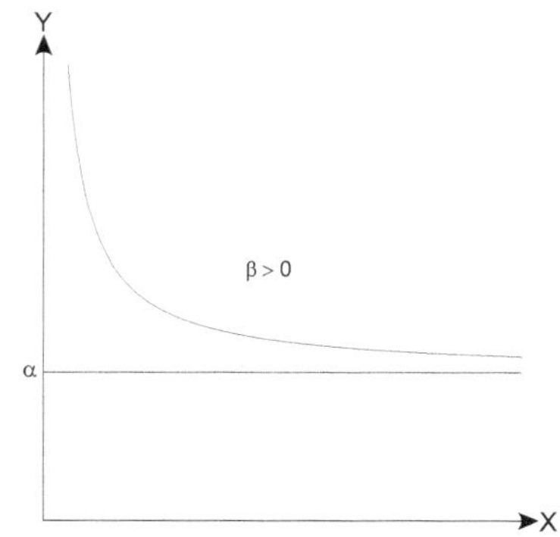

Fig. 2.11 Reciprocal model

Example 2.6 For example, suppose that, for a given country, the regression of the inflation rate (π_t) on the inverse of the unemployment rate (u_t) leads to the following results:

$$\widehat{\pi_t} = -2.3030 + 20.0103\left(\frac{1}{u_t}\right) \tag{2.242}$$

These results show that even if the unemployment rate rises indefinitely, the largest change in prices will be a drop in the inflation rate of about 2.30 points.

Remark 2.15 The reciprocal model corresponds to the case where $\lambda_Y = 1$ and $\lambda_X = -1$ in the Box-Cox transformation.

2.6.4 Log-Inverse or Log-Reciprocal Model

The **log-inverse** or **log-reciprocal** model is given by:

$$\log Y_t = \alpha - \beta\left(\frac{1}{X_t}\right) + \varepsilon_t \tag{2.243}$$

Ignoring the error term, this model can still be written:

$$Y_t = \exp\left(\alpha - \beta\left(\frac{1}{X_t}\right)\right) \tag{2.244}$$

When X_t tends to zero by positive values $\left(X_t \to 0^+\right)$, Y_t tends to zero. Furthermore, the slope is given by:

$$\frac{dY_t}{dX_t} = \left(\frac{\beta}{X_t^2}\right)\exp\left(\alpha - \beta\left(\frac{1}{X_t}\right)\right) \tag{2.245}$$

Fig. 2.12 Log-inverse model

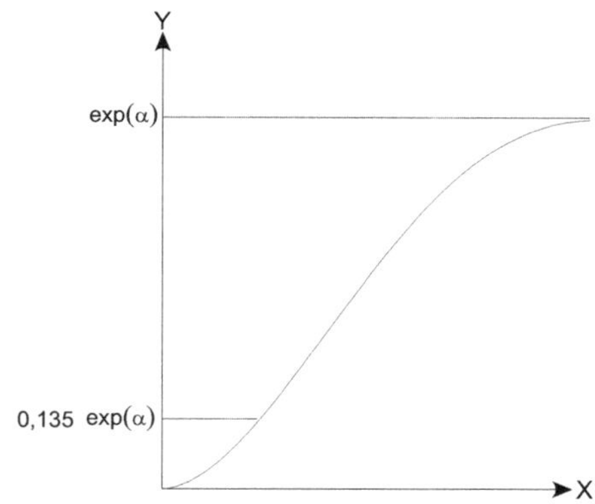

It is positive if $\beta > 0$. Moreover, the second derivative is written:

$$\frac{d^2Y_t}{dX_t^2} = \left(\frac{\beta^2}{X_t^4} - \frac{2\beta}{X_t^3}\right)\exp\left(\alpha - \beta\left(\frac{1}{X_t}\right)\right) \tag{2.246}$$

The cancellation of this second derivative, i.e., $\left(\frac{\beta}{X_t^4} - \frac{2\beta}{X_t^3}\right) = \beta^2 - 2\beta X_t = 0$, shows that there is an inflection point for $X_t = \beta/2$. Moreover, when X_t tends to infinity, Y_t tends to $exp\,(\alpha)$ by virtue of (2.244). By replacing X_t with $\beta/2$ in Eq. (2.244), the value of Y_t at the inflection point is given by:

$$Y_t = \exp(\alpha - 2) = 0.135\exp(\alpha) \tag{2.247}$$

As shown in Fig. 2.12, we see that, initially, Y_t grows at an increasing rate (the curve is convex), then, after the inflection point, the variable grows at a decreasing rate.

Remark 2.16 The log-reciprocal model corresponds to the case where $\lambda_Y = 0$ and $\lambda_X = -1$ in the Box-Cox transformation.

Conclusion

This chapter has presented the basic model of econometrics, namely, the simple regression model. In this model, only one explanatory variable is introduced. In practice, however, it is rare that a single variable can explain the behavior of the dependent variable. It is possible, then, to refine the study of the dynamics of the dependent variable by adding explanatory variables to the model. This is known as a multiple regression model. This model is the subject of the next chapter.

The Gist of the Chapter

Simple regression model	$Y_t = \alpha + \beta X_t + \varepsilon_t$
Variables	Explained (dependent) variable: Y_t
	Explanatory (independent) variable: X_t
	Error: ε_t
Hypotheses	Zero mean error: $E(\varepsilon_t) = 0 \ \forall t$
	Non-autocorrelation and homoskedasticity:
	$E(\varepsilon_t \varepsilon_{t'}) = \begin{cases} 0 & \forall t \neq t' \\ \sigma_\varepsilon^2 & \forall t = t' \end{cases}$
	Normality: $\varepsilon_t \sim N(0, \sigma_\varepsilon^2)$
Regression line	$\hat{Y}_t = \hat{\alpha} + \hat{\beta} X_t$
Residuals	$e_t = Y_t - \hat{Y}_t = Y_t - \hat{\alpha} - \hat{\beta} X_t$
OLS estimators	$\hat{\alpha} = \bar{Y} - \hat{\beta}\bar{X}$
	$\hat{\beta} = \frac{Cov(X_t, Y_t)}{V(X_t)}$
	$\hat{\sigma}_\varepsilon^2 = \frac{1}{T-2} \sum_{t=1}^{T} e_t^2$
t-Statistic	$t_{\hat{\beta}} = \frac{\hat{\beta}}{\hat{\sigma}_{\hat{\beta}}}$
Coefficient of determination	$R^2 = \frac{V(\hat{Y}_t)}{V(Y_t)} = \frac{\sum_{t=1}^{T} (\hat{Y}_t - \bar{Y})^2}{\sum_{t=1}^{T} (Y_t - \bar{Y})^2} = 1 - \frac{\sum_{t=1}^{T} e_t^2}{\sum_{t=1}^{T} (Y_t - \bar{Y})^2}$,
	$0 \leq R^2 \leq 1$

Further Reading

Developments on the linear regression model and the ordinary least squares method can be found in any econometrics textbook (see the references cited at the end of the book), including Johnston and Dinardo (1996), Davidson and MacKinnon (1993), or Greene (2020). For a more mathematical presentation, see, for example, Florens et al. (2007).

For further developments related to tests and laws, readers may refer to Lehnan (1959), Rao (1965), Kmenta (1971), Mood et al. (1974), or Hurlin and Mignon (2022).

For extensions of the linear regression model, interested readers can refer to Davidson and MacKinnon (1993) or Gujarati et al. (2017). Nonlinear regression models are discussed in Goldfeld and Quandt (1972), Gallant (1987), Pindyck and Rubinfeld (1991), Davidson and MacKinnon (1993), or Gujarati et al. (2017).

Appendix 2.1: Demonstrations

Appendix 2.1.1: Demonstration of the Linearity of the OLS Estimators

In order to demonstrate the linearity of the OLS estimators and in particular of $\hat{\beta}$, let us consider the centered variables:

$$x_t = X_t - \bar{X} \tag{2.248}$$

and

$$y_t = Y_t - \bar{Y} \tag{2.249}$$

In this case, the estimator $\hat{\beta}$ is given by:

$$\hat{\beta} = \frac{\sum_{t=1}^{T} x_t y_t}{\sum_{t=1}^{T} x_t^2} = \frac{\sum_{t=1}^{T} x_t (Y_t - \bar{Y})}{\sum_{t=1}^{T} x_t^2} = \frac{\sum_{t=1}^{T} x_t Y_t}{\sum_{t=1}^{T} x_t^2} - \bar{Y} \times \frac{\sum_{t=1}^{T} x_t}{\sum_{t=1}^{T} x_t^2} \tag{2.250}$$

Thus:

$$\sum_{t=1}^{T} x_t = \sum_{t=1}^{T} (X_t - \bar{X}) = \sum_{t=1}^{T} X_t - T\bar{X} = 0$$

Hence:[8]

$$\hat{\beta} = \frac{\sum_{t=1}^{T} x_t Y_t}{\sum_{t=1}^{T} x_t^2} = \sum_{t=1}^{T} w_t Y_t \tag{2.251}$$

with:

$$w_t = \frac{x_t}{\sum_{t=1}^{T} x_t^2} \tag{2.252}$$

The expression (2.251) reflects the fact that $\hat{\beta}$ is a **linear estimator** of β: $\hat{\beta}$ appears as a linear function of the dependent variable Y_t. We can also highlight

[8] Since X_t is nonrandom, so is x_t.

a certain number of characteristics of the weighting coefficients w_t that can be grouped under the following property.

Property 2.12 By virtue of the definition of w_t (Eq. (2.252)), we can write:

$$\sum_{t=1}^{T} w_t = \frac{\sum_{t=1}^{T} x_t}{\sum_{t=1}^{T} x_t^2} = 0 \tag{2.253}$$

In addition:

$$\sum_{t=1}^{T} w_t x_t = \sum_{t=1}^{T} w_t \left(X_t - \bar{X}\right) = \sum_{t=1}^{T} w_t X_t - \bar{X} \sum_{t=1}^{T} w_t = \sum_{t=1}^{T} w_t X_t \tag{2.254}$$

And:

$$\sum_{t=1}^{T} w_t x_t = \sum_{t=1}^{T} \frac{x_t}{\sum_{t=1}^{T} x_t^2} x_t = \frac{\sum_{t=1}^{T} x_t^2}{\sum_{t=1}^{T} x_t^2} = 1 \tag{2.255}$$

So:

$$\sum_{t=1}^{T} w_t x_t = \sum_{t=1}^{T} w_t X_t = 1 \tag{2.256}$$

We also have:

$$\sum_{t=1}^{T} w_t^2 = \sum_{t=1}^{T} \left(\frac{x_t}{\sum_{t=1}^{T} x_t^2} \right)^2 = \sum_{t=1}^{T} \frac{x_t^2}{\left(\sum_{t=1}^{T} x_t^2\right)^2} = \frac{1}{\sum_{t=1}^{T} x_t^2} \tag{2.257}$$

The linearity of the estimator $\hat{\alpha}$ can also be demonstrated by noting that:

$$\hat{\alpha} = \bar{Y} - \hat{\beta}\bar{X} = \frac{1}{T}\sum_{t=1}^{T} Y_t - \bar{X}\sum_{t=1}^{T} w_t Y_t \tag{2.258}$$

and using relation (2.251). We can therefore write:

$$\hat{\alpha} = \sum_{t=1}^{T} \left(\frac{1}{T} - \bar{X} w_t \right) Y_t \tag{2.259}$$

which shows that $\hat{\alpha}$ is a linear function of Y_t: $\hat{\alpha}$ is a linear estimator of α.

Appendix 2.1.2: Demonstration of the Unbiasedness Property of the OLS Estimators

Let us prove that $\hat{\beta}$ is an unbiased estimator of β, that is, $E\left(\hat{\beta}\right) = \beta$. From Eq. (2.251), we have:

$$\hat{\beta} = \sum_{t=1}^{T} w_t Y_t = \sum_{t=1}^{T} w_t \left(\alpha + \beta X_t + \varepsilon_t \right) = \alpha \sum_{t=1}^{T} w_t + \beta \sum_{t=1}^{T} w_t X_t + \sum_{t=1}^{T} w_t \varepsilon_t \tag{2.260}$$

By virtue of Eqs. (2.253) and (2.256), we can write:

$$\hat{\beta} = \beta + \sum_{t=1}^{T} w_t \varepsilon_t \tag{2.261}$$

Let us calculate the mathematical expectation of this expression:

$$E\left(\hat{\beta}\right) = E\left(\beta + \sum_{t=1}^{T} w_t \varepsilon_t \right) = \beta + E\left(\sum_{t=1}^{T} w_t \varepsilon_t \right) \tag{2.262}$$

This can also be written, noting that w_t is nonrandom:

$$E\left(\hat{\beta}\right) = \beta + \sum_{t=1}^{T} w_t E\left(\varepsilon_t\right) \tag{2.263}$$

Since $E\left(\varepsilon_t\right) = 0$, we deduce:

$$E\left(\hat{\beta}\right) = \beta \tag{2.264}$$

It follows that $\hat{\beta}$ is an unbiased estimator of β.

In order to show that $\hat{\alpha}$ is also an unbiased estimator of α, let us start again from the linearity property:

$$\hat{\alpha} = \sum_{t=1}^{T} \left(\frac{1}{T} - \bar{X} w_t \right) Y_t \tag{2.265}$$

that is:

$$\hat{\alpha} = \sum_{t=1}^{T} \left(\frac{1}{T} - \bar{X} w_t \right) (\alpha + \beta X_t + \varepsilon_t) \tag{2.266}$$

$$\hat{\alpha} = \alpha - \bar{X}\alpha \sum_{t=1}^{T} w_t + \beta \sum_{t=1}^{T} \frac{X_t}{T} - \bar{X}\beta \sum_{t=1}^{T} w_t X_t + \sum_{t=1}^{T} \left(\frac{1}{T} - \bar{X} w_t \right) \varepsilon_t \tag{2.267}$$

Given that, in accordance with Property 2.12, $\sum_{t=1}^{T} w_t = 0$ and $\sum_{t=1}^{T} w_t X_t = 1$, we deduce:

$$\hat{\alpha} = \alpha + \sum_{t=1}^{T} \left(\frac{1}{T} - \bar{X} w_t \right) \varepsilon_t \tag{2.268}$$

Let us take the mathematical expectation of this expression:

$$E\left(\hat{\alpha}\right) = E\left(\alpha + \sum_{t=1}^{T} \left(\frac{1}{T} - \bar{X} w_t \right) \varepsilon_t \right) \tag{2.269}$$

which can also be written as:

$$E\left(\hat{\alpha}\right) = \alpha + \sum_{t=1}^{T} \left(\frac{1}{T} - \bar{X} w_t \right) E\left(\varepsilon_t\right) \tag{2.270}$$

Since $E\left(\varepsilon_t\right) = 0$, we obtain the result we are looking for:

$$E\left(\hat{\alpha}\right) = \alpha \tag{2.271}$$

Appendix 2.1.3: Demonstration of the Consistency and Minimum Variance Property of the OLS Estimators

Let us start by showing that the OLS estimators $\hat{\alpha}$ and $\hat{\beta}$ are **consistent** estimators, that is, their variance tends to zero when T tends to infinity, i.e.:

$$\lim_{T\to\infty} V(\hat{\alpha}) = 0 \text{ and } \lim_{T\to\infty} V(\hat{\beta}) = 0 \tag{2.272}$$

Let us calculate $V(\hat{\beta})$. We have, by definition:

$$V(\hat{\beta}) = E\left[\hat{\beta} - E\left(\hat{\beta}\right)\right]^2 = E\left[\hat{\beta} - \beta\right]^2 \tag{2.273}$$

since $\hat{\beta}$ is an unbiased estimator of β. Using (2.262), we can write:

$$V(\hat{\beta}) = E\left(\sum_{t=1}^{T} w_t \varepsilon_t\right)^2 = E\left(\sum_{t=1}^{T} w_t^2 \varepsilon_t^2 + 2\sum_{t<t'} w_t w_{t'} \varepsilon_t \varepsilon_{t'}\right) \tag{2.274}$$

$$V(\hat{\beta}) = \sum_{t=1}^{T} w_t^2 E\left(\varepsilon_t^2\right) + 2\sum_{t<t'} w_t w_{t'} E\left(\varepsilon_t \varepsilon_{t'}\right) \tag{2.275}$$

We know that:

$$E\left(\varepsilon_t^2\right) = \sigma_\varepsilon^2 \tag{2.276}$$

where σ_ε^2 denotes the variance of the error, and:

$$E\left(\varepsilon_t \varepsilon_{t'}\right) = 0 \quad \forall t \neq t' \tag{2.277}$$

We deduce:

$$V(\hat{\beta}) = \sigma_\varepsilon^2 \sum_{t=1}^{T} w_t^2 \tag{2.278}$$

which can be written using (2.257):

$$V(\hat{\beta}) = \frac{\sigma_\varepsilon^2}{\sum_{t=1}^{T} x_t^2} \tag{2.279}$$

Given that $\sum_{t=1}^{T} x_t^2 = \sum_{t=1}^{T} \left(X_t - \bar{X}\right)^2 = TV\left(X_t\right)$, we deduce the following relationship:

$$V(\hat{\beta}) = \frac{\sigma_\varepsilon^2}{TV\left(X_t\right)} \tag{2.280}$$

Thus, if $T \to \infty$, then $V(\hat{\beta}) \to 0$, which implies that $\hat{\beta}$ is a consistent estimator. We also note that the variance of the estimator $\hat{\beta}$ is smaller when the variance of the explanatory variable is larger.

Let us now calculate $V(\hat{\alpha})$ to show that $\hat{\alpha}$ is also a consistent estimator. As before, using the fact that $\hat{\alpha}$ is an unbiased estimator, we have by definition:

$$V(\hat{\alpha}) = E\left[\hat{\alpha} - E\left(\hat{\alpha}\right)\right]^2 = E\left[\hat{\alpha} - \alpha\right]^2 \tag{2.281}$$

Using (2.268), we can write:

$$V(\hat{\alpha}) = E\left(\sum_{t=1}^{T}\left(\frac{1}{T} - \bar{X}w_t\right)\varepsilon_t\right)^2 \tag{2.282}$$

$$V(\hat{\alpha}) = E\left(\sum_{t=1}^{T}\left(\frac{1}{T} - \bar{X}w_t\right)^2 \varepsilon_t^2 + 2\sum_{t<t'}\left(\frac{1}{T} - \bar{X}w_t\right)\left(\frac{1}{T} - \bar{X}w_{t'}\right)\varepsilon_t\varepsilon_{t'}\right) \tag{2.283}$$

Or:

$$V(\hat{\alpha}) = \sum_{t=1}^{T}\left(\frac{1}{T} - \bar{X}w_t\right)^2 E\left(\varepsilon_t^2\right) + 2\sum_{t<t'}\left(\frac{1}{T} - \bar{X}w_t\right)\left(\frac{1}{T} - \bar{X}w_{t'}\right)E\left(\varepsilon_t\varepsilon_{t'}\right) \tag{2.284}$$

Hence:

$$V(\hat{\alpha}) = \sigma_\varepsilon^2 \sum_{t=1}^{T}\left(\frac{1}{T} - \bar{X}w_t\right)^2 = \sigma_\varepsilon^2 \sum_{t=1}^{T}\left(\frac{1}{T^2} - 2\frac{1}{T}\bar{X}w_t + \bar{X}^2 w_t^2\right) \tag{2.285}$$

$$V(\hat{\alpha}) = \sigma_\varepsilon^2\left(\frac{1}{T} + \bar{X}^2\sum_{t=1}^{T} w_t^2 - 2\frac{1}{T}\bar{X}\sum_{t=1}^{T} w_t\right) \tag{2.286}$$

So, using (2.257) and noting that $\sum_{t=1}^{T} x_t^2 = TV(X_t)$:

$$V(\hat{\alpha}) = \sigma_\varepsilon^2 \left(\frac{1}{T} + \frac{\bar{X}^2}{\sum_{t=1}^{T} x_t^2} \right) = \sigma_\varepsilon^2 \frac{\sum_{t=1}^{T} x_t^2 + T\bar{X}^2}{T \sum_{t=1}^{T} x_t^2} = \sigma_\varepsilon^2 \frac{\sum_{t=1}^{T} X_t^2}{T^2 V(X_t)} \tag{2.287}$$

Thus, if $T \to \infty$, then $V(\hat{\alpha}) \to 0$, which implies that $\hat{\alpha}$ is a consistent estimator.

It remains to be shown that the OLS estimators $\hat{\alpha}$ and $\hat{\beta}$ are estimators of minimum variance among the class of linear unbiased estimators. Let us start by treating the case of the estimator $\hat{\beta}$. In general and by definition, a linear estimator of β can be written as:

$$\beta^* = \sum_{t=1}^{T} \gamma_t Y_t \tag{2.288}$$

where the γ_t are weighting coefficients that must be determined. Given that $Y_t = \alpha + \beta X_t + \varepsilon_t$, we have:

$$\beta^* = \sum_{t=1}^{T} \gamma_t (\alpha + \beta X_t + \varepsilon_t) \tag{2.289}$$

that is:

$$\beta^* = \alpha \sum_{t=1}^{T} \gamma_t + \beta \sum_{t=1}^{T} \gamma_t X_t + \sum_{t=1}^{T} \gamma_t \varepsilon_t \tag{2.290}$$

Let us determine the mathematical expectation of this expression:

$$E(\beta^*) = E\left(\alpha \sum_{t=1}^{T} \gamma_t + \beta \sum_{t=1}^{T} \gamma_t X_t + \sum_{t=1}^{T} \gamma_t \varepsilon_t \right) \tag{2.291}$$

By distributing the expectation operator and using the fact that $E(\varepsilon_t) = 0$, we get:

$$E(\beta^*) = \alpha \sum_{t=1}^{T} \gamma_t + \beta \sum_{t=1}^{T} \gamma_t X_t \tag{2.292}$$

β^* is an unbiased estimator of β if $E(\beta^*) = \beta$, i.e., if:

$$\sum_{t=1}^{T} \gamma_t = 0 \tag{2.293}$$

and

$$\sum_{t=1}^{T} \gamma_t X_t = 1 \tag{2.294}$$

If these two conditions hold, we have according to (2.290):

$$\beta^* = \beta + \sum_{t=1}^{T} \gamma_t \varepsilon_t \tag{2.295}$$

and the variance is given by:

$$V(\beta^*) = E(\beta^* - \beta)^2 = E\left(\sum_{t=1}^{T} \gamma_t \varepsilon_t\right)^2 \tag{2.296}$$

By applying the same reasoning as that used to demonstrate the consistency of the estimator $\hat{\beta}$, we obtain:

$$V(\beta^*) = \sigma_\varepsilon^2 \sum_{t=1}^{T} \gamma_t^2 \tag{2.297}$$

We must therefore compare this variance with that of the OLS estimator, i.e., $V(\hat{\beta})$ given by:

$$V(\hat{\beta}) = \sigma_\varepsilon^2 \sum_{t=1}^{T} w_t^2 \tag{2.298}$$

To this end, let us posit:

$$\gamma_t = w_t + (\gamma_t - w_t) \tag{2.299}$$

We have:

$$\sum_{t=1}^{T} \gamma_t^2 = \sum_{t=1}^{T} (w_t + (\gamma_t - w_t))^2 = \sum_{t=1}^{T} w_t^2 + \sum_{t=1}^{T} (\gamma_t - w_t)^2 + 2\sum_{t=1}^{T} w_t (\gamma_t - w_t) \tag{2.300}$$

According to (2.257):

$$\sum_{t=1}^{T} w_t^2 = \frac{1}{\sum_{t=1}^{T} x_t^2} \tag{2.301}$$

and, in line with (2.252) and using the fact that $\sum_{t=1}^{T} \gamma_t X_t = \sum_{t=1}^{T} \gamma_t x_t = 1$:

$$\sum_{t=1}^{T} w_t \gamma_t = \frac{\sum_{t=1}^{T} x_t \gamma_t}{\sum_{t=1}^{T} x_t^2} = \frac{1}{\sum_{t=1}^{T} x_t^2} \tag{2.302}$$

According to (2.300), we have:

$$\begin{aligned} 2\sum_{t=1}^{T} w_t\left(\gamma_t - w_t\right) &= \sum_{t=1}^{T} \gamma_t^2 - \sum_{t=1}^{T} w_t^2 - \sum_{t=1}^{T} \left(\gamma_t - w_t\right)^2 \\ &= -2\sum_{t=1}^{T} w_t^2 + 2\sum_{t=1}^{T} w_t \gamma_t \end{aligned} \tag{2.303}$$

Hence:

$$\sum_{t=1}^{T} w_t\left(\gamma_t - w_t\right) = -\frac{1}{\sum_{t=1}^{T} x_t^2} + \frac{1}{\sum_{t=1}^{T} x_t^2} = 0 \tag{2.304}$$

We have:

$$V(\beta^*) = \sigma_\varepsilon^2 \sum_{t=1}^{T} \gamma_t^2 = \sigma_\varepsilon^2 \sum_{t=1}^{T} \left(w_t + \left(\gamma_t - w_t\right)\right)^2 = \sigma_\varepsilon^2 \left(\sum_{t=1}^{T} w_t^2 + \sum_{t=1}^{T} \left(\gamma_t - w_t\right)^2\right) \tag{2.305}$$

So, using (2.278):

$$V(\beta^*) = V(\hat{\beta}) + \sigma_\varepsilon^2 \left(\sum_{t=1}^{T} \left(\gamma_t - w_t\right)^2\right) \tag{2.306}$$

Since $\sum_{t=1}^{T} (\gamma_t - w_t)^2 \geq 0$, we have:

$$V(\beta^*) \geq V(\hat{\beta}) \tag{2.307}$$

It follows that, among the class of unbiased estimators, the OLS estimator $\hat{\beta}$ is the one with the lowest variance.

By applying similar reasoning, it is possible to show that the OLS estimator $\hat{\alpha}$ also satisfies the same property.

Appendix 2.1.4: Calculation of the Estimator of the Variance of the Error Term

We seek to determine an estimator $\hat{\sigma}_\varepsilon^2$ of the variance of the error term σ_ε^2. By definition, the residuals are given by:

$$e_t = Y_t - \hat{Y}_t = \alpha + \beta X_t + \varepsilon_t - \hat{\alpha} - \hat{\beta} X_t \tag{2.308}$$

$$e_t = \varepsilon_t - (\hat{\alpha} - \alpha) - \left(\hat{\beta} - \beta\right) X_t \tag{2.309}$$

We also have:

$$\bar{Y} = \frac{1}{T}\sum_{t=1}^{T} Y_t = \frac{1}{T}\sum_{t=1}^{T} (\alpha + \beta X_t + \varepsilon_t) = \alpha + \beta \bar{X} + \bar{\varepsilon} \tag{2.310}$$

and:

$$y_t = Y_t - \bar{Y} = \alpha + \beta X_t + \varepsilon_t - \left(\alpha + \beta \bar{X} + \bar{\varepsilon}\right) = \beta\left(X_t - \bar{X}\right) + (\varepsilon_t - \bar{\varepsilon}) \tag{2.311}$$

hence:

$$y_t = \beta x_t + (\varepsilon_t - \bar{\varepsilon}) \tag{2.312}$$

Given that:

$$e_t = y_t - \hat{y}_t = y_t - \hat{\beta} x_t = \beta x_t + (\varepsilon_t - \bar{\varepsilon}) - \hat{\beta} x_t \tag{2.313}$$

we deduce:

$$e_t = (\varepsilon_t - \bar{\varepsilon}) - \left(\hat{\beta} - \beta\right) x_t \tag{2.314}$$

In order to introduce the sum of squared residuals, which is necessary to calculate the estimator of the variance of the error term, this expression is squared and summed:

$$\sum_{t=1}^{T} e_t^2 = \left[\sum_{t=1}^{T} (\varepsilon_t - \bar{\varepsilon}) - \left(\hat{\beta} - \beta\right) x_t\right]^2 \tag{2.315}$$

which gives:

$$\sum_{t=1}^{T} e_t^2 = \sum_{t=1}^{T} (\varepsilon_t - \bar{\varepsilon})^2 + \left(\hat{\beta} - \beta\right)^2 \sum_{t=1}^{T} x_t^2 - 2\left(\hat{\beta} - \beta\right) \sum_{t=1}^{T} (\varepsilon_t - \bar{\varepsilon})\, x_t \tag{2.316}$$

Let us calculate the mathematical expectation of the different terms of this equation:

$$\begin{aligned} E\left(\sum_{t=1}^{T} (\varepsilon_t - \bar{\varepsilon})^2\right) &= E\left(\sum_{t=1}^{T} \varepsilon_t^2 - T\bar{\varepsilon}^2\right) \\ &= \sum_{t=1}^{T} E\left(\varepsilon_t^2\right) - \frac{1}{T}\left(\sum_{t=1}^{T} \varepsilon_t\right)^2 \\ &= \sum_{t=1}^{T} E\left(\varepsilon_t^2\right) - \frac{1}{T}\left(\sum_{t=1}^{T} E\left(\varepsilon_t^2\right) + 2\sum_{t \neq t'} E\left(\varepsilon_t \varepsilon_{t'}\right)\right)^2 \end{aligned} \tag{2.317}$$

Given that $E\left(\varepsilon_t \varepsilon_{t'}\right) = 0$ for $t \neq t'$, we have:

$$E\left(\sum_{t=1}^{T} (\varepsilon_t - \bar{\varepsilon})^2\right) = (T-1)\,\sigma_\varepsilon^2 \tag{2.318}$$

Now consider the second term of (2.316):

$$E\left(\left(\hat{\beta} - \beta\right)^2 \sum_{t=1}^{T} x_t^2\right) = \sum_{t=1}^{T} x_t^2 E\left(\hat{\beta} - \beta\right)^2 \tag{2.319}$$

We know that:

$$E\left(\hat{\beta} - \beta\right)^2 = V(\hat{\beta}) = \frac{\sigma_\varepsilon^2}{\sum\limits_{t=1}^{T} x_t^2} \tag{2.320}$$

Hence:

$$E\left(\left(\hat{\beta}-\beta\right)^2\sum_{t=1}^{T}x_t^2\right)=\sigma_\varepsilon^2 \tag{2.321}$$

In addition:

$$\left(\hat{\beta}-\beta\right)\sum_{t=1}^{T}\left(\varepsilon_t-\bar{\varepsilon}\right)x_t=\sum_{t=1}^{T}w_t\varepsilon_t\left(\sum_{t=1}^{T}\varepsilon_t x_t-\bar{\varepsilon}\sum_{t=1}^{T}x_t\right) \tag{2.322}$$

because $\hat{\beta}-\beta=\sum\limits_{t=1}^{T}w_t\varepsilon_t$ according to (2.261). Furthermore, by virtue of (2.252) and since $\sum\limits_{t=1}^{T}x_t=0$, we have:

$$\left(\hat{\beta}-\beta\right)\sum_{t=1}^{T}\left(\varepsilon_t-\bar{\varepsilon}\right)x_t=\sum_{t=1}^{T}\frac{x_t\varepsilon_t}{\sum\limits_{t=1}^{T}x_t^2}\sum_{t=1}^{T}\varepsilon_t x_t=\frac{\left(\sum\limits_{t=1}^{T}\varepsilon_t x_t\right)^2}{\sum\limits_{t=1}^{T}x_t^2} \tag{2.323}$$

Let us take the expectation of this expression:

$$\begin{aligned}E\left(\left(\hat{\beta}-\beta\right)\sum_{t=1}^{T}\left(\varepsilon_t-\bar{\varepsilon}\right)x_t\right)&=E\left(\frac{\left(\sum\limits_{t=1}^{T}\varepsilon_t x_t\right)^2}{\sum\limits_{t=1}^{T}x_t^2}\right)\\&=\frac{1}{\sum\limits_{t=1}^{T}x_t^2}E\left(\sum_{t=1}^{T}\varepsilon_t^2x_t^2+2\sum_{t\neq t'}\varepsilon_t\varepsilon_{t'}x_tx_{t'}\right)\end{aligned} \tag{2.324}$$

Noting that $E\left(\sum\limits_{t=1}^{T}\varepsilon_t^2x_t^2\right)=\sum\limits_{t=1}^{T}x_t^2E\left(\varepsilon_t^2\right)=\sigma_\varepsilon^2\sum\limits_{t=1}^{T}x_t^2$, we deduce:

$$E\left(\left(\hat{\beta}-\beta\right)\sum_{t=1}^{T}\left(\varepsilon_t-\bar{\varepsilon}\right)x_t\right)=\sigma_\varepsilon^2 \tag{2.325}$$

If we take Eq. (2.316), using relations (2.318), (2.321), and (2.325), we obtain:

$$E\left(\sum_{t=1}^{T} e_t^2\right) = (T-1)\,\sigma_\varepsilon^2 + \sigma_\varepsilon^2 - 2\sigma_\varepsilon^2 = (T-2)\,\sigma_\varepsilon^2 \tag{2.326}$$

We finally deduce the estimator $\hat{\sigma}_\varepsilon^2$ of the variance of the error term:

$$\hat{\sigma}_\varepsilon^2 = \frac{1}{T-2}\sum_{t=1}^{T} e_t^2 \tag{2.327}$$

This is an unbiased estimator of σ_ε^2.

Appendix 2.1.5: Calculation of the Standard Deviation of the Forecast Error and Prediction Interval

In order to determine a **prediction interval**, the variance of the forecast error must be calculated. We have:

$$\begin{aligned} V\left(e_{T+h}\right) &= E\left(\varepsilon_{T+h} - (\hat{\alpha}-\alpha) - \left(\hat{\beta}-\beta\right)X_{T+h}\right)^2 \\ &= E\left(\varepsilon_{T+h} - \left[(\hat{\alpha}-\alpha) + \left(\hat{\beta}-\beta\right)X_{T+h}\right]\right)^2 \end{aligned} \tag{2.328}$$

$$\begin{aligned} V\left(e_{T+h}\right) = {} & E\left(\varepsilon_{T+h}\right)^2 - 2E\left(\varepsilon_{T+h}\left[(\hat{\alpha}-\alpha) + \left(\hat{\beta}-\beta\right)X_{T+h}\right]\right) \\ & + E\left((\hat{\alpha}-\alpha) + \left(\hat{\beta}-\beta\right)X_{T+h}\right)^2 \end{aligned} \tag{2.329}$$

Knowing that $E\left(\varepsilon_{T+h}X_{T+h}\right) = 0$, $E\left(\hat{\alpha}-\alpha\right)^2 = V\left(\hat{\alpha}\right)$, $E\left(\hat{\beta}-\beta\right)^2 = V\left(\hat{\beta}\right)$ and $E\left((\hat{\alpha}-\alpha)\left(\hat{\beta}-\beta\right)\right) = E\left((\hat{\alpha}-E(\hat{\alpha}))\left(\hat{\beta}-E\left(\hat{\beta}\right)\right)\right) = cov\left(\hat{\alpha},\hat{\beta}\right)$, we have:

$$V\left(e_{T+h}\right) = V\left(\varepsilon_{T+h}\right) + V\left(\hat{\alpha}\right) + X_{T+h}^2 V\left(\hat{\beta}\right) + 2X_{T+h}cov\left(\hat{\alpha},\hat{\beta}\right) \tag{2.330}$$

Let us calculate the covariance between $\hat{\alpha}$ and $\hat{\beta}$:

$$cov\left(\hat{\alpha},\hat{\beta}\right) = E\left((\hat{\alpha}-\alpha)\left(\hat{\beta}-\beta\right)\right) = E\left(\left(\bar{Y} - \hat{\beta}\bar{X} - \bar{Y} + \beta\bar{X} + \bar{\varepsilon}\right)\left(\hat{\beta}-\beta\right)\right) \tag{2.331}$$

that is:

$$cov\left(\hat{\alpha},\hat{\beta}\right)=E\left(\left(\bar{\varepsilon}-\left(\hat{\beta}-\beta\right)\bar{X}\right)\left(\hat{\beta}-\beta\right)\right) \tag{2.332}$$

Since $E\left(\bar{\varepsilon}\left(\hat{\beta}-\beta\right)\right)=0$, we have:

$$cov\left(\hat{\alpha},\hat{\beta}\right)=-\bar{X}E\left(\hat{\beta}-\beta\right)^2 \tag{2.333}$$

From (2.279), we know that:

$$E\left(\hat{\beta}-\beta\right)^2=V(\hat{\beta})=\frac{\sigma_\varepsilon^2}{\sum_{t=1}^{T}x_t^2} \tag{2.334}$$

Hence:

$$cov\left(\hat{\alpha},\hat{\beta}\right)=-\bar{X}\frac{\sigma_\varepsilon^2}{\sum_{t=1}^{T}x_t^2} \tag{2.335}$$

Furthermore, according to (2.287):

$$V(\hat{\alpha})=\sigma_\varepsilon^2\left(\frac{1}{T}+\frac{\bar{X}^2}{\sum_{t=1}^{T}x_t^2}\right) \tag{2.336}$$

Transferring these various expressions into (2.330), we obtain:

$$V\left(e_{T+h}\right)=\sigma_\varepsilon^2\left(1+\frac{1}{T}+\frac{\bar{X}^2}{\sum_{t=1}^{T}x_t^2}+X_{T+h}^2\frac{1}{\sum_{t=1}^{T}x_t^2}-2X_{T+h}\bar{X}\frac{1}{\sum_{t=1}^{T}x_t^2}\right) \tag{2.337}$$

or:

$$V\left(e_{T+h}\right)=\sigma_\varepsilon^2\left(1+\frac{1}{T}+\frac{\left(X_{T+h}-\bar{X}\right)^2}{\sum_{t=1}^{T}x_t^2}\right)=\sigma_\varepsilon^2\left(1+\frac{1}{T}+\frac{\left(X_{T+h}-\bar{X}\right)^2}{\sum_{t=1}^{T}\left(X_t-\bar{X}\right)^2}\right) \tag{2.338}$$

The relationship (2.338) shows that the variance of the forecast error is an increasing function of the squared deviation between the value of the explanatory variable in $T + h$ and its mean. In other words, the larger the variance, i.e., the more the value of X in $T + h$ deviates from the mean, the higher the variance of the forecast error. The forecast error being a linear function of variables following normal distributions (relation (2.205)), it is normally distributed:

$$\frac{e_{T+h}}{\sigma_\varepsilon \sqrt{1 + \frac{1}{T} + \frac{(X_{T+h} - \bar{X})^2}{\sum_{t=1}^{T} x_t^2}}} \sim N(0, 1) \tag{2.339}$$

By replacing σ_ε with its estimator $\hat{\sigma}_\varepsilon$, we obtain:

$$\frac{e_{T+h}}{\hat{\sigma}_\varepsilon \sqrt{1 + \frac{1}{T} + \frac{(X_{T+h} - \bar{X})^2}{\sum_{t=1}^{T} x_t^2}}} \sim t(T - 2) \tag{2.340}$$

that is:

$$\frac{Y_{T+h} - \hat{Y}_{T+h}}{\hat{\sigma}_\varepsilon \sqrt{1 + \frac{1}{T} + \frac{(X_{T+h} - \bar{X})^2}{\sum_{t=1}^{T} x_t^2}}} \sim t(T - 2) \tag{2.341}$$

with $\hat{Y}_{T+h} = \hat{\alpha} + \hat{\beta} X_{T+h}$. We deduce a $100(1 - p)\%$ prediction interval for Y_{T+h}:

$$\left(\hat{\alpha} + \hat{\beta} X_{T+h}\right) \pm t_{p/2} \hat{\sigma}_\varepsilon \sqrt{1 + \frac{1}{T} + \frac{(X_{T+h} - \bar{X})^2}{\sum_{t=1}^{T} x_t^2}} \tag{2.342}$$

Appendix 2.2: Normal Distribution and Normality Test

The **normal distribution** is the most widespread statistical distribution. The density function of a random variable x following a general normal distribution is given by:

$$p(x) = \frac{1}{\sqrt{2\pi}\sigma} \exp\left(-\frac{1}{2}\left(\frac{x - m}{\sigma}\right)^2\right) \tag{2.343}$$

where exp is the exponential, m is the mean of the variable x, and σ its standard deviation. We note:

$$x \sim N\left(m, \sigma^2\right) \tag{2.344}$$

The "bell curve" that represents it passes through a maximum at $x = m$ and is symmetrical with respect to the vertical line passing through this point. In practice, a particularly important case is that of the **standard normal distribution**. The density function of a random variable z following a standard normal distribution is given by:

$$p(z) = \frac{1}{\sqrt{2\pi}} \exp\left(-\frac{1}{2}z^2\right) \tag{2.345}$$

It is always possible to return to the case of the standard normal distribution by transforming the variable: $z = \frac{x-m}{\sigma}$. We note:

$$z \sim N(0, 1) \tag{2.346}$$

Consider a series X_t, $t = 1, \ldots, T$. The normal distribution is characterized by a **skewness** coefficient equal to zero and a **kurtosis** coefficient of 3. Skewness is a measure of the asymmetry of a distribution, and kurtosis is a measure that quantifies the shape of a distribution by providing information about its tails and peakedness compared to a normal distribution. The skewness coefficient, noted S, associated with the series X_t is given by:

$$S = \frac{\left[\frac{1}{T}\sum_{t=1}^{T}\left(X_t - \bar{X}\right)^3\right]^2}{\left[\frac{1}{T}\sum_{t=1}^{T}\left(X_t - \bar{X}\right)^2\right]^3} = \frac{\mu_3^2}{\mu_2^3} \tag{2.347}$$

and the kurtosis coefficient, denoted K, is written:

$$K = \frac{\frac{1}{T}\sum_{t=1}^{T}\left(X_t - \bar{X}\right)^4}{\left[\frac{1}{T}\sum_{t=1}^{T}\left(X_t - \bar{X}\right)^2\right]^2} = \frac{\mu_4}{\mu_2^2} \tag{2.348}$$

where $\bar{X}$ is the mean of the series X_t, $t = 1, \ldots, T$, and the μ_i are the centered moments of order i. For a normal distribution, we have:

$$\begin{cases} \mu_3 = 0 \\ \mu_4 = 3\mu_2^2 \end{cases} \tag{2.349}$$

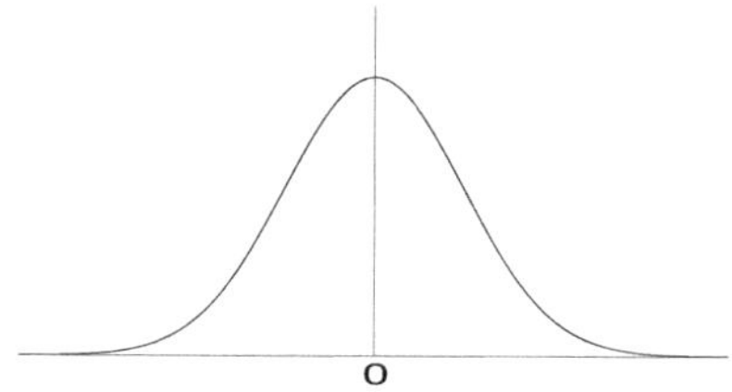

Fig. 2.13 Density function of the normal distribution

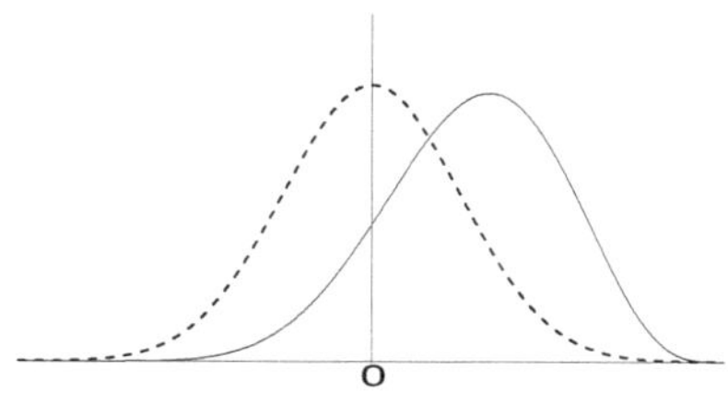

Fig. 2.14 Asymmetric distribution (spread to the left)

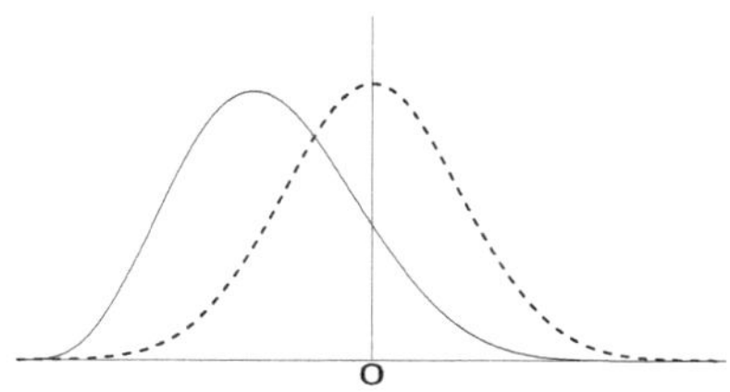

Fig. 2.15 Asymmetric distribution (spread to the right)

that is:

$$\begin{cases} S = 0 \\ K = 3 \end{cases} \tag{2.350}$$

When $S \neq 0$, the distribution of the series is said to be **asymmetric**. More specifically, when $S < 0$, the distribution is skewed to the right (or spread to the left), and when $S < 0$ it is skewed to the left (or spread to the right). Figures 2.13, 2.14, and 2.15 represent the case of a series following a normal distribution (symmetric distribution), a series whose distribution is skewed to the right, and a series with a left-skewed distribution, respectively. On these last two graphs, the dotted curve is the one relating to the normal distribution.

Concerning the kurtosis coefficient, if $K < 3$, the distribution is said to be flattened or **platykurtic**. If $K > 3$, the distribution presents an excess of kurtosis; we speak about **leptokurtic** distribution. Figure 2.13 shows the normal distribution ($K = 3$), Fig. 2.16 the case where $K < 3$, and Fig. 2.17 the case where $K > 3$. On these last two graphs, the dotted curve represents the normal distribution.

The **Jarque and Bera test** (Jarque and Bera, 1980) is based on the definition of skewness and kurtosis coefficients and allows us to test the null hypothesis of

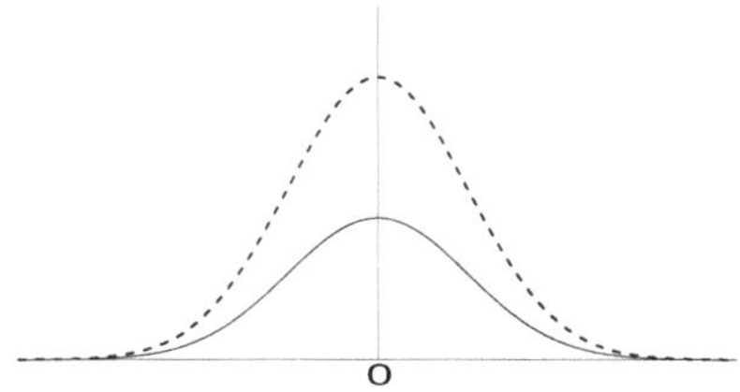

Fig. 2.16 Platykurtic distribution

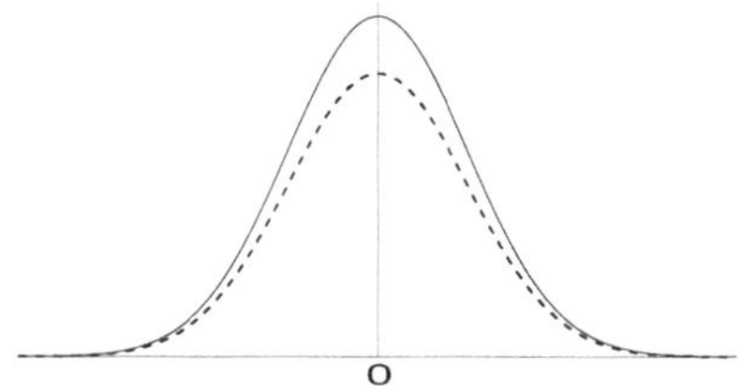

Fig. 2.17 Leptokurtic distribution

normality of a distribution. The test statistic, denoted JB, is written as:

$$JB = \frac{T}{6}\left[S^2 + \frac{1}{4}(K-3)^2\right] \tag{2.351}$$

Under the null hypothesis of normality, the test statistic follows a Chi-squared distribution with 2 degrees of freedom. Therefore, if the calculated value of the JB test statistic is lower than the theoretical value of the Chi-squared distribution with 2 degrees of freedom, the null hypothesis of normality is not rejected. On the other hand, if JB is greater than the critical value, the null hypothesis of normality is rejected.

This Jarque and Bera test is used to test the normality of the residuals.

Appendix 2.3: The Maximum Likelihood Method

The **maximum likelihood method** is another technique for estimating a regression model. It is assumed that the error ε_t is normally distributed with mean zero and variance σ_ε^2. In the regression model:

$$Y_t = \alpha + \beta X_t + \varepsilon_t \tag{2.352}$$

Y_t is a linear function of the error term. Consequently, Y_t is also normally distributed with mean:

$$E(Y_t) = \alpha + \beta X_t \tag{2.353}$$

and variance:

$$V(Y_t) = \sigma_\varepsilon^2 \tag{2.354}$$

We note $f\left(Y_1, Y_2, \ldots, Y_T \left| \alpha + \beta X_t, \sigma_\varepsilon^2\right.\right)$ the **joint probability density function** of $(Y_1, Y_2, \ldots, Y_T)$. Since the Y_t are supposed to be independent, we can write:

$$f\left(Y_1, Y_2, \ldots, Y_T \left| \alpha + \beta X_t, \sigma_\varepsilon^2\right.\right) = f\left(Y_1 \left| \alpha + \beta X_t, \sigma_\varepsilon^2\right.\right) \times f\left(Y_2 \left| \alpha + \beta X_t, \sigma_\varepsilon^2\right.\right) \times \ldots \times f\left(Y_T \left| \alpha + \beta X_t, \sigma_\varepsilon^2\right.\right) \tag{2.355}$$

with:

$$f(Y_t) = \frac{1}{\sqrt{2\pi}\sigma_\varepsilon} \exp\left(-\frac{1}{2}\left(\frac{Y_t - \alpha - \beta X_t}{\sigma_\varepsilon}\right)^2\right) \tag{2.356}$$

which is the density function of the general normal distribution.

By substituting (2.356) into (2.355), we get:

$$f\left(Y_1, Y_2, \ldots, Y_T \left| \alpha + \beta X_t, \sigma_\varepsilon^2\right.\right) = \frac{1}{\left(\sqrt{2\pi}\right)^T \sigma_\varepsilon^T} \exp\left(-\frac{1}{2}\sum_{t=1}^{T}\left(\frac{Y_t - \alpha - \beta X_t}{\sigma_\varepsilon}\right)^2\right) \tag{2.357}$$

Assuming that the $(Y_1, Y_2, \ldots, Y_T)$ are known, expression (2.357) is called the **likelihood function**. It is noted:

$$L\left(\alpha, \beta, \sigma_\varepsilon^2\right) = \frac{1}{\left(\sqrt{2\pi}\right)^T \sigma_\varepsilon^T} \exp\left(-\frac{1}{2}\sum_{t=1}^{T}\left(\frac{Y_t - \alpha - \beta X_t}{\sigma_\varepsilon}\right)^2\right) \tag{2.358}$$

The maximum likelihood method consists in maximizing the latter expression in order to obtain the estimators of α, β, and σ_ε^2. To this end, we write the log-likelihood function in logarithms:

$$\ln L = -T \ln \sigma_\varepsilon - \frac{T}{2} \ln (2\pi) - \frac{1}{2}\sum_{t=1}^{T}\left(\frac{Y_t - \alpha - \beta X_t}{\sigma_\varepsilon}\right)^2 \tag{2.359}$$

which can be expressed as follows:

$$\ln L = -\frac{T}{2}\ln\sigma_{\varepsilon}^{2} - \frac{T}{2}\ln(2\pi) - \frac{1}{2}\sum_{t=1}^{T}\left(\frac{Y_t - \alpha - \beta X_t}{\sigma_{\varepsilon}}\right)^{2} \tag{2.360}$$

This expression is called the **log-likelihood**. We differentiate Eq. (2.360) with respect to α, β, and σ_{ε}^{2}, i.e.:

$$\frac{\partial \ln L}{\partial \alpha} = -\sum_{t=1}^{T}\left(\frac{Y_t - \alpha - \beta X_t}{\sigma_{\varepsilon}^{2}}\right)(-1) \tag{2.361}$$

$$\frac{\partial \ln L}{\partial \beta} = -\sum_{t=1}^{T}\left(\frac{Y_t - \alpha - \beta X_t}{\sigma_{\varepsilon}^{2}}\right)(-X_t) \tag{2.362}$$

$$\frac{\partial \ln L}{\partial \sigma_{\varepsilon}^{2}} = -\frac{T}{2\sigma_{\varepsilon}^{2}} + \frac{1}{2}\sum_{t=1}^{T}\left(\frac{Y_t - \alpha - \beta X_t}{\sigma_{\varepsilon}^{2}}\right)^{2} \tag{2.363}$$

Equalizing each of these equations to zero (first-order optimization condition) and denoting by ML the maximum likelihood estimators, we obtain:

$$\sum_{t=1}^{T}\left(Y_t - \hat{\alpha}_{ML} - \hat{\beta}_{ML}X_t\right) = 0 \tag{2.364}$$

$$\sum_{t=1}^{T}\left(Y_t - \hat{\alpha}_{ML} - \hat{\beta}_{ML}X_t\right)X_t = 0 \tag{2.365}$$

$$-T + \frac{1}{\hat{\sigma}_{\varepsilon,ML}^{2}}\sum_{t=1}^{T}\left(Y_t - \hat{\alpha}_{ML} - \hat{\beta}_{ML}X_t\right)^{2} = 0 \tag{2.366}$$

Equations (2.364) and (2.365) can be written:

$$\sum_{t=1}^{T}Y_t = T\hat{\alpha}_{ML} + \hat{\beta}_{ML}\sum_{t=1}^{T}X_t \tag{2.367}$$

$$\sum_{t=1}^{T}X_tY_t = \hat{\alpha}_{ML}\sum_{t=1}^{T}X_t + \hat{\beta}_{ML}\sum_{t=1}^{T}X_t^{2} \tag{2.368}$$

Equations (2.367) and (2.368) correspond exactly to Eqs. (2.44) and (2.48). The maximum likelihood estimators of the coefficients α and β are therefore identical to the OLS estimators.

Let us now determine the estimator of the variance of the error term:

$$\begin{aligned} \hat{\sigma}^2_{\varepsilon,ML} &= \frac{1}{T}\sum_{t=1}^{T}\left(Y_t - \hat{\alpha}_{ML} - \hat{\beta}_{ML}X_t\right)^2 \\ &= \frac{1}{T}\sum_{t=1}^{T}\left(Y_t - \hat{\alpha} - \hat{\beta}X_t\right)^2 \\ &= \frac{1}{T}\sum_{t=1}^{T} e_t^2 \end{aligned} \tag{2.369}$$

The maximum likelihood estimator of the variance of the error is therefore different from that obtained by OLS (Eq. (2.327)):

$$\hat{\sigma}^2_{\varepsilon} = \frac{1}{T-2}\sum_{t=1}^{T} e_t^2 \tag{2.370}$$

Since the OLS estimator is an unbiased estimator, it follows that the maximum likelihood estimator is a biased estimator. This is a consistent estimator and the bias decreases as the sample size increases.

The Multiple Regression Model 3

The simple regression model studied in Chap. 2 had only one explanatory variable. In practice, however, it is common for an (explained) variable to depend on several explanatory variables.

As in the previous chapter, let us take a few examples to illustrate the questions to be answered in this chapter. Does a family's consumption expenditure depend more on its income or its size? To what extent is the quantity demanded of a particular good a function of the price of that good, the price of other goods, and consumer income? Do wages depend more on the level of education or work experience?

In these different cases where several explanatory variables come into play, we speak of a **multiple regression model**. This chapter proposes an in-depth study of this model.[1]

3.1 Writing the Model in Matrix Form

The multiple regression model with k explanatory variables is written as:

$$Y_t = \alpha + \beta_1 X_{1t} + \beta_2 X_{2t} + \ldots + \beta_k X_{kt} + \varepsilon_t \tag{3.1}$$

where $t = 1, \ldots, T$, Y_t is the explained (or dependent) variable, $X_{1t}, X_{2t}, \ldots, X_{kt}$ are the k explanatory variables, and ε_t is the error term.

This model thus corresponds to an extension of the simple regression model to the case of k explanatory variables with $k > 1$.

The coefficients $\beta_1, \ldots, \beta_k$ are called **partial regression coefficients** or **partial slope coefficients.** β_1 measures the change in the mean value of Y, with the value of the other explanatory variables remaining constant (i.e., all other things being

[1] This chapter calls upon various notions of matrix algebra. In Appendix 3.1, readers will find the elements of matrix algebra necessary to understand the various developments here.

V. Mignon, *Principles of Econometrics*, Classroom Companion: Economics,
https://doi.org/10.1007/978-3-031-52535-3_3

equal). The regression coefficient β_1 therefore measures the effect of a 1-point variation in X_{1t} on the mean value of Y_t, this effect being net of any potential influence of the other explanatory variables on the mean value of Y_t. The same type of reasoning obviously applies to the other regression coefficients.

We can write Eq. (3.1) for each value of t, $t = 1, \ldots, T$:

$$\begin{aligned} Y_1 &= \alpha + \beta_1 X_{11} + \beta_2 X_{21} + \ldots + \beta_k X_{k1} + \varepsilon_1 \\ Y_2 &= \alpha + \beta_1 X_{12} + \beta_2 X_{22} + \ldots + \beta_k X_{k2} + \varepsilon_2 \\ &\vdots \\ Y_t &= \alpha + \beta_1 X_{1t} + \beta_2 X_{2t} + \ldots + \beta_k X_{kt} + \varepsilon_t \\ &\vdots \\ Y_T &= \alpha + \beta_1 X_{1T} + \beta_2 X_{2T} + \ldots + \beta_k X_{kT} + \varepsilon_T \end{aligned} \tag{3.2}$$

Or, in matrix form:

$$\begin{pmatrix} Y_1 \\ Y_2 \\ \vdots \\ Y_t \\ \vdots \\ Y_T \end{pmatrix} = \begin{pmatrix} 1 & X_{11} & X_{21} & \cdots & X_{k1} \\ 1 & X_{12} & X_{22} & \cdots & X_{k2} \\ \vdots & \vdots & \vdots & \ldots & \vdots \\ 1 & X_{1t} & X_{2t} & \cdots & X_{kt} \\ \vdots & \vdots & \vdots & \ldots & \vdots \\ 1 & X_{1T} & X_{2T} & \cdots & X_{kT} \end{pmatrix} \begin{pmatrix} \alpha \\ \beta_1 \\ \beta_2 \\ \vdots \\ \beta_k \end{pmatrix} + \begin{pmatrix} \varepsilon_1 \\ \varepsilon_2 \\ \vdots \\ \varepsilon_t \\ \vdots \\ \varepsilon_T \end{pmatrix} \tag{3.3}$$

The multiple regression model can then be written:[2]

$$\underset{(T,1)}{\boldsymbol{Y}} = \underset{(T,k+1)}{\boldsymbol{X}} \underset{(k+1,1)}{\boldsymbol{\beta}} + \underset{(T,1)}{\boldsymbol{\varepsilon}} \tag{3.4}$$

where:

- $\boldsymbol{Y}$ is the column vector containing the values of the dependent variable: $\boldsymbol{Y} = \begin{pmatrix} Y_1 \\ Y_2 \\ \vdots \\ Y_t \\ \vdots \\ Y_T \end{pmatrix}$

[2] Matrices and vectors are written in **bold** characters. This notation convention will be used throughout the book.

– $\boldsymbol{X}$ is the matrix whose first column is composed of 1 and each of whose subsequent columns contains the values of each of the k explanatory variables:
$\boldsymbol{X} = \begin{pmatrix} 1 & X_{11} & X_{21} & \cdots & X_{k1} \\ 1 & X_{12} & X_{22} & \cdots & X_{k2} \\ \vdots & \vdots & \vdots & \cdots & \vdots \\ 1 & X_{1t} & X_{2t} & \cdots & X_{kt} \\ \vdots & \vdots & \vdots & \cdots & \vdots \\ 1 & X_{1T} & X_{2T} & \cdots & X_{kT} \end{pmatrix}$. Each explanatory variable, X_{it}, $i = 1, \ldots, k$, is thus represented as a column vector of T observations of the variable X_i.

– $\boldsymbol{\beta}$ is the vector of parameters to be estimated: $\boldsymbol{\beta} = \begin{pmatrix} \alpha \\ \beta_1 \\ \beta_2 \\ \vdots \\ \beta_k \end{pmatrix}$

– $\boldsymbol{\varepsilon}$ is the column vector containing the values of the error term: $\boldsymbol{\varepsilon} = \begin{pmatrix} \varepsilon_1 \\ \varepsilon_2 \\ \vdots \\ \varepsilon_t \\ \vdots \\ \varepsilon_T \end{pmatrix}$

The first column of the matrix $\boldsymbol{X}$ contains only 1s in order to take into account the constant α. This allows us to keep a compact matrix form, making it easier to present the developments linked to the multiple regression model.

As in the case of the simple regression model, the aim is to obtain an estimate of the parameter vector $\boldsymbol{\beta}$. To this end, the ordinary least squares (OLS) method is applied.

3.2 The OLS Estimators

3.2.1 Assumptions of the Multiple Regression Model

Like the simple regression model, the multiple regression model is based on a number of assumptions, which we present below.

Hypothesis 1: The Matrix X Is Nonrandom

For the sake of simplification, it is frequently assumed that the values of the variables in the matrix $\boldsymbol{X}$ are observed without error, i.e., that the matrix $\boldsymbol{X}$ is nonrandom. This assumption may seem strong,[3] but it amounts to assuming that the explanatory variables are controlled, which considerably simplifies the derivation of some fundamental statistical results. It is thus a purely technical assumption, allowing us to consider each vector of the matrix $\boldsymbol{X}$ as a known constant for the probability distribution of Y_t.

It is possible to relax this assumption about the nonrandom character of the matrix $\boldsymbol{X}$ and assume that it is independent of each value of the error term. Such an assumption can be written as:

$$E(\varepsilon_t | \boldsymbol{X}) = \begin{pmatrix} E(\varepsilon_1 | \boldsymbol{X}) \\ E(\varepsilon_2 | \boldsymbol{X}) \\ \vdots \\ E(\varepsilon_T | \boldsymbol{X}) \end{pmatrix} = 0 \tag{3.5}$$

where the notation $E(a | b)$ denotes the expectation of a conditional on b. Thus, the expectation of the error term conditional on the values of the explanatory variables is zero. This is equivalent to assuming that the explanatory variables do not provide any information about the mean value of the error term. They are not involved in predicting it.

Hypothesis 2: The Matrix X Is of Full Rank

Assuming that the matrix $\boldsymbol{X}$ is of **full rank** is equivalent to writing:

$$Rank(\boldsymbol{X}) = k + 1 \tag{3.6}$$

This assumption states that the explanatory variables are linearly independent.[4] Such an assumption of independence among the explanatory variables is necessary for estimating the parameter vector $\boldsymbol{\beta}$.

If the number of observations T is less than $k + 1$, then the matrix $\boldsymbol{X}$ cannot be of full rank. For this reason, we assume that the number of observations is greater than the number of explanatory variables, i.e.:

$$T > k + 1 \tag{3.7}$$

Hypothesis 3: The Expectation of the Error Term Is Zero

As in the case of the simple regression model, the error term is assumed to have zero expectation. The reasoning is the same: the error term can take on both negative and positive values, and there is no bias in favor of positive values, nor in favor of

[3] In the sense that the matrix of explanatory variables is assumed to be unchanged whatever the sample of observations.

[4] We will see later that such an assumption implies that there is no **collinearity** between the explanatory variables.

negative ones. We deduce that the mathematical expectation of the errors is zero:

$$E(\boldsymbol{\varepsilon}) = \begin{pmatrix} E(\varepsilon_1) \\ E(\varepsilon_2) \\ \vdots \\ E(\varepsilon_T) \end{pmatrix} = 0 \tag{3.8}$$

Hence:

$$E(\boldsymbol{Y}) = \boldsymbol{X\beta} \tag{3.9}$$

Hypothesis 4: Homoskedasticity and the Absence of Autocorrelation of Errors

Let $E(\boldsymbol{\varepsilon\varepsilon}')$ be the variance-covariance matrix of the error term with $\boldsymbol{\varepsilon}'$ denoting the transpose of $\boldsymbol{\varepsilon}$. The assumption of homoskedasticity—i.e., constancy of variance—and of absence of autocorrelation is written:

$$E(\boldsymbol{\varepsilon\varepsilon}') = \sigma_\varepsilon^2 \boldsymbol{I} \tag{3.10}$$

where $\boldsymbol{I}$ denotes the identity matrix and σ_ε^2 the variance of the error term.

To understand expression (3.10), let us write the variance-covariance matrix of the error term:

$$E(\boldsymbol{\varepsilon\varepsilon}') = \begin{pmatrix} V(\varepsilon_1) & Cov(\varepsilon_1, \varepsilon_2) & \cdots & Cov(\varepsilon_1, \varepsilon_T) \\ Cov(\varepsilon_2, \varepsilon_1) & V(\varepsilon_2) & \cdots & Cov(\varepsilon_2, \varepsilon_T) \\ \vdots & \vdots & \ddots & \vdots \\ Cov(\varepsilon_T, \varepsilon_1) & Cov(\varepsilon_T, \varepsilon_2) & \cdots & V(\varepsilon_T) \end{pmatrix} \tag{3.11}$$

$$= \begin{pmatrix} E(\varepsilon_1^2) & E(\varepsilon_1\varepsilon_2) & \cdots & E(\varepsilon_1\varepsilon_T) \\ E(\varepsilon_2\varepsilon_1) & E(\varepsilon_2^2) & \cdots & E(\varepsilon_2\varepsilon_T) \\ \vdots & \vdots & \ddots & \vdots \\ E(\varepsilon_T\varepsilon_1) & E(\varepsilon_T\varepsilon_2) & \cdots & E(\varepsilon_T^2) \end{pmatrix}$$

Under the assumption of no autocorrelation, all terms off the diagonal are zero. In accordance with the assumption of homoskedasticity, the terms on the diagonal are constant and equal to σ_ε^2. We therefore have:

$$E(\boldsymbol{\varepsilon\varepsilon}') = \begin{pmatrix} \sigma_\varepsilon^2 & 0 & \cdots & 0 \\ 0 & \sigma_\varepsilon^2 & \cdots & 0 \\ \vdots & \vdots & \ddots & \vdots \\ 0 & 0 & \cdots & \sigma_\varepsilon^2 \end{pmatrix} = \sigma_\varepsilon^2 \begin{pmatrix} 1 & 0 & \cdots & 0 \\ 0 & 1 & \cdots & 0 \\ \vdots & \vdots & \ddots & \vdots \\ 0 & 0 & \cdots & 1 \end{pmatrix} = \sigma_\varepsilon^2 \boldsymbol{I} \tag{3.12}$$

Recall that non-autocorrelated and homoskedastic errors are said to be spherical.

Hypothesis 5: Normality of Errors
Errors are assumed to follow a normal distribution of zero mean and constant variance, i.e.:

$$\boldsymbol{\varepsilon} \sim N\left(0, \sigma_{\varepsilon}^{2}\boldsymbol{I}\right) \tag{3.13}$$

As in the case of the simple regression model, the normality assumption is not necessary to establish the results of the multiple regression model. However, it allows us to derive statistical results and construct test statistics (see below).

3.2.2 Estimation of Coefficients

Consider the multiple regression model:

$$\boldsymbol{Y} = \boldsymbol{X\beta} + \boldsymbol{\varepsilon} \tag{3.14}$$

Our objective is to estimate the vector $\boldsymbol{\beta}$ of parameters by the OLS method. This vector $\hat{\boldsymbol{\beta}}$ of estimated parameters is given by:

$$\hat{\boldsymbol{\beta}} = \left(\boldsymbol{X}'\boldsymbol{X}\right)^{-1}\boldsymbol{X}'\boldsymbol{Y} \tag{3.15}$$

Let us demonstrate this relationship. Consider the vector $\boldsymbol{e}$ of residuals:

$$\boldsymbol{e} = \boldsymbol{Y} - \boldsymbol{X}\hat{\boldsymbol{\beta}} \tag{3.16}$$

where $\hat{\boldsymbol{\beta}}$ denotes the vector of parameters estimated by OLS.

Recall that the OLS method consists in minimizing the sum of squared errors. Since the errors are not observable, we minimize the sum of squared residuals:

$$Min \sum_{t=1}^{T} \varepsilon_t^2 \equiv Min\left(\boldsymbol{e}'\boldsymbol{e}\right) \tag{3.17}$$

$$\begin{aligned} \boldsymbol{e}'\boldsymbol{e} &= \left(\boldsymbol{Y} - \boldsymbol{X}\hat{\boldsymbol{\beta}}\right)'\left(\boldsymbol{Y} - \boldsymbol{X}\hat{\boldsymbol{\beta}}\right) \\ &= \boldsymbol{Y}'\boldsymbol{Y} - \hat{\boldsymbol{\beta}}'\boldsymbol{X}'\boldsymbol{Y} - \boldsymbol{Y}'\boldsymbol{X}\hat{\boldsymbol{\beta}} + \hat{\boldsymbol{\beta}}'\boldsymbol{X}'\boldsymbol{X}\hat{\boldsymbol{\beta}} \end{aligned} \tag{3.18}$$

$\hat{\boldsymbol{\beta}}'\boldsymbol{X}'\boldsymbol{Y}$ and $\boldsymbol{Y}'\boldsymbol{X}\hat{\boldsymbol{\beta}}$ are scalars. Knowing that a scalar is equal to its transpose:

$$\hat{\boldsymbol{\beta}}'\boldsymbol{X}'\boldsymbol{Y} = \left(\hat{\boldsymbol{\beta}}'\boldsymbol{X}'\boldsymbol{Y}\right)' = \boldsymbol{Y}'\boldsymbol{X}\hat{\boldsymbol{\beta}} \tag{3.19}$$

we deduce:

$$\boldsymbol{e}'\boldsymbol{e} = \boldsymbol{Y}'\boldsymbol{Y} - 2\hat{\boldsymbol{\beta}}'\boldsymbol{X}'\boldsymbol{Y} + \hat{\boldsymbol{\beta}}'\boldsymbol{X}'\boldsymbol{X}\hat{\boldsymbol{\beta}} \tag{3.20}$$

We seek to minimize this expression, i.e.:

$$\frac{\partial\left(\boldsymbol{e}'\boldsymbol{e}\right)}{\partial\hat{\boldsymbol{\beta}}} = 0 \tag{3.21}$$

$$\frac{\partial\left(\boldsymbol{e}'\boldsymbol{e}\right)}{\partial\hat{\boldsymbol{\beta}}} = -2\frac{\partial\left(\hat{\boldsymbol{\beta}}'\boldsymbol{X}'\boldsymbol{Y}\right)}{\partial\hat{\boldsymbol{\beta}}} + \frac{\partial\left(\hat{\boldsymbol{\beta}}'\boldsymbol{X}'\boldsymbol{X}\hat{\boldsymbol{\beta}}\right)}{\partial\hat{\boldsymbol{\beta}}} \tag{3.22}$$

We have:

$$\frac{\partial\left(\hat{\boldsymbol{\beta}}'\boldsymbol{X}'\boldsymbol{Y}\right)}{\partial\hat{\boldsymbol{\beta}}} = \boldsymbol{X}'\boldsymbol{Y} \tag{3.23}$$

because $\frac{\partial(\boldsymbol{A}'\boldsymbol{B})}{\partial\boldsymbol{B}} = \frac{\partial(\boldsymbol{B}'\boldsymbol{A})}{\partial\boldsymbol{B}} = \boldsymbol{A}$ where $\boldsymbol{A}$ and $\boldsymbol{B}$ denote vectors (of dimension $(k+1, 1)$ in our case).

Moreover:

$$\frac{\partial\left(\hat{\boldsymbol{\beta}}'\boldsymbol{X}'\boldsymbol{X}\hat{\boldsymbol{\beta}}\right)}{\partial\hat{\boldsymbol{\beta}}} = 2\left(\boldsymbol{X}'\boldsymbol{X}\right)\hat{\boldsymbol{\beta}} \tag{3.24}$$

since $\frac{\partial(\boldsymbol{B}'\boldsymbol{C}\boldsymbol{B})}{\partial\boldsymbol{B}} = 2\boldsymbol{C}\boldsymbol{B}$ where $\boldsymbol{C}$ denotes a symmetric matrix and $\boldsymbol{B}$ a vector. Relationship (3.22) is therefore written as:

$$\frac{\partial\left(\boldsymbol{e}'\boldsymbol{e}\right)}{\partial\hat{\boldsymbol{\beta}}} = -2\boldsymbol{X}'\boldsymbol{Y} + 2\left(\boldsymbol{X}'\boldsymbol{X}\right)\hat{\boldsymbol{\beta}} = 0 \tag{3.25}$$

Hence:

$$\left(\boldsymbol{X}'\boldsymbol{X}\right)\hat{\boldsymbol{\beta}} = \boldsymbol{X}'\boldsymbol{Y} \tag{3.26}$$

Insofar as the matrix $\boldsymbol{X}$ is of full rank, it follows that the matrix $\left(\boldsymbol{X}'\boldsymbol{X}\right)$ is also of rank $k+1$, implying that its inverse exists. Consequently, we deduce from Eq. (3.26) the expression giving the vector $\hat{\boldsymbol{\beta}}$ of parameters estimated by OLS:

$$\hat{\boldsymbol{\beta}} = \left(\boldsymbol{X}'\boldsymbol{X}\right)^{-1}\boldsymbol{X}'\boldsymbol{Y} \tag{3.27}$$

Remark 3.1 As in the case of the simple regression model, it is possible to estimate the multiple regression model by the **maximum likelihood method**. It can be shown that the maximum likelihood estimator of the parameter vector $\boldsymbol{\beta}$ is identical to that obtained with OLS. Furthermore, as in the simple regression model, the maximum likelihood estimator of the error variance is biased, unlike that of OLS. For a detailed presentation of the maximum likelihood method, see Greene (2020).

3.2.3 Properties of OLS Estimators

Linearity of the Estimator

We have, according to relationship (3.27):

$$\hat{\boldsymbol{\beta}} = \left(\boldsymbol{X}'\boldsymbol{X}\right)^{-1} \boldsymbol{X}'\boldsymbol{Y} = \left(\boldsymbol{X}'\boldsymbol{X}\right)^{-1} \boldsymbol{X}' \left(\boldsymbol{X}\boldsymbol{\beta} + \boldsymbol{\varepsilon}\right) \tag{3.28}$$

that is:

$$\hat{\boldsymbol{\beta}} = \left(\boldsymbol{X}'\boldsymbol{X}\right)^{-1} \boldsymbol{X}'\boldsymbol{X}\boldsymbol{\beta} + \left(\boldsymbol{X}'\boldsymbol{X}\right)^{-1} \boldsymbol{X}'\boldsymbol{\varepsilon} \tag{3.29}$$

Hence:

$$\hat{\boldsymbol{\beta}} = \boldsymbol{\beta} + \left(\boldsymbol{X}'\boldsymbol{X}\right)^{-1} \boldsymbol{X}'\boldsymbol{\varepsilon} \tag{3.30}$$

This last equation shows that $\hat{\boldsymbol{\beta}}$ is a linear estimator of $\boldsymbol{\beta}$.

Unbiased Estimator

From Eq. (3.30), we can write:

$$E\left(\hat{\boldsymbol{\beta}}\right) = E\left(\boldsymbol{\beta}\right) + \left(\boldsymbol{X}'\boldsymbol{X}\right)^{-1} \boldsymbol{X}'E\left(\boldsymbol{\varepsilon}\right) \tag{3.31}$$

Given that $E\left(\boldsymbol{\beta}\right) = \boldsymbol{\beta}$ and $E\left(\boldsymbol{\varepsilon}\right) = 0$, we deduce:

$$E\left(\hat{\boldsymbol{\beta}}\right) = \boldsymbol{\beta} \tag{3.32}$$

which means that $\hat{\boldsymbol{\beta}}$ is an unbiased estimator of $\boldsymbol{\beta}$.

Variance-Covariance Matrix of Coefficients

Let $\boldsymbol{\Omega}_{\hat{\beta}}$ be the variance-covariance matrix of OLS coefficients, i.e., of $\hat{\boldsymbol{\beta}}$, given by:

$$\boldsymbol{\Omega}_{\hat{\beta}} = E\left[\left(\hat{\boldsymbol{\beta}} - \boldsymbol{\beta}\right)\left(\hat{\boldsymbol{\beta}} - \boldsymbol{\beta}\right)'\right] \tag{3.33}$$

that is:

$$\boldsymbol{\Omega}_{\hat{\boldsymbol{\beta}}} = \begin{pmatrix} V(\hat{\alpha}) & Cov(\hat{\alpha}, \hat{\beta}_1) & Cov(\hat{\alpha}, \hat{\beta}_2) & \cdots & Cov(\hat{\alpha}, \hat{\beta}_k) \\ Cov(\hat{\beta}_1, \hat{\alpha}) & V(\hat{\beta}_1) & Cov(\hat{\beta}_1, \hat{\beta}_2) & \cdots & Cov(\hat{\beta}_1, \hat{\beta}_k) \\ Cov(\hat{\beta}_2, \hat{\alpha}) & Cov(\hat{\beta}_2, \hat{\beta}_1) & V(\hat{\beta}_2) & \cdots & \vdots \\ \vdots & \vdots & \vdots & \ddots & \vdots \\ Cov(\hat{\beta}_k, \hat{\alpha}) & \cdots & \cdots & \cdots & V(\hat{\beta}_k) \end{pmatrix} \tag{3.34}$$

This matrix is symmetric. Furthermore, using relation (3.30), we can write (3.33) as follows:

$$\boldsymbol{\Omega}_{\hat{\boldsymbol{\beta}}} = E\left[\left(\left(\boldsymbol{X}'\boldsymbol{X}\right)^{-1}\boldsymbol{X}'\boldsymbol{\varepsilon}\right)\left(\left(\boldsymbol{X}'\boldsymbol{X}\right)^{-1}\boldsymbol{X}'\boldsymbol{\varepsilon}\right)'\right] \tag{3.35}$$

or:

$$\boldsymbol{\Omega}_{\hat{\boldsymbol{\beta}}} = E\left[\left(\boldsymbol{X}'\boldsymbol{X}\right)^{-1}\boldsymbol{X}'\boldsymbol{\varepsilon}\boldsymbol{\varepsilon}'\boldsymbol{X}\left(\boldsymbol{X}'\boldsymbol{X}\right)^{-1}\right] \tag{3.36}$$

because $\left(\boldsymbol{X}'\boldsymbol{X}\right)^{-1}$ is a symmetric matrix, implying: $\left(\left(\boldsymbol{X}'\boldsymbol{X}\right)^{-1}\right)' = \left(\boldsymbol{X}'\boldsymbol{X}\right)^{-1}$.

Knowing that $E\left(\boldsymbol{\varepsilon}\boldsymbol{\varepsilon}'\right) = \sigma_{\varepsilon}^2\boldsymbol{I}$, we deduce:

$$\boldsymbol{\Omega}_{\hat{\boldsymbol{\beta}}} = \left(\boldsymbol{X}'\boldsymbol{X}\right)^{-1}\boldsymbol{X}'\sigma_{\varepsilon}^2\boldsymbol{I}\boldsymbol{X}\left(\boldsymbol{X}'\boldsymbol{X}\right)^{-1} \tag{3.37}$$

Hence:

$$\boldsymbol{\Omega}_{\hat{\boldsymbol{\beta}}} = \sigma_{\varepsilon}^2\left(\boldsymbol{X}'\boldsymbol{X}\right)^{-1} \tag{3.38}$$

This determination of the variance-covariance matrix will allow us to show that $\hat{\boldsymbol{\beta}}$ is of minimum variance.

Minimum Variance Estimator

It can be shown (see Appendix 3.2.1) that, among the class of linear unbiased estimators, the OLS estimator $\hat{\boldsymbol{\beta}}$ has the smallest variance. We can therefore state the following property, as in the case of the simple regression model.

Property 3.1 The OLS estimator $\hat{\boldsymbol{\beta}}$ of $\boldsymbol{\beta}$ is the best linear unbiased estimator (**BLUE**).

3.2.4 Error Variance Estimation

The error is, of course, unknown in the model. In order to estimate the variance σ_ε^2 of the errors, we need to use the residuals $\boldsymbol{e}$:

$$\boldsymbol{e} = \boldsymbol{Y} - \boldsymbol{X}\hat{\boldsymbol{\beta}} \tag{3.39}$$

From this expression and after calculating $E\left(\boldsymbol{e}'\boldsymbol{e}\right)$, we can show that the estimator $\hat{\sigma}_\varepsilon^2$ of the error variance is written (see Appendix 3.2.2):

$$\hat{\sigma}_\varepsilon^2 = \frac{\boldsymbol{e}'\boldsymbol{e}}{T-k-1} \equiv \frac{1}{T-k-1}\sum_{t=1}^{T} e_t^2 \tag{3.40}$$

This is an unbiased estimator of σ_ε^2.

3.2.5 Example

Consider the following example comprising one explained variable Y_t and three explanatory variables X_{1t}, X_{2t}, and X_{3t} for $t = 1, \ldots, 6$ (see Table 3.1). This example is only given for illustrative purposes to put into practice the concepts previously presented, since it obviously makes little sense to carry out a 6-point regression.

The model is written as:

$$Y_t = \alpha + \beta_1 X_{1t} + \beta_2 X_{2t} + \beta_3 X_{3t} + \varepsilon_t \tag{3.41}$$

or, in matrix form:

$$\boldsymbol{Y} = \boldsymbol{X}\boldsymbol{\beta} + \boldsymbol{\varepsilon} \tag{3.42}$$

Table 3.1 Illustrative example

t	Y_t	X_{1t}	X_{2t}	X_{3t}
1	4	3	5	4
2	2	5	6	8
3	1	7	3	9
4	3	2	2	5
5	6	9	1	2
6	8	4	7	3
Sum	24	30	24	31

with:

$$Y = \begin{pmatrix} 4 \\ 2 \\ 1 \\ 3 \\ 6 \\ 8 \end{pmatrix}, \boldsymbol{X} = \begin{pmatrix} 1\,3\,5\,4 \\ 1\,5\,6\,8 \\ 1\,7\,3\,9 \\ 1\,2\,2\,5 \\ 1\,9\,1\,2 \\ 1\,4\,7\,3 \end{pmatrix}, \boldsymbol{\beta} = \begin{pmatrix} \alpha \\ \beta_1 \\ \beta_2 \\ \beta_3 \end{pmatrix} \text{ and } \boldsymbol{\varepsilon} = \begin{pmatrix} \varepsilon_1 \\ \varepsilon_2 \\ \varepsilon_3 \\ \varepsilon_4 \\ \varepsilon_5 \\ \varepsilon_6 \end{pmatrix}$$

Determination of OLS Estimators

By writing the transpose of $\boldsymbol{X}$:

$$\boldsymbol{X}' = \begin{pmatrix} 1 & 1 & 1 & 1 & 1 & 1 \\ 3 & 5 & 7 & 2 & 9 & 4 \\ 5 & 6 & 3 & 2 & 1 & 7 \\ 4 & 58 & 9 & 5 & 2 & 3 \end{pmatrix} \tag{3.43}$$

we can calculate the matrix product $\boldsymbol{X}'\boldsymbol{X}$:

$$\boldsymbol{X}'\boldsymbol{X} = \begin{pmatrix} 6 & 30 & 24 & 31 \\ 30 & 184 & 107 & 155 \\ 24 & 107 & 124 & 128 \\ 31 & 155 & 128 & 199 \end{pmatrix} \tag{3.44}$$

We deduce the inverse matrix:

$$\left(\boldsymbol{X}'\boldsymbol{X}\right)^{-1} = \begin{pmatrix} 2.87 & -0.24 & -0.24 & -0.11 \\ -0.24 & 0.04 & 0.02 & -0.002 \\ -0.24 & 0.02 & 0.04 & -0.004 \\ -0.11 & -0.002 & -0.004 & 0.03 \end{pmatrix} \tag{3.45}$$

Calculating the product $\boldsymbol{X}'\boldsymbol{Y}$ gives:

$$\boldsymbol{X}'\boldsymbol{Y} = \begin{pmatrix} 24 \\ 121 \\ 103 \\ 92 \end{pmatrix} \tag{3.46}$$

Finally:

$$\left(\boldsymbol{X}'\boldsymbol{X}\right)^{-1}\boldsymbol{X}'\boldsymbol{Y} = \begin{pmatrix} 5.57 \\ 0.21 \\ 0.47 \\ -0.87 \end{pmatrix} = \begin{pmatrix} \hat{\alpha} \\ \hat{\beta}_1 \\ \hat{\beta}_2 \\ \hat{\beta}_3 \end{pmatrix} \tag{3.47}$$

The estimated model is therefore:

$$\hat{Y}_t = 5.57 + 0.21X_{1t} + 0.47X_{2t} - 0.87X_{3t} \tag{3.48}$$

Practical Calculation

In practice, it is possible to use certain matrix algebra results to simplify calculations. Thus, in a model with three explanatory variables, such as the one in this example, the matrix $\left(\boldsymbol{X}'\boldsymbol{X}\right)$ is given by:

$$\boldsymbol{X}'\boldsymbol{X} = \begin{pmatrix} T & \sum X_{1t} & \sum X_{2t} & \sum X_{3t} \\ \sum X_{1t} & \sum X_{1t}^2 & \sum X_{1t}X_{2t} & \sum X_{1t}X_{3t} \\ \sum X_{2t} & \sum X_{1t}X_{2t} & \sum X_{2t}^2 & \sum X_{2t}X_{3t} \\ \sum X_{3t} & \sum X_{1t}X_{3t} & \sum X_{2t}X_{3t} & \sum X_{3t}^2 \end{pmatrix} \tag{3.49}$$

and the matrix $\boldsymbol{X}'\boldsymbol{Y}$ is written as:

$$\boldsymbol{X}'\boldsymbol{Y} = \begin{pmatrix} \sum Y_t \\ \sum Y_t X_{1t} \\ \sum Y_t X_{2t} \\ \sum Y_t X_{3t} \end{pmatrix} \tag{3.50}$$

Calculating these various sums gives the following matrices:

$$\boldsymbol{X}'\boldsymbol{X} = \begin{pmatrix} 6 & 30 & 24 & 31 \\ 30 & 184 & 107 & 155 \\ 24 & 107 & 124 & 128 \\ 31 & 155 & 128 & 199 \end{pmatrix} \tag{3.51}$$

and:

$$\boldsymbol{X}'\boldsymbol{Y} = \begin{pmatrix} 24 \\ 121 \\ 103 \\ 92 \end{pmatrix} \tag{3.52}$$

3.3 Tests on the Regression Coefficients

3.3.1 Distribution of Estimators

Assume that the errors are normally distributed:

$$\boldsymbol{\varepsilon} \sim N\left(0, \sigma_\varepsilon^2 \boldsymbol{I}\right) \tag{3.53}$$

By virtue of Relationship (3.30):

$$\hat{\boldsymbol{\beta}} = \boldsymbol{\beta} + \left(\boldsymbol{X}'\boldsymbol{X}\right)^{-1} \boldsymbol{X}'\boldsymbol{\varepsilon} \tag{3.54}$$

$\hat{\boldsymbol{\beta}}$ is a linear function of $\boldsymbol{\varepsilon}$ and therefore also follows a normal distribution with $(k+1)$ dimensions. Furthermore, we have shown that the variance-covariance matrix is given by:

$$\boldsymbol{\Omega}_{\hat{\boldsymbol{\beta}}} = \sigma_\varepsilon^2 \left(\boldsymbol{X}'\boldsymbol{X}\right)^{-1} \tag{3.55}$$

So we have:

$$\hat{\boldsymbol{\beta}} \sim N\left(\boldsymbol{\beta}, \sigma_\varepsilon^2 \left(\boldsymbol{X}'\boldsymbol{X}\right)^{-1}\right) \tag{3.56}$$

Consider the coefficient $\hat{\beta}_i$ associated with the ith explanatory variable X_{it}. We have:

$$\hat{\beta}_i \sim N\left(\beta_i, \sigma_\varepsilon^2 a_{i+1,i+1}\right) \tag{3.57}$$

where $a_{i+1,i+1}$ denotes the $(i+1)$th element[5] of the diagonal of $\left(\boldsymbol{X}'\boldsymbol{X}\right)^{-1}$. We can write:

$$\frac{\hat{\beta}_i - \beta_i}{\sigma_\varepsilon \sqrt{a_{i+1,i+1}}} \sim N(0, 1) \tag{3.58}$$

We have also seen that:

$$\hat{\sigma}_\varepsilon^2 = \frac{\boldsymbol{e}'\boldsymbol{e}}{T-k-1} \tag{3.59}$$

[5] We consider the $(i+1)$th element and not the ith since the first element of the matrix $\left(\boldsymbol{X}'\boldsymbol{X}\right)^{-1}$ relates to the constant term. Obviously, if the variables are centered, it is appropriate to choose the ith element.

Using the result that if $\boldsymbol{w} \sim N\left(0, \sigma_w^2 \boldsymbol{I}\right)$, $\frac{\boldsymbol{w}'\boldsymbol{w}}{\sigma_w^2}$ follows a Chi-squared distribution, we have:

$$\frac{\boldsymbol{e}'\boldsymbol{e}}{\sigma_\varepsilon^2} \sim \chi^2_{T-k-1} \tag{3.60}$$

that is:

$$\frac{(T-k-1)\hat{\sigma}_\varepsilon^2}{\sigma_\varepsilon^2} \sim \chi^2_{T-k-1} \tag{3.61}$$

Remembering the property that (see Box 2.2 in Chap. 2) if $z \sim N(0, 1)$ and $v \sim \chi_r^2$, the quantity:

$$t = \frac{z\sqrt{r}}{\sqrt{v}} \tag{3.62}$$

follows a Student's t distribution with r degrees of freedom, we have:

$$t = \frac{\frac{\hat{\beta}_i - \beta_i}{\sigma_\varepsilon \sqrt{a_{i+1,i+1}}}\sqrt{T-k-1}}{\sqrt{\frac{(T-k-1)\hat{\sigma}_\varepsilon^2}{\sigma_\varepsilon^2}}} \tag{3.63}$$

or finally:

$$t = \frac{\hat{\beta}_i - \beta_i}{\hat{\sigma}_\varepsilon \sqrt{a_{i+1,i+1}}} \sim t(T-k-1) \tag{3.64}$$

for $i = 1, \ldots, k+1$. This result allows us to proceed with tests on β_i.

3.3.2 Tests on a Regression Coefficient

From relationship (3.64), we can construct a $100(1-p)\%$ confidence interval for β_i, i.e.:

$$\hat{\beta}_i \pm t_{p/2}\hat{\sigma}_\varepsilon \sqrt{a_{i+1,i+1}} \tag{3.65}$$

As in the case of the simple regression model, it is possible to test the null hypothesis that β_i is equal to a certain value β_0, i.e.:

$$H_0 : \beta_i = \beta_0 \tag{3.66}$$

against the alternative hypothesis:

$$H_1 : \beta_i \neq \beta_0 \tag{3.67}$$

If the null hypothesis is true, then:

$$\frac{\hat{\beta}_i - \beta_0}{\hat{\sigma}_\varepsilon \sqrt{a_{i+1,i+1}}} \sim t(T-k-1) \tag{3.68}$$

and the decision rule is given by:

- If $\left|\frac{\hat{\beta}_i - \beta_0}{\hat{\sigma}_\varepsilon \sqrt{a_{i+1,i+1}}}\right| \leq t_{p/2}$: the null hypothesis is not rejected at the $100p\%$ significance level, so $\beta_i = \beta_0$.
- If $\left|\frac{\hat{\beta}_i - \beta_0}{\hat{\sigma}_\varepsilon \sqrt{a_{i+1,i+1}}}\right| > t_{p/2}$: the null hypothesis is rejected at the $100p\%$ significance level, so $\beta_i \neq \beta_0$.

In practice, the most commonly used test consists in testing the null hypothesis:

$$H_0 : \beta_i = 0 \tag{3.69}$$

against the alternative hypothesis:

$$H_0 : \beta_i \neq 0 \tag{3.70}$$

This is a **coefficient significance test** (or ***t*-test**): under the null hypothesis, the coefficient associated with the variable X_{it} is not significant, i.e., this variable plays no role in determining the dependent variable Y_t. In practical terms, we proceed as in the case of the simple regression model, i.e., we calculate the t-statistic of the coefficient $\hat{\beta}_i$:

$$t_{\hat{\beta}_i} = \frac{\hat{\beta}_i}{\widehat{\sigma_{\hat{\beta}_i}}} \tag{3.71}$$

where $\widehat{\sigma_{\hat{\beta}_i}}$ denotes the estimate of the standard deviation of the coefficient $\hat{\beta}_i$, i.e.:

$$\widehat{\sigma_{\hat{\beta}_i}} = \hat{\sigma}_\varepsilon \sqrt{a_{i+1,i+1}} \tag{3.72}$$

The decision rule associated with the significance test of the coefficient β_i is written:

- If $\left|t_{\hat{\beta}_i}\right| \leq t_{p/2}$: the null hypothesis is not rejected at the $100p\%$ significance level, so $\beta_i = 0$—the variable X_{it} does not contribute to the explanation of Y_t.
- If $\left|t_{\hat{\beta}_i}\right| > t_{p/2}$: the null hypothesis is rejected at the $100p\%$ significance level, so $\beta_i \neq 0$—the coefficient associated with the variable X_{it} is significantly different from zero, indicating that X_{it} contributes to the explanation of the dependent variable Y_t.

3.3.3 Significance Tests of Several Coefficients

We have previously presented the t-test, which allows us to test the significance of each of the regression coefficients in isolation. It is also possible to simultaneously test the significance of several or even all the coefficients of the estimated model. For this purpose, we use a Fisher test.

Assume that the elements of the vector $\boldsymbol{\beta}$ are subject to q constraints:

$$\boldsymbol{R}\boldsymbol{\beta} = \boldsymbol{r} \tag{3.73}$$

where $\boldsymbol{R}$ is a given matrix of size $(q, k+1)$ and $\boldsymbol{r}$ is the vector of constraints of dimension $(q, 1)$. It is further assumed that $q \leq k+1$ and that the matrix $\boldsymbol{R}$ is of full rank, meaning that the q constraints are linearly independent.

Equation (3.73) is very general. We will show that various tests can be carried out based on this expression.

Test on a Particular Regression Coefficient

The case previously presented of the test of significance of one of the coefficients β_i corresponds to:

$$\boldsymbol{R} = [0 \cdots 0\ 1\ 0 \cdots 0] \text{ and } \boldsymbol{r} = 0 \tag{3.74}$$

The matrix $\boldsymbol{R}$ contains a 1 in the $(i+1)$th place[6] and $\boldsymbol{r}$ is null, which amounts to testing the significance of the coefficient β_i. If we set $\boldsymbol{r} = \beta_0$, we find the test of equality of β_i at a certain value β_0 presented earlier.

[6] Recall that the first element of the matrix corresponds to the constant term, which explains why we consider the $(i+1)$th element and not the ith.

Test of Equality of Coefficients

From Eq. (3.73), it is possible to test the equality of two (or more) coefficients. For example, writing:

$$\boldsymbol{R} = [0\,0\,1\,0\; -1\,0 \cdots 0] \text{ and } \boldsymbol{r} = 0 \tag{3.75}$$

allows us to test:[7]

$$\beta_2 - \beta_4 = 0 \tag{3.76}$$

that is:

$$\beta_2 = \beta_4 \tag{3.77}$$

Similarly, the relationship:

$$\boldsymbol{R} = [0\,0\,1\,0\,1\,0 \cdots 0] \text{ and } \boldsymbol{r} = 0 \tag{3.78}$$

is equivalent to performing the test:

$$\beta_2 + \beta_4 = 0 \Longleftrightarrow \beta_2 = -\beta_4 \tag{3.79}$$

Significance Test for All Coefficients

It is also possible, starting from (3.73), to perform a test of significance of all the coefficients of the explanatory variables of the regression. Such a test is called **test of regression significance**. It corresponds to the following case:

$$\boldsymbol{R} = \begin{pmatrix} 0 & 1 & 0 & \cdots & 0 \\ 0 & 0 & 1 & \cdots & 0 \\ \vdots & \vdots & \vdots & \ddots & \vdots \\ 0 & \cdots & \cdots & 0 & 1 \end{pmatrix} \text{ and } \boldsymbol{r} = \begin{pmatrix} 0 \\ 0 \\ \vdots \\ 0 \end{pmatrix} \tag{3.80}$$

This gives:

$$\begin{pmatrix} 0 & 1 & 0 & \cdots & 0 \\ 0 & 0 & 1 & \cdots & 0 \\ \vdots & \vdots & \vdots & \ddots & \vdots \\ 0 & \cdots & \cdots & 0 & 1 \end{pmatrix} \begin{pmatrix} \alpha \\ \beta_1 \\ \vdots \\ \vdots \\ \beta_k \end{pmatrix} = \begin{pmatrix} 0 \\ 0 \\ \vdots \\ 0 \end{pmatrix} \tag{3.81}$$

[7] The first element of the matrix corresponding to the constant α.

Here we test $\beta_1 = \beta_2 = \cdots = \beta_k = 0$. This involves testing the null hypothesis that no coefficient is significant, i.e., that none of the k explanatory variables contributes to explaining the dependent variable Y_t. This test is particularly helpful and frequently used.

Significance Test of a Subset of Coefficients

It is possible to test the significance of a subset of explanatory variables, which corresponds to the following case:

$$\boldsymbol{R} = [\,\boldsymbol{0}\ \boldsymbol{I}_s\,] \text{ and } \boldsymbol{r} = \boldsymbol{0} \tag{3.82}$$

where $\boldsymbol{r}$ is a column vector with s elements and $\boldsymbol{0}$ is a null matrix of size $(s, k+1-s)$. We test $\beta_{k-s+2} = \beta_{k-s+3} = \cdots = \beta_k = 0$. This involves testing the null hypothesis that the last s elements of the vector $\boldsymbol{\beta}$ are insignificant.

We can thus see that expression (3.73) brings together a large number of tests, which are presented in detail in Appendix 3.2.3. We now propose a synthesis.

Synthesis

The various tests mentioned above are **Fisher tests**. They consist in considering two regression models: an **unconstrained model** and a **constrained model**. The unconstrained model involves regressing the dependent variable on all the explanatory variables. The constrained model involves regressing the dependent variable on just some of the explanatory variables. It is called a constrained model insofar as a constraint is imposed on one or more coefficients included in the regression. Generally speaking, Fisher tests are written:

$$F = \frac{(RSS_c - RSS_{nc})\,/q}{RSS_{nc}/\,(T-k-1)} \sim F\,(q, T-k-1) \tag{3.83}$$

where RSS_{nc} is the residual sum of squares of the unconstrained model and RSS_c denotes the residual sum of squares of the constrained model, q being the number of constraints.

Remark 3.2 The t-test, which consists in testing the significance of a single coefficient, can also be interpreted as a Fisher test. Indeed, the test of the null hypothesis $\beta_i = 0$ amounts to:

- Regressing the dependent variable on all the explanatory variables: unconstrained model
- Regressing the dependent variable on all the explanatory variables except X_i: constrained model

3.4 Analysis of Variance (ANOVA) and Adjusted Coefficient of Determination

After estimating the regression parameters and testing the significance of the coefficients, it is necessary—as in the case of the simple regression model—to assess the goodness of fit. In other words, we study whether the scatter plot is well represented by the regression line, by analyzing whether the scatter is concentrated or, on the contrary, dispersed around the line. This can be done by using the analysis of variance of the regression and calculating the coefficient of determination.

3.4.1 Analysis-of-Variance Equation

Case of Centered Variables

Consider the centered variables:

$$\begin{cases} y_t = Y_t - \bar{Y} \\ x_{1t} = X_{1t} - \bar{X}_1 \\ \dots \\ x_{it} = X_{it} - \bar{X}_i \\ \dots \\ x_{kt} = X_{kt} - \bar{X}_k \end{cases} \tag{3.84}$$

The multiple regression model is then written as:

$$\boldsymbol{y} = \boldsymbol{x\beta} + \boldsymbol{\varepsilon} \tag{3.85}$$

where:

- $\boldsymbol{y} = \begin{pmatrix} y_1 \\ \vdots \\ y_T \end{pmatrix}$
- $\boldsymbol{x} = \begin{pmatrix} x_{11} & \cdots & x_{k1} \\ \vdots & \ddots & \\ x_{1T} & & x_{kT} \end{pmatrix}$
- $\boldsymbol{\beta} = \begin{pmatrix} \beta_1 \\ \vdots \\ \beta_k \end{pmatrix}$
- and $\boldsymbol{\varepsilon} = \begin{pmatrix} \varepsilon_1 \\ \vdots \\ \varepsilon_T \end{pmatrix}$

As in the case of the simple regression model, we will establish the analysis-of-variance equation by starting from the expression of the residuals:

$$\boldsymbol{e} = \boldsymbol{y} - \boldsymbol{x}\hat{\boldsymbol{\beta}} \tag{3.86}$$

Hence:

$$\boldsymbol{e}'\ \boldsymbol{e} = \left(\boldsymbol{y} - \boldsymbol{x}\hat{\boldsymbol{\beta}}\right)'\left(\boldsymbol{y} - \boldsymbol{x}\hat{\boldsymbol{\beta}}\right) = \boldsymbol{y}'\boldsymbol{y} - 2\hat{\boldsymbol{\beta}}'\boldsymbol{x}'\boldsymbol{y} + \hat{\boldsymbol{\beta}}'\boldsymbol{x}'\boldsymbol{x}\hat{\boldsymbol{\beta}} \tag{3.87}$$

Knowing that $\boldsymbol{x}'\boldsymbol{y} = \boldsymbol{x}'\boldsymbol{x}\hat{\boldsymbol{\beta}}$, this relationship can be written as:

$$\boldsymbol{e}'\boldsymbol{e} = \boldsymbol{y}'\boldsymbol{y} - \hat{\boldsymbol{\beta}}'\boldsymbol{x}'\boldsymbol{x}\hat{\boldsymbol{\beta}} \tag{3.88}$$

that is:

$$\boldsymbol{y}'\boldsymbol{y} = \hat{\boldsymbol{\beta}}'\boldsymbol{x}'\boldsymbol{x}\hat{\boldsymbol{\beta}} + \boldsymbol{e}'\boldsymbol{e} \tag{3.89}$$

or:

$$\boldsymbol{y}'\boldsymbol{y} = \hat{\boldsymbol{\beta}}'\boldsymbol{x}'\boldsymbol{y} + \boldsymbol{e}'\boldsymbol{e} \tag{3.90}$$

Equation (3.90) is the **analysis-of-variance** equation according to which the total sum of squares is equal to the explained sum of squares and the residual sum of squares where:

- The total sum of squares (TSS) is given by: $\boldsymbol{y}'\ \boldsymbol{y} = \sum_t y_t^2 = \sum_t \left(Y_t - \bar{Y}\right)^2$
- The explained sum of squares (ESS) is given by: $\hat{\boldsymbol{\beta}}'\boldsymbol{x}'\boldsymbol{y}$ or $\hat{\boldsymbol{\beta}}'\boldsymbol{x}'\boldsymbol{x}\hat{\boldsymbol{\beta}}$
- The residual sum of squares (RSS) is given by: $\boldsymbol{e}'\ \boldsymbol{e} = \sum_t e_t^2$

Case of Noncentered Variables

Of course, it is also possible to determine the analysis-of-variance equation in the case where the variables are not centered. We proceed in the same way, starting from the expression:

$$\boldsymbol{e}'\ \boldsymbol{e} = \left(\boldsymbol{Y} - \boldsymbol{X}\hat{\boldsymbol{\beta}}\right)'\left(\boldsymbol{Y} - \boldsymbol{X}\hat{\boldsymbol{\beta}}\right) = \boldsymbol{Y}'\boldsymbol{Y} - 2\hat{\boldsymbol{\beta}}'\boldsymbol{X}'\boldsymbol{Y} + \hat{\boldsymbol{\beta}}'\boldsymbol{X}'\boldsymbol{X}\hat{\boldsymbol{\beta}} \tag{3.91}$$

Knowing that $\boldsymbol{X}'\boldsymbol{Y} = \boldsymbol{X}'\boldsymbol{X}\hat{\boldsymbol{\beta}}$, this relationship can still be written as:

$$\boldsymbol{Y}'\boldsymbol{Y} = \hat{\boldsymbol{\beta}}'\boldsymbol{X}'\boldsymbol{X}\hat{\boldsymbol{\beta}} + \boldsymbol{e}'\boldsymbol{e} \tag{3.92}$$

or also:

$$\boldsymbol{Y}'\boldsymbol{Y} = \hat{\boldsymbol{\beta}}'\boldsymbol{X}'\boldsymbol{Y} + \boldsymbol{e}'\boldsymbol{e} \tag{3.93}$$

If we compare this equation with (3.90), we see that the residual sum of squares is the same, but the left-hand side is different. Specifically:

- In (3.93), the left-hand side is given by: $\boldsymbol{Y}'\ \boldsymbol{Y} = \sum_t Y_t^2$.
- In (3.90), the left-hand side is written: $\boldsymbol{y}'\ \boldsymbol{y} = \sum_t y_t^2 = \sum_t \left(Y_t - \bar{Y}\right)^2 = \sum_t Y_t^2 - T\bar{Y}^2$.

We can therefore write the analysis-of-variance Eq. (3.93) as follows:

$$\underbrace{\boldsymbol{Y}'\boldsymbol{Y} - T\bar{Y}^2}_{TSS} = \underbrace{\left(\hat{\boldsymbol{\beta}}'\boldsymbol{X}'\boldsymbol{Y} - T\bar{Y}^2\right)}_{ESS} + \underbrace{\boldsymbol{e}'\boldsymbol{e}}_{RSS} \tag{3.94}$$

It is then possible to give the definition of the coefficient of determination.

3.4.2 Coefficient of Determination

The **coefficient of determination**, denoted R^2 (R-squared), of the multiple regression model is written as in the case of the simple regression model:

$$R^2 = \frac{ESS}{TSS} = 1 - \frac{RSS}{TSS} \tag{3.95}$$

That is, in the case of centered variables:

$$R^2 = \frac{\hat{\boldsymbol{\beta}}'\boldsymbol{x}'\boldsymbol{y}}{\boldsymbol{y}'\boldsymbol{y}} = 1 - \frac{\boldsymbol{e}'\boldsymbol{e}}{\boldsymbol{y}'\boldsymbol{y}} \tag{3.96}$$

and, in the case of noncentered variables:

$$R^2 = \frac{\hat{\boldsymbol{\beta}}'\boldsymbol{X}'\boldsymbol{Y} - T\bar{Y}^2}{\boldsymbol{Y}'\boldsymbol{Y} - T\bar{Y}^2} = 1 - \frac{\boldsymbol{e}'\boldsymbol{e}}{\boldsymbol{Y}'\boldsymbol{Y} - T\bar{Y}^2} \tag{3.97}$$

As in the case of the simple regression model, we have:

$$0 \le R^2 \le 1 \tag{3.98}$$

Remark 3.3 $\sqrt{R^2}$ is called the **multiple correlation coefficient**. However, it is not used in practice.

The interpretation of the coefficient of determination is similar to that presented for the simple regression model. It measures the percentage of the total variance explained by the explanatory variables and thus enables us to judge the quality of the model's fit.

Remark 3.4 (Coefficient of Determination and Fisher Test) Considering the Fisher test of regression significance, it is possible to rewrite expression (3.83) as follows:

$$F = \frac{ESS/k}{RSS/(T-k-1)} \tag{3.99}$$

or, alternatively, by introducing the coefficient of determination:

$$F = \frac{R^2/k}{\left(1-R^2\right)/(T-k-1)} \sim F\,(k, T-k-1) \tag{3.100}$$

The latter expression is used to test the significance of the coefficient of determination, i.e., the null hypothesis that $R^2 = 0$.

3.4.3 Adjusted Coefficient of Determination

The coefficient of determination is a nondecreasing function of the number of explanatory variables included in the model. Thus, when the number of explanatory variables in the model increases, the value of the coefficient of determination also increases. This is because introducing an additional explanatory variable cannot increase the residual sum of squares, so it cannot decrease the value of the coefficient of determination. To overcome this problem, the number of explanatory variables in the model must be taken into account when calculating the coefficient of determination. In other words, the coefficient of determination must be corrected by the number of degrees of freedom. This is done by calculating the **adjusted (or corrected) coefficient of determination**. The latter is noted $\bar{R}^2$ and given by:

$$\bar{R}^2 = 1 - \frac{e'e/(T-k-1)}{y'y/(T-1)} \tag{3.101}$$

that is:

$$\bar{R}^2 = 1 - \frac{T-1}{T-k-1}\left(1-R^2\right) \tag{3.102}$$

Remark 3.5 The determination and adjusted determination coefficients make it possible to compare models. Of course, these models must have the same dependent variable and the same number of observations. When the models to be compared involve the same number of explanatory variables, we can use the R^2. On the other hand, whenever the models to be compared differ in the number of explanatory variables introduced, the $\bar{R}^2$ should be used. The model with the highest coefficient of determination—or adjusted coefficient of determination—is then selected.

3.4.4 Partial Correlation Coefficient

We saw in the first chapter that the correlation coefficient is an indicator of the link between two variables. For a model with k explanatory variables, it is possible to calculate several correlation coefficients. For example, if the model has two explanatory variables X_1 and X_2, three correlation coefficients can be calculated: r_{YX_1}, r_{YX_2}, and $r_{X_1X_2}$.

However, it is questionable whether r_{YX_1} measures the true link between Y and X_1 in the presence of X_2. For this to be the case, it is necessary to calculate a correlation coefficient r_{YX_1} that is independent of the influence that X_2 may have on Y and on X_1. Such a coefficient is called the **partial correlation coefficient**. It is denoted r_{YX_1,X_2} and is given by:

$$r_{YX_1,X_2} = \frac{r_{YX_1} - r_{YX_2}r_{X_1X_2}}{\sqrt{\left(1 - r_{YX_2}^2\right)\left(1 - r_{X_1X_2}^2\right)}} \tag{3.103}$$

Similarly, we can define:

$$r_{YX_2,X_1} = \frac{r_{YX_2} - r_{YX_1}r_{X_1X_2}}{\sqrt{\left(1 - r_{YX_1}^2\right)\left(1 - r_{X_1X_2}^2\right)}} \tag{3.104}$$

and:

$$r_{X_1X_2,Y} = \frac{r_{X_1X_2} - r_{YX_1}r_{YX_2}}{\sqrt{\left(1 - r_{YX_1}^2\right)\left(1 - r_{YX_2}^2\right)}} \tag{3.105}$$

r_{YX_1,X_2} (respectively r_{YX_2,X_1}) is the partial correlation coefficient between Y and X_1 (respectively between Y and X_2), the influence of X_2 (respectively X_1) having been removed. Similarly, $r_{X_1X_2,Y}$ is the partial correlation coefficient between the two explanatory variables X_1 and X_2, the influence of Y having been removed. A partial correlation coefficient therefore measures the link between two variables, the influence of one or more other explanatory variables having been removed. The three partial correlation coefficients presented above are first-order coefficients in the sense that only the influence of one variable is removed.

It is also possible to calculate **second-order partial correlation coefficients**. Consider, for example, a model with three explanatory variables: X_1, X_2, and X_3.

Three second-order partial correlation coefficients can be calculated: $r_{YX_1.X_2X_3}$, $r_{YX_2.X_1X_3}$, and $r_{YX_3.X_1X_2}$. In a model with four explanatory variables, third-order partial correlation coefficients can be calculated, and so on.

The coefficient $r^2_{YX_1.X_2}$ is sometimes called the **partial coefficient of determination**: it measures the proportion of the variation in Y, which is not explained by X_2, but by X_1. The notion of partial correlation is thus very important as it enables us to judge the relevance of including one (or more) explanatory variable(s) in a model. The correlation and partial correlation coefficients are also linked to the coefficient of determination through the following relationships:

$$R^2 = \frac{r^2_{YX_1} + r^2_{YX_2} - 2r_{YX_1}r_{YX_2}r_{X_1X_2}}{1 - r^2_{X_1X_2}} \tag{3.106}$$

$$R^2 = r^2_{YX_1} + \left(1 - r^2_{YX_1}\right) r_{YX_2.X_1} \tag{3.107}$$

$$R^2 = r^2_{YX_2} + \left(1 - r^2_{YX_2}\right) r_{YX_1.X_2} \tag{3.108}$$

3.4.5 Example

Let us return to the previous simple example to illustrate the various concepts presented (see Sect. 3.2.5). Recall that we had a model with three explanatory variables whose values are shown in Table 3.2.

We have seen that the application of the OLS method led to the following estimated model:

$$\hat{Y}_t = 5.57 + 0.21X_{1t} + 0.47X_{2t} - 0.87X_{3t} \tag{3.109}$$

Analysis-of-Variance Equation: Case of Centered Variables

To determine the centered variables, we first calculate the means of each of the variables: $\bar{Y} = 4$, $\bar{X}_1 = 5$, $\bar{X}_2 = 4$, and $\bar{X}_3 = 5.17$. The values of the centered variables are shown in Table 3.3. The necessary calculations are shown in Table 3.4.

Table 3.2 Illustrative example

t	Y_t	X_{1t}	X_{2t}	X_{3t}
1	4	3	5	4
2	2	5	6	8
3	1	7	3	9
4	3	2	2	5
5	6	9	1	2
6	8	4	7	3

Table 3.3 Illustrative example: centered variables

t	y_t	x_{1t}	x_{2t}	x_{3t}
1	0	−2	1	−1.17
2	−2	0	2	2.83
3	−3	2	−1	3.83
4	−1	−3	−2	−0.17
5	2	4	−3	−3.17
6	4	−1	3	−2.17
Sum	0	0	0	0
Mean	0	0	0	0

Table 3.4 Illustrative example: calculations

t	y_t^2	$\hat{y}_t$	$\hat{y}_t^2$	e_t	e_t^2
1	0	1.07	1.15	−1.07	1.15
2	4	−1.53	2.34	−0.47	0.25
3	9	−3.40	11.54	0.40	0.16
4	1	−1.43	2.04	0.43	0.18
5	4	2.19	4.78	−0.19	0.03
6	16	3.10	9.59	0.90	0.81
Sum	34	0	31.44	0	2.56
Mean	5.67	0	5.24	0	

These calculations give us:

- The total sum of squares: $TSS = 34$
- The explained sum of squares: $ESS = 31.44$
- The sum of squared residuals: $RSS = 2.56$

The analysis-of-variance equation is therefore written: $34 = 31.44 + 2.56$. It is then possible to determine the value of the coefficients of determination and adjusted determination:

- $R^2 = \frac{ESS}{TSS} = 0.92$ or $R^2 = 1 - \frac{RSS}{TSS} = 0.92$
- $\bar{R}^2 = 1 - \frac{6-1}{6-3-1}(1 - 0.92) = 0.81$

We can see that we have $\bar{R}^2 \leq R^2$. The model explains around 90% of the variance of Y_t according to R^2 and 80% according to $\bar{R}^2$.

Analysis-of-Variance Equation: Case of Noncentered Variables

Let us check that we obtain the same results by reasoning directly from the raw data. The necessary calculations are presented in Table 3.5.

Table 3.5 Example illustration: calculations

t	Y_t^2	$\hat{Y}_t$	$\hat{Y}_t^2$	e_t	e_t^2
1	16	5.07	25.71	−1.07	1.14
2	4	2.47	6.11	−0.48	0.23
3	1	0.60	0.36	0.38	0.14
4	9	2.57	6.61	0.42	0.18
5	36	6.19	38.28	−0.19	0.04
6	64	7.10	50.37	0.91	0.83
Sum	130	24	127.44	0	2.56
Mean	21.67	4	21.24	0	

These calculations give us:

- The total sum of squares: $TSS = 130 - 6 \times 4^2 = 34$
- The explained sum of squares: $ESS = 127.44 - 6 \times 4^2 = 31.44$
- The sum of squared residuals: $RSS = 2.56$

The analysis-of-variance equation is of course similar to that obtained in the case of centered variables: $34 = 31.44 + 2.56$. As before, let us calculate the value of the coefficients of determination and adjusted determination:

- $R^2 = \frac{ESS}{TSS} = 0.92$ or $R^2 = 1 - \frac{SCR}{SCT} = 0.92$
- $\bar{R}^2 = 1 - \frac{6-1}{6-3-1}(1 - 0.92) = 0.81$

We find the result that the model explains about 90% of the variance of Y_t according to R^2 and 80% according to $\bar{R}^2$.

Tests on the Regression Coefficients

Recall that we found:

$$\begin{pmatrix} \hat{\alpha} \\ \hat{\beta}_1 \\ \hat{\beta}_2 \\ \hat{\beta}_3 \end{pmatrix} = \begin{pmatrix} 5.57 \\ 0.21 \\ 0.47 \\ -0.87 \end{pmatrix} \tag{3.110}$$

The significance of the coefficients can then be tested.

Significance Test of a Coefficient For example, let us look at the test of the null hypothesis:

$$H_0 : \beta_2 = 0 \tag{3.111}$$

We know that:

$$\frac{\hat{\beta}_2}{\hat{\sigma}_\varepsilon \sqrt{a_{33}}} \sim t(6-3-1) \tag{3.112}$$

where a_{33} is the third element of the diagonal of the matrix $\left(\boldsymbol{X}'\boldsymbol{X}\right)^{-1}$. Recall that we consider the third element and not the second insofar as the matrix $\left(\boldsymbol{X}'\boldsymbol{X}\right)^{-1}$ takes the constant into account. We have seen that $a_{33} = 0.04$, so $\sqrt{a_{33}} = 0.2$. Furthermore, we know that $\sum_t e_t^2 = 2.56$, so: $\hat{\sigma}_\varepsilon^2 = \frac{1}{6-3-1} 2.56 = 1.28$. The test statistic is given by:

$$t = \frac{0.47}{\sqrt{1.28} \times 0.2} = 2.07 \tag{3.113}$$

This statistic follows a Student's t distribution with 2 degrees of freedom. The Student's law table for a 5% significance level gives us $t(2) = 4.303$. Since $2.07 < 4.303$, we do not reject the null hypothesis of nonsignificance of the coefficient β_2: the variable X_{2t} does not contribute to the explanation of the dependent variable Y_t.

We previously mentioned that the t-test can also be interpreted as a Fisher test:

$$F = \frac{\left(RSS_c - RSS_{nc}\right)/q}{RSS_{nc}/\left(T-k-1\right)} \sim F\left(q, T-k-1\right) \tag{3.114}$$

where RSS_{nc} is the sum of squared residuals of the unconstrained model and RSS_c denotes the sum of squared residuals of the constrained model, q being the number of constraints.

We know that: $RSS_{nc} = 2.56$. The constrained model consists of the regression of Y_t on X_{1t} and X_{3t}. The estimation of this model leads to a residual sum of squares equal to $RSS_c = 7.60$. The Fisher test statistic is written as:

$$F = \frac{\left(7.60 - 2.56\right)/1}{2.56/\left(6-3-1\right)} = 3.93 \sim F\left(q, T-k-1\right) \tag{3.115}$$

Under the null hypothesis, $F \sim F\left(1, 2\right) = 18.513$ at the 5% significance level. The calculated value of the statistic being lower than the critical value, we do not reject the null hypothesis. We obtain the same result as with the Student's t-test.

Significance Test of the Whole Regression Let us now consider the test of the null hypothesis:

$$H_0 : \beta_1 = \beta_2 = \beta_3 = 0 \tag{3.116}$$

This is a Fisher test. We can use relationship (3.99):

$$F = \frac{ESS/k}{RSS/(T-k-1)} = \frac{31.44/3}{2.56/2} = 8.18 \tag{3.117}$$

The Fisher table gives us $F(3, 2) = 19.164$ at the 5% significance level. Since the calculated value of the F statistic is lower than the critical value, we do not reject the null hypothesis that $\beta_1 = \beta_2 = \beta_3 = 0$. The fact that the coefficient of determination is large even though the variables are not significant may be due to the small number of observations. Remember that this example is for illustrative purposes only.

We can also use the relationship involving the coefficient of determination:

$$F = \frac{R^2/k}{\left(1-R^2\right)/(T-k-1)} \tag{3.118}$$

which gives:

$$F = \frac{0.92/3}{(1-0.92)/(6-3-1)} = 8.18 \tag{3.119}$$

The interpretation is similar to the previous one, namely, that the explanatory variables do not contribute to the explanation of the dependent variable.

Significance Test of a Subset of Explanatory Variables Consider the test of the null hypothesis:

$$H_0 : \beta_2 = \beta_3 = 0 \tag{3.120}$$

This involves testing whether or not the coefficients associated with the variables X_{2t} and X_{3t} are significant. We perform a Fisher test:

- The unconstrained model has already been estimated. The corresponding sum of squared residuals is: $RSS_{nc} = 2.56$.
- The constrained model is estimated by regressing Y_t on X_{1t}. The estimation of this model leads to a sum of squared residuals equal to: $RSS_c = 33.97$.
- We calculate the test statistic:

$$F = \frac{(33.97-2.56)/2}{2.56/(6-3-1)} = 12.27 \tag{3.121}$$

- The Fisher table gives, at the 5% significance level: $F(2, 2) = 19$. Since the calculated value of the statistic is lower than the critical value, we cannot reject the null hypothesis that the coefficients associated with the variables X_{2t} and X_{3t} are not significant.

Calculation of the Partial Correlation Coefficients

Suppose we want to calculate the following two partial correlation coefficients:

- r_{YX_1,X_2}, measuring the influence of X_1 on Y, the influence of X_2 having been removed
- r_{YX_3,X_1X_2}, measuring the influence of X_3 on Y, the influences of X_1 and X_2 having been removed

In order to calculate r_{YX_1,X_2}, we start by regressing Y on a constant and X_2, which gives:

$$\hat{Y}_t = \underset{(1.23)}{3.00} + \underset{(0.46)}{0.25} X_{2t} \tag{3.122}$$

We derive the residual series:

$$e_{1t} = \hat{Y}_t - 3.00 - 0.25X_{2t} \tag{3.123}$$

We then regress X_1 on a constant and X_2:

$$\hat{X}_{1t} = \underset{(3.02)}{6.86} - \underset{(-0.93)}{0.46} X_{2t} \tag{3.124}$$

and we deduce the residual series:

$$e_{2t} = \hat{X}_{1t} - 6.86 + 0.46X_{2t} \tag{3.125}$$

The partial correlation coefficient r_{YX_1,X_2} is equal to the correlation coefficient between e_1 and e_2; hence:

$$r_{YX_1,X_2} = r_{e_1e_2} = 0.14 \tag{3.126}$$

The correlation between Y and X_1 is equal to 0.14, when the influence of X_2 is removed.

To determine r_{YX_3,X_1X_2}, we first regress Y on X_1 and X_2, i.e.:

$$\hat{Y}_t = \underset{(0.38)}{1.95} + \underset{(0.25)}{0.15} X_{1t} + \underset{(0.47)}{0.32} X_{2t} \tag{3.127}$$

and we note e_{3t} the residual series:

$$e_{3t} = \hat{Y}_t - 1.95 - 0.15X_{1t} - 0.32X_{2t} \tag{3.128}$$

We then regress X_3 on X_1 and X_2 and obtain:

$$\hat{X}_{3t} = \underset{(0.75)}{4.14} + \underset{(0.09)}{0.07} X_{1t} + \underset{(0.23)}{0.17} X_{2t} \tag{3.129}$$

from which we deduce the residual series:

$$e_{4t} = \hat{X}_{3t} - 4.14 - 0.07X_{1t} - 0.17X_{2t} \tag{3.130}$$

We then obtain the desired result:

$$r_{YX_3,X_1X_2} = r_{e_3e_4} = -0.96 \tag{3.131}$$

The correlation between Y and X_3 is -0.96, when the influence of the variables X_1 and X_2 is removed.

3.5 Some Examples of Cross-Sectional Applications

As in the previous chapter, we offer a few examples from the literature to highlight the utility of the multiple regression model for cross-sectional studies.

3.5.1 Determinants of Crime

The first example concerns the determinants of crime in developing countries. In this context, the study by Puech (2005) aims to investigate the explanatory factors of violent crime rates in 2000 for 723 municipalities in the Brazilian state of Minas Gerais. Two categories of crimes are considered: crimes against property and crimes against persons. The explanatory variables selected by Puech (2005) include:

- Level of development (DEV), measured through the municipality's average income
- Income inequality, measured by the Gini coefficient of income distribution ($GINI$)
- Changes in economic conditions, captured by the income growth rate between 1991 and 2000 ($GROWTH$)
- Demographic and sociological variables such as:
  - The number of police officers per capita (POL)
  - The urbanization rate (URB)
  - The proportion of young people aged between 15 and 24 in the municipality's population ($YOUNG$)
  - Family instability, measured by the proportion of households headed by single women ($WOMEN$)

As previously mentioned, two dependent variables are considered:

- The aggregate rate of violent property crimes: sum of theft, armed robbery, and auto theft

Table 3.6 Determinants of crime

	Property	Persons
DEV	9.12 (2.43)	−0.03 (−0.10)
DEV^2	−0.68 (−1.91)	
$GINI$	3.13 (3.84)	2.12 (2.70)
$GROWTH$	−0.50 (−1.99)	0.20 (0.82)
POL	75.01 (1.54)	−41.61 (−0.89)
URB	−0.18 (−0.35)	0.53 (1.07)
$YOUNG$	39.29 (5.04)	25.11 (3.39)
$WOMEN$	14.71 (2.40)	20.51 (3.49)
$Constant$	−33.13 (−3.19)	−6.64 (−2.24)

Source: Puech (2005). Figures in parentheses are t-statistics of the estimated coefficients. In the column "Property" (resp. "Persons"), the explained variable is the rate of crimes against property (resp. persons).

– The aggregate rate of violent crimes against persons: sum of attempted homicides, homicides, and assaults

Table 3.6 reports Puech (2005) estimation results. Schematically, if we consider property crimes, it emerges that both average income and income inequality have a positive impact on this type of crime. We also note that property crimes increase with the proportion of young people and vary with family vulnerability in the expected direction. Turning now to crimes against persons, we see that average income has no impact, while inequality has an expected positive effect. Like property crimes, crimes against people also vary positively with the proportion of young people in the population and with family instability. In summary, the three main determinants of crime in Minas Gerais are income inequality, the proportion of young people in the municipality's population, and family instability.

3.5.2 Health Econometrics

The second example falls within the field of health econometrics. It is based on the work of Thuilliez (2007) and highlights the relationship between malaria and primary education. Primary education is apprehended in terms of school results through repetition and completion rates in primary school. With regard to malaria ($MALA$), the author uses an index measured as the proportion of the population "at risk" (i.e., likely to contract malaria) in a country in 1994. The analysis covers a set of 80 countries. Thuilliez (2007) also considers the following explanatory variables:

- Per capita income (INC): this variable takes into account the fact that countries with higher income levels offer a better quality of education than others. Per capita income is measured by GDP per capita at the purchasing power parity level (in logarithms).
- Level of urbanization (URB): this variable is expected to have a positive impact on educational outcomes. The variable is expressed in logarithms.
- Public expenditure on primary education, expressed as a percentage of GDP (EXP) and in logarithms.
- Public expenditure management efficiency (GEI). The measure used is the government effectiveness index proposed by Kaufmann et al. (2006).
- Geographical location variables: percentage of regions belonging to tropical zones ($TROP$), on the one hand, and subtropical zones ($SUBTROP$), on the other hand.
- Infant mortality rate (MOR), in logarithms.

The results obtained by Thuilliez (2007) are displayed in Table 3.7. They show a relationship between malaria and primary school repetition and completion rates. Concerning the repetition rate, the coefficient assigned to the variable $MALA$ is positive, meaning that malaria tends to increase the repetition rate, all other things being equal. The value of 0.096 shows that children living in malaria-risk countries

Table 3.7 Education and malaria

	Repetition	Completion
$MALA$	0.096 (2.46)	−0.295 (−3.40)
INC	0.023 (1.53)	−0.003 (−0.06)
URB	0.031 (0.97)	0.072 (1.36)
EXP	−0.019 (−1.36)	0.018 (0.60)
GEI	−0.001 (−0.08)	−0.016 (−0.42)
$TROP$	0.017 (0.39)	−0.031 (−0.40)
$SUBTROP$	0.038 (1.31)	0.008 (0.10)
MOR	0.053 (3.12)	−0.095 (−1.51)
$Constant$	−0.530 (−2.83)	0.991 (1.77)

Source: Thuilliez (2007). Numbers in parentheses are t-statistics of the estimated coefficients. In the "Repetition" (resp. "Completion") column, the dependent variable is the repetition (resp. completion) rate in primary school.

have repetition rates 9.6% higher than those living in noninfested countries, all other things being equal. Similarly, if we consider the regression with completion rate as the explained variable, it appears that the coefficient is negative: malaria has a negative impact on the completion rate. The estimated value of the coefficient also shows that high-risk countries have primary school completion rates 29.5% lower than those of non-risk countries, all other things being equal. On the other hand, we find that per capita income has no significant impact on repetition and completion rates, nor do geographic location variables and public expenditure variables. The fact that public expenditure does not appear to be significant simply illustrates that increasing school resources does not imply children get better results. Only the infant mortality rate has a positive effect on the primary school repetition rate. In summary, this study by Thuilliez (2007) shows that malaria has a negative impact on children's school performance, which is in line with expectations.

3.5.3 Inequalities and Financial Openness

The third example comes from the work of Bénassy-Quéré and Salins (2005) and examines the impact of financial openness on income inequalities for 42 developing and emerging countries in 2001. Four types of inequalities are distinguished: inter-individual inequalities, geographical disparities, urban/rural disparities, and regional disparities. The financial openness variable (denoted $OPENFI$) is defined as the sum of the absolute values of the country's capital inflows and outflows, as a percentage of its GDP. The explanatory variables selected, which are control variables, differ according to the type of inequality considered.

For the study of inter-individual income inequalities, three control variables are used:

- Social mobility ($SOCMOB$): this variable can take four values, ranging from 1 (for the worst institutional environment) to 4 (for the best environment). Social mobility corresponds to recruitment and promotion in the public and private sectors. A value of 1 corresponds to recruitment or promotion based on social position, and a value of 4 corresponds to recruitment or promotion based on merit.
- Trade openness ($TRADE$): this represents the scale of tariff and nontariff barriers. It ranges from 1 (high barriers) to 4 (low barriers).
- The scale of structural reforms undertaken in the country under consideration following financial and trade openness (REF). The variable ranges from 1 (no reform) to 4 (very extensive reforms).

Concerning geographical disparities, the following three control variables are considered:

- GDP per capita (in logarithms), denoted GDP.
- The level of infrastructure development ($INFRA$). The variable is 1 when the level of infrastructure is low and 4 when it is high.
- The geographical mobility of a country's inhabitants ($GEOMOB$). It varies from 1 (low mobility) to 4 (high mobility).

The results concerning the effect of financial openness on income equality are given in the equation below:

$$\widehat{EQUALITY} = \underset{(3.43)}{1.730} - \underset{(0.43)}{0.226}OPENFI + \underset{(2.19)}{0.436}SOCMOB - \underset{(-2.37)}{0.403}TRADE + \underset{(1.89)}{0.262}REF \quad (3.132)$$

The dependent variable is an indicator of inter-individual income equality within a given population. Consequently, when an explanatory variable has a positive coefficient, it tends to increase equality and therefore reduce income inequality. The figures in parentheses below the values of the estimated coefficients are t-statistics of these coefficients. This estimation shows that financial openness has no significant impact on income inequalities: the t-statistics assigned to the coefficient associated with the variable $OPENFI$ is indeed lower than the critical value at the 5% significance level, implying that $OPENFI$ does not contribute to the explanation of income inequalities. Concerning the other three variables, it turns out that social mobility and reforms reduce inequalities, since their associated coefficients are positive, whereas trade openness tends to increase them.

Table 3.8 presents the results relating to geographical disparities. As shown, financial openness tends to reduce geographical income inequalities, as well as urban-rural inequalities. However, the impact of financial openness on regional inequalities is not significant. The level of infrastructure development tends to

Table 3.8 Impact of financial openness on geographic disparities

	Geographic equality	Urban-rural equality	Regional equality
$Constant$	0.524 (0.46)	0.720 (0.47)	0.251 (0.27)
$OPENFI$	1.146 (4.92)	2.172 (6.19)	−0.283 (−0.78)
GDP	0.130 (0.98)	0.089 (0.51)	0.186 (1.71)
$INFRA$	−0.198 (−2.33)	−0.049 (−0.37)	−0.406 (−2.9)
$GEOMOB$	0.033 (0.39)	−0.036 (−0.32)	0.128 (1.36)

Source: Bénassy-Quéré and Salins (2005). Figures in parentheses are t-statistics of the estimated coefficients.

increase regional inequalities, while the level of GDP per capita tends to reduce them.

3.5.4 Inequality and Voting Behavior

The last cross-sectional example we present is based on the article by Farvaque et al. (2007), and concerns the impact of environmental variables on the electoral behavior of residents in France. The sample consists of 560 French municipalities with more than 10,000 inhabitants and covers the 2001 municipal elections.

The dependent variable is the outcome of the incumbent mayor's party ($PARTY$), expressed as a percentage of total votes cast. The explanatory variables are:

- Incumbent bonus ($SORT$): this is the percentage of votes obtained by the incumbent party in the previous municipal elections.
- A binary variable that is 1 if the incumbent party is the same as the one to which the President of the Republic belongs (Jacques Chirac in 2001). This variable, noted $PRES$, equals 0 when the party is different.
- A variable representing the merger ($MERGER$) of lists between the two rounds of voting. The value of this variable is 1 when a merger occurs between the two rounds of voting, 0 otherwise.
- Two ecological variables are also used:
  - An air pollution indicator representing the ozone concentration ($OZONE$) for each region considered
  - A soil pollution indicator representing the number of polluted sites and soils in the county ($POLL$)
- Finally, a control variable, noted POP, is introduced to take into account the size of the cities: the variable POP is 1 if the city counts more than 100,000 inhabitants, 0 otherwise.

The results obtained by Farvaque et al. (2007) are:

$$\widehat{PARTY} = \underset{(6.23)}{33.43} + \underset{(6.21)}{0.47}SORT + \underset{(4.19)}{6.10}PRES - \underset{(-3.25)}{12.24}MERGER \qquad (3.133)$$

$$- \underset{(-2.82)}{0.02}\ OZONE - \underset{(-2.75)}{0.02}\ POLL - \underset{(-3.51)}{4.40}\ POP$$

Among the political variables, the incumbent's bonus has a positive and highly significant effect. Thus, all other things being equal, almost half of the votes obtained by the party in the previous elections are transferred to the explained variable. The variable $PRES$ also has a significant positive impact, illustrating significant regional and national interactions. Finally, the variable $MERGER$ has a negative sign, indicating that the merger of the incumbent party with another list between the two rounds of voting has a negative effect on the incumbent party. Also noteworthy is the significant impact of the variable POP showing that the size of the municipality has an unfavorable effect on the incumbent party.

Turning now to the ecological variables, their impact is negative. The variable $OZONE$ has a significant and negative coefficient: air pollution has a negative impact on the election of the incumbent mayor. The same is true for the variable $POLL$, indicating that voters tend to punish incumbent mayors from municipalities with the most polluted sites and soils. Overall, this study shows that ecological inequalities, as reflected in environmental variables, have a significant impact on electoral behavior.

3.6 Prediction

One of the practical interests of the regression model lies in forecasting. Thus, once the model has been estimated, it can be used to predict the evolution of the dependent variable.

3.6.1 Determination of Predicted Value and Prediction Interval

The following model was estimated:

$$Y_t = \alpha + \beta_1 X_{1t} + \beta_2 X_{2t} + \ldots + \beta_k X_{kt} + \varepsilon_t \tag{3.134}$$

for $t = 1, \ldots, T$, i.e.:

$$\hat{Y}_t = \hat{\alpha} + \hat{\beta}_1 X_{1t} + \hat{\beta}_2 X_{2t} + \ldots + \hat{\beta}_k X_{kt} \tag{3.135}$$

We seek to determine the forecast of the dependent variable for a horizon h, i.e., $\hat{Y}_{T+h}$, as well as the associated prediction interval. The latter is given by:

$$\hat{Y}_{T+h} \pm t_{p/2} \hat{\sigma}_\varepsilon \sqrt{\boldsymbol{R} \left(\boldsymbol{X}'\boldsymbol{X}\right)^{-1} \boldsymbol{R}' + 1} \tag{3.136}$$

where $\boldsymbol{R}$ is the matrix containing the values of the explanatory variables at date $T + h$ and whose first element is 1.

Let us explain this expression and the value taken by $\hat{Y}_{T+h}$. Assuming that the relationship generating the explained variable remains identical and that the values of the explanatory variables are known in $T + h$, we have:

$$\hat{Y}_{T+h} = \hat{\alpha} + \hat{\beta}_1 X_{1T+h} + \hat{\beta}_2 X_{2T+h} + \ldots + \hat{\beta}_k X_{kT+h} \tag{3.137}$$

The prediction error is:

$$e_{T+h} = Y_{T+h} - \hat{Y}_{T+h} \tag{3.138}$$
$$= \varepsilon_{T+h} - (\hat{\alpha} - \alpha) - \left(\hat{\beta}_1 - \beta_1\right) X_{1T+h} - \ldots - \left(\hat{\beta}_k - \beta_k\right) X_{kT+h}$$

Since the OLS estimators of the coefficients are unbiased estimators and given that $E(\varepsilon_{T+h}) = 0$, we deduce:

$$E(e_{T+h}) = 0 \tag{3.139}$$

In order to determine a prediction interval, we need to calculate the variance of the forecast error. To do this, let us express the forecast in matrix form. We can write:

$$Y_{T+h} = \boldsymbol{R\beta} + \varepsilon_{T+h} \tag{3.140}$$

where, as before, $\boldsymbol{R}$ is the matrix (row vector) containing the values of the explanatory variables at date $T + h$ and whose first element is 1. The forecast is then given by:

$$\hat{Y}_{T+h} = \boldsymbol{R}\hat{\boldsymbol{\beta}} \tag{3.141}$$

and the forecast error is written as:

$$e_{T+h} = Y_{T+h} - \hat{Y}_{T+h} = \varepsilon_{T+h} - \boldsymbol{R}\left(\hat{\boldsymbol{\beta}} - \boldsymbol{\beta}\right) \tag{3.142}$$

Knowing that $E\left(\hat{\boldsymbol{\beta}}\right) = \boldsymbol{\beta}$ and $E(\varepsilon_{T+h}) = 0$, we have:

$$E(e_{T+h}) = 0 \tag{3.143}$$

The variance of the forecast error is:

$$V(e_{T+h}) = E\left[\left(\varepsilon_{T+h} - \boldsymbol{R}\left(\hat{\boldsymbol{\beta}} - \boldsymbol{\beta}\right)\right)\left(\varepsilon_{T+h} - \boldsymbol{R}\left(\hat{\boldsymbol{\beta}} - \boldsymbol{\beta}\right)\right)'\right] \tag{3.144}$$

Using (3.265), we obtain:

$$V(e_{T+h}) = \sigma_{\varepsilon}^2\left(\boldsymbol{R}\left(\boldsymbol{X}'\boldsymbol{X}\right)^{-1}\boldsymbol{R}' + 1\right) \tag{3.145}$$

Knowing that the forecast error is normally distributed, we have:

$$\frac{e_{T+h}}{\sigma_{\varepsilon}\sqrt{\boldsymbol{R}\left(\boldsymbol{X}'\boldsymbol{X}\right)^{-1}\boldsymbol{R}' + 1}} \sim N(0, 1) \tag{3.146}$$

Replacing σ_ε by its estimator (see Eq. (3.40)):

$$\hat{\sigma}_\varepsilon = \sqrt{\frac{\boldsymbol{e}'\boldsymbol{e}}{T-k-1}} \tag{3.147}$$

we get:

$$\frac{Y_{T+h} - \hat{Y}_{T+h}}{\hat{\sigma}_\varepsilon \sqrt{\boldsymbol{R}\left(\boldsymbol{X}'\boldsymbol{X}\right)^{-1}\boldsymbol{R}' + 1}} \sim t(T-k-1) \tag{3.148}$$

It is then possible to construct a $100(1-p)\%$ prediction interval for Y_{T+h}:

$$\hat{Y}_{T+h} \pm t_{p/2}\hat{\sigma}_\varepsilon \sqrt{\boldsymbol{R}\left(\boldsymbol{X}'\boldsymbol{X}\right)^{-1}\boldsymbol{R}' + 1} \tag{3.149}$$

Taking the usual value $p = 5\%$, the 95% interval is thus written:

$$\hat{Y}_{T+h} \pm t_{0.025}\hat{\sigma}_\varepsilon \sqrt{\boldsymbol{R}\left(\boldsymbol{X}'\boldsymbol{X}\right)^{-1}\boldsymbol{R}' + 1} \tag{3.150}$$

3.6.2 Example

Let us take the example studied throughout this chapter (see Sect. 3.2.5) and suppose that we wish to predict the value of Y for the date $t = 7$. Also assume that the values of the three explanatory variables are known in $t = 7$ and are given by: $X_{17} = 6$, $X_{27} = 8$, and $X_{37} = 1$. The matrix $\boldsymbol{R}$ is written as:

$$\boldsymbol{R} = \begin{pmatrix} 1 & 6 & 8 & 1 \end{pmatrix} \tag{3.151}$$

The expected value of Y_7 is given by:

$$\hat{Y}_7 = \hat{\alpha} + \hat{\beta}_1 X_{17} + \hat{\beta}_2 X_{27} + \hat{\beta}_3 X_{37} \tag{3.152}$$

that is:

$$\hat{Y}_7 = 5.57 + 0.21 \times 6 + 0.47 \times 8 - 0.87 \times 1 \tag{3.153}$$

Hence:

$$\hat{Y}_7 = 9.72 \tag{3.154}$$

Now let us determine the 95% prediction interval for Y_7. We had found:

$$\left(X'X\right)^{-1} = \begin{pmatrix} 2.87 & -0.24 & -0.24 & -0.11 \\ -0.24 & 0.04 & 0.02 & -0.002 \\ -0.24 & 0.02 & 0.04 & -0.004 \\ -0.11 & -0.002 & -0.004 & 0.03 \end{pmatrix} \tag{3.155}$$

Therefore:

$$R\left(X'X\right)^{-1}R' = \begin{pmatrix} 1 & 6 & 8 & 1 \end{pmatrix} \begin{pmatrix} 2.87 & -0.24 & -0.24 & -0.11 \\ -0.24 & 0.04 & 0.02 & -0.002 \\ -0.24 & 0.02 & 0.04 & -0.004 \\ -0.11 & -0.002 & -0.004 & 0.03 \end{pmatrix} \begin{pmatrix} 1 \\ 6 \\ 8 \\ 1 \end{pmatrix} \tag{3.156}$$

Hence:

$$R\left(X'X\right)^{-1}R' = 1.79 \tag{3.157}$$

We had also obtained: $\hat{\sigma}_{\varepsilon}^2 = 1.28$, i.e., $\hat{\sigma}_{\varepsilon} = 1.13$. Using (3.150) and knowing that $t(T-k-1) = t(2) = 4.303$, the 95% prediction interval for Y_7 is:

$$9.72 \pm 4.303 \times 1.13\sqrt{1.79+1} = 9.72 \pm 8.12 \tag{3.158}$$

which corresponds to an interval ranging from 1.60 to 17.84.

3.7 Model Comparison Criteria

We have previously presented the coefficient of determination and the adjusted coefficient of determination. These two statistics can be used to compare models. There are also other criteria, which we describe below.

3.7.1 Explanatory Power/Predictive Power of a Model

It is worth distinguishing the **explanatory power** of a model from its **predictive power**, or, in other words, **in-sample prediction** from **out-of-sample prediction**. In-sample prediction indicates how the estimated model fits the observations in a given sample. It thus concerns the explanatory power of the model. Out-of-sample prediction refers to how an estimated model predicts the future values of the explained variable, given the future values of the explanatory variables. Out-of-sample forecasting therefore concerns the predictive power of a model.

3.7.2 Coefficient of Determination and Adjusted Coefficient of Determination

These coefficients have already been presented. Let us just recall some essential points.

The coefficient of determination lies between 0 and 1; the closer it is to 1, the better the quality of the fit. It measures the quality of the fit within the sample and is therefore a measure of a model's explanatory power. It is not because a model has a high coefficient of determination that it will perform well in out-of-sample forecasting; the coefficient of determination is not a measure of a model's predictive power. If several models—with the same explained variable and the same number of explanatory variables—are compared on the basis of the coefficient of determination, the one with the highest R^2 value should be selected.

It is important to remember, however, that there is a nondecreasing relationship between the value of the coefficient of determination and the number of explanatory variables introduced into a model. For this reason, the adjusted (or corrected) coefficient of determination has been proposed and can be used to compare models with a different number of explanatory variables; the best model being the one with the highest adjusted coefficient of determination. Like the usual coefficient of determination, the adjusted coefficient of determination only allows us to judge the explanatory power of a model, not its predictive power.

3.7.3 Information Criteria

The **information criteria** are based on information theory and are intended to assess the loss of information—called the amount of Kullback information[8]—when an estimated model is thought to represent the true data-generating process. Since the aim is to minimize this loss of information, the model to be preferred, among all the models estimated, is the one that will minimize the information criteria.

The information criteria (IC) are based on the use of the maximum likelihood method (see Appendix 2.3 to Chap. 2). They are of the form:

$$IC = -\frac{2\ell}{T} + \frac{p(T)}{T} \tag{3.159}$$

where ℓ is the **log-likelihood function**:

$$\ell = -\frac{T}{2}\left(1 + \log(2\pi) + \log\left(\frac{e'e}{T}\right)\right) \tag{3.160}$$

[8] Strictly speaking, this is known as Kullback-Leibler information (see Kullback and Leibler, 1951).

and $p(T)$ is a **penalty function** that increases with the model's complexity, i.e., with the number of explanatory variables introduced into the model. In other words, the information criteria penalize the addition of variables to guard against the risk of over-fitting or over-parameterization. The various information criteria proposed in the literature are distinguished by the penalty function adopted. The most frequently used criteria are those introduced by Akaike (1973),[9] Schwarz (1978), and, to a lesser extent, Hannan and Quinn (1979), which we present below.

Akaike Information Criterion (AIC)

The Akaike information criterion is written:

$$AIC = \frac{-2\ell}{T} + \frac{2(k+1)}{T} \tag{3.161}$$

which can still be expressed, under the assumption of error normality,[10] as follows:

$$AIC = \log \hat{\sigma}_{\varepsilon}^2 + \frac{2(k+1)}{T} \tag{3.162}$$

where $\hat{\sigma}_{\varepsilon}^2$ is the error variance estimator, k is the number of explanatory variables included in the model, and T is the number of observations. If several models are compared on the basis of this criterion, the model with the lowest AIC will be selected. This is thus a criterion to be minimized. Unlike the usual and adjusted coefficients of determination, the AIC can be used to assess the explanatory power of a model, but also its predictive power.

Remark 3.6 When the number of explanatory variables k is large relative to the number of observations—which may happen in the case of a small sample—it is possible to use the corrected AIC (see Hurvich and Tsai, 1989), denoted AIC_c, given by:

$$AIC_c = AIC + \frac{2(k+1)(k+2)}{T-k-2} \tag{3.163}$$

Schwarz Information Criterion (SIC)

The SIC is given by:

$$SIC = \frac{-2\ell}{T} + \frac{(k+1)}{T}\log(T) \tag{3.164}$$

---

[9] See also Akaike (1969, 1974).

[10] In this case, the OLS and maximum likelihood estimators are equivalent.

or, under the assumption of error normality:

$$SIC = \log \hat{\sigma}_{\varepsilon}^2 + \frac{(k+1)}{T} \log T \tag{3.165}$$

where $\hat{\sigma}_{\varepsilon}^2$ is the error variance estimator, k is the number of explanatory variables included in the model, and T is the number of observations. As with the AIC, the SIC must be minimized. Thus, the best model will be the one with the lowest SIC value. Like the AIC, the SIC criterion can be used to compare predictive performance both within and out of the sample.

The SIC is more parsimonious than the AIC, as it penalizes the number of variables in the model more heavily. In other words, it penalizes more the over-parameterization. As a result, the SIC tends to select models with either the same number of variables or fewer variables than those selected by the AIC.

Hannan-Quinn Information Criterion (HQ)

The Hannan-Quinn information criterion, also used to assess the explanatory and predictive power of a model, is given by:

$$HQ = \frac{-2\ell}{T} + 2c \frac{(k+1) \log(\log T)}{T} \tag{3.166}$$

or, under the assumption of error normality:

$$HQ = \log \hat{\sigma}_{\varepsilon}^2 + 2c \frac{(k+1) \log(\log T)}{T} \tag{3.167}$$

where $\hat{\sigma}_{\varepsilon}^2$ is the error variance estimator, k is the number of explanatory variables included in the model, T is the number of observations, and c is a constant term for which a value of 1 is very frequently used. Like the other two information criteria, the HQ criterion must be minimized.

3.7.4 The Mallows Criterion

Suppose we want to compare a model with k explanatory variables to a model with h explanatory variables, with $h \leq k$. We denote RSS_h the sum of squared residuals associated with the model with h regressors and $\hat{\sigma}_{\varepsilon}^2$ the error variance estimate associated with the model with k regressors. The Mallows statistic is given by:

$$C_h = \frac{RSS_h}{\hat{\sigma}_{\varepsilon}^2} + (2(h+1) - T) \tag{3.168}$$

It can be shown that:

$$E(C_h) \simeq h \tag{3.169}$$

In choosing a model based on this statistic, we should keep the model whose statistic C_h is the closest to h.

3.8 Empirical Application

We propose to study the relationship between the following three series of stock market returns:

- Returns of the US Dow Jones Industrial Average index: $RDJIND$
- Returns of the UK, $FTSE$ 100 index: $RFTSE$
- Returns of the Japanese, $NIKKEI$ 225 index: $RNIKKEI$

The data are shown in Table 3.9 and are taken from the Macrobond database. The series are monthly and cover the period from February 1984 to June 2021, i.e., a number of observations $T = 449$. Suppose we wish to explain the returns of the UK index by the returns of the Japanese and US stock indexes. The dependent variable is therefore $RFTSE$ and the explanatory variables are $RNIKKEI$ and $RDJIND$. We seek to estimate the following model:

$$RFTSE_t = \alpha + \beta_1 RNIKKEI_t + \beta_2 RDJIND_t + \varepsilon_t \tag{3.170}$$

Table 3.9 Data on stock market returns

t	$RFTSE$	$RNIKKEI$	$RDJIND$
1984.02	−0.0216	−0.0164	−0.0555
1984.03	0.0671	0.0858	0.0088
1984.04	0.0229	0.0048	0.0050
...	...	...	...
2021.04	0.0374	−0.0126	0.0267
2021.05	0.0075	0.0016	0.0191
2021.06	0.0021	−0.0024	−0.0008
Sum	1.8902	1.0381	3.3417

Data source: Macrobond

3.8.1 Practical Calculation of the OLS Estimators

We aim to estimate the vector $\boldsymbol{\beta} = \begin{pmatrix} \alpha \\ \beta_1 \\ \beta_2 \end{pmatrix}$ using the OLS method. The matrix $\boldsymbol{X}$ is written as:

$$\boldsymbol{X} = \begin{pmatrix} 1 & -0.0164 & -0.0555 \\ 1 & 0.0858 & 0.0088 \\ 1 & 0.0048 & 0.0050 \\ \vdots & \vdots & \vdots \\ 1 & -0.0126 & 0.0267 \\ 1 & 0.0016 & 0.0191 \\ 1 & -0.0024 & -0.0008 \end{pmatrix} \quad (3.171)$$

In practice, as we have previously seen, we can calculate the matrix $\boldsymbol{X}'\boldsymbol{X}$ as follows:

$$\boldsymbol{X}'\boldsymbol{X} = \begin{pmatrix} T & \sum RNIKKEI & \sum RDJIND \\ \sum RNIKKEI & \sum RNIKKEI^2 & \sum RNIKKEI \times RDJIND \\ \sum RDJIND & \sum RNIKKEI \times RDJIND & \sum RDJIND^2 \end{pmatrix} \quad (3.172)$$

and the matrix $\boldsymbol{X}'\boldsymbol{Y}$ as follows:

$$\boldsymbol{X}'\boldsymbol{Y} = \begin{pmatrix} \sum RFTSE \\ \sum RFTSE \times RNIKKEI \\ \sum RFTSE \times RDJIND \end{pmatrix} \quad (3.173)$$

Calculating the products between the variables gives:

- $\sum_{t=1}^{449} RFTSE_t \times RNIKKEI_t = 0.5788$
- $\sum_{t=1}^{449} RFTSE_t \times RDJIND_t = 0.6902$
- $\sum_{t=1}^{449} RNIKKEI_t \times RDJIND_t = 0.6178$

and calculating the sums of squares leads to:

- $\sum_{t=1}^{449} RFTSE_t^2 = 0.8937$
- $\sum_{t=1}^{449} RNIKKEI_t^2 = 1.5665$
- $\sum_{t=1}^{449} RDJIND_t^2 = 0.8890$

We deduce the matrices:

$$\boldsymbol{X}'\boldsymbol{X} = \begin{pmatrix} 449 & 1.0381 & 3.3417 \\ 1.0381 & 1.5665 & 0.6178 \\ 3.3417 & 0.6178 & 0.8890 \end{pmatrix} \tag{3.174}$$

and:

$$\boldsymbol{X}'\boldsymbol{Y} = \begin{pmatrix} 1.8902 \\ 0.5788 \\ 0.6902 \end{pmatrix} \tag{3.175}$$

Calculating the inverse of $(\boldsymbol{X}'\boldsymbol{X})$ gives:

$$(\boldsymbol{X}'\boldsymbol{X})^{-1} = \begin{pmatrix} 0.0023 & 0.0026 & -0.0104 \\ 0.0026 & 0.8823 & -0.6230 \\ -0.0104 & -0.6230 & 1.5971 \end{pmatrix} \tag{3.176}$$

Hence:

$$\hat{\boldsymbol{\beta}} = \begin{pmatrix} 0.0023 & 0.0026 & -0.0104 \\ 0.0026 & 0.8823 & -0.6230 \\ -0.0104 & -0.6230 & 1.5971 \end{pmatrix} \begin{pmatrix} 1.8902 \\ 0.5788 \\ 0.6902 \end{pmatrix} \tag{3.177}$$

or finally:

$$\hat{\boldsymbol{\beta}} = \begin{pmatrix} -0.0014 \\ 0.0856 \\ 0.7221 \end{pmatrix} \tag{3.178}$$

The estimated model is therefore:

$$\widehat{RFTSE}_t = -0.0014 + 0.0856 RNIKKEI_t + 0.7221 RDJIND_t \tag{3.179}$$

Table 3.10 Series of stock market returns. OLS estimation

Dependent variable: RFTSE				
Variable	Coefficient	Std. error	t-Statistic	Prob.
C	−0.001362	0.001340	−1.016558	0.3099
RNIKKEI	0.085603	0.026253	3.260758	0.0012
RDJIND	0.722065	0.035320	20.44346	0.0000
R-squared	0.606707	Mean dependent var	0.004210	
Adjusted R-squared	0.604944	S.D. dependent var	0.044466	
S.E. of regression	0.027948	Akaike info criterion	−4.310264	
Sum squared resid	0.348373	Schwarz criterion	−4.282823	
Log likelihood	970.6542	Hannan-Quinn criterion	−4.299447	
F-statistic	344.0076	Durbin-Watson stat	2.246332	
Prob(F-statistic)	0.000000			

3.8.2 Software Estimation

Using *Eviews* software leads to the results shown in Table 3.10. In addition to the estimation of the three coefficients, we have several results.

In particular, Table 3.10 gives us the standard deviations and t-statistics of the estimated coefficients. We thus have:

$$t_{\hat{\alpha}} = \frac{\hat{\alpha}}{\widehat{\sigma_{\hat{\alpha}}}} = \frac{-0.0014}{0.0013} = -1.0166 \quad (3.180)$$

$$t_{\hat{\beta}_1} = \frac{\hat{\beta}_1}{\widehat{\sigma_{\hat{\beta}1}}} = \frac{0.0856}{0.0263} = 3.2608 \quad (3.181)$$

$$t_{\hat{\beta}_2} = \frac{\hat{\beta}_2}{\widehat{\sigma_{\hat{\beta}2}}} = \frac{0.7221}{0.0353} = 20.4435 \quad (3.182)$$

It is then possible to perform significance tests on each of these coefficients. Each of the t-statistics has a Student's t distribution with $(T - k - 1) = (449 - 2 - 1) = 446$ degrees of freedom. At the 5% significance level, the critical value is equal to 1.96. It can be seen that:

- $|t_{\hat{\alpha}}| = 1.0166 < 1.96$: we do not reject the null hypothesis that $\alpha = 0$. The constant is not significantly different from zero.
- $\left|t_{\hat{\beta}_1}\right| = 3.2608 > 1.96$: we reject the null hypothesis that $\beta_1 = 0$. The coefficient associated with the Japanese variable is significant, meaning that $RNIKKEI$ contributes to the explanation of $RFTSE$.

– $\left|t_{\hat{\beta}_2}\right| = 20.4435 > 1.96$: we reject the null hypothesis that $\beta_2 = 0$. The coefficient associated with the US variable is significant, meaning that $RDJIND$ contributes to the explanation of $RFTSE$.

Furthermore, Table 3.10 gives us the value of the F-statistic allowing us to test the significance of the regression as a whole, i.e., to test the null hypothesis that $\beta_1 = \beta_2 = 0$. The statistic F follows a Fisher distribution with $(q, T - k - 1) = (2449 - 2 - 1) = (2446)$ degrees of freedom. The Fisher table, for a 5% significance level, gives us: $F(2146) = 2.997$. Given that $344.0076 > 2.997$, the null hypothesis that the coefficients are both equal to zero is rejected. This conclusion was expected in view of the results from the t-tests.

Table 3.10 provides the value of the coefficient of determination $\left(R^2 = 0.6067\right)$ and the adjusted coefficient of determination $\left(\bar{R}^2 = 0.6049\right)$. Thus, about 60% of the variation in UK returns is explained by the model, i.e., by Japanese and US returns.

If we compare the t-statistics of the coefficients assigned to $RNIKKEI$ and $RDJIND$, we can see that $\left|t_{\hat{\beta}_2}\right| > \left|t_{\hat{\beta}_1}\right|$. Therefore, the variable $RDJIND$ has more influence than $RNIKKEI$ on $RFTSE$. We can then look at the simpler model consisting of regressing UK returns on US returns only. The results of this estimation are given in Table 3.11.

Table 3.11 Regression of RFTSE on RDJIND

Dependent variable: RFTSE				
Variable	Coefficient	Std. error	t-Statistic	Prob.
C	−0.001614	0.001352	−1.193763	0.2332
RDJIND	0.782505	0.030388	25.75060	0.0000
R-squared	0.597331	Mean dependent var	0.004210	
Adjusted R-squared	0.596430	S.D. dependent var	0.044466	
S.E. of regression	0.028248	Akaike info criterion	−4.291158	
Sum squared resid	0.356678	Schwarz criterion	−4.272864	
Log likelihood	965.3650	Hannan-Quinn criterion	−4.283947	
F-statistic	663.0935	Durbin-Watson stat	2.224700	
Prob(F-statistic)	0.0000			

Since the two estimated models (Tables 3.10 and 3.11) have the same dependent variable and cover the same period, they can be compared using the adjusted coefficient of determination. We see that: $0.5964 < 0.6049$. The first model, which also incorporates Japanese returns, is to be preferred, as it explains a higher percentage of the variation in $RFTSE$. We can also note that, as expected, the sum of squared residuals of the first model is lower than that associated with the second model: $0.3484 < 0.3567$, corroborating the superiority of the first regression. These results are confirmed by the values taken by the Akaike, Schwarz, and Hannan-Quinn information criteria. Indeed, the model minimizing these three criteria is the one that includes the returns of the Japanese stock index. These results were expected since, even if the explanatory power of Japanese returns is lower than that of US returns, the Japanese variable contributes to the explanation of the UK series.

The model in Table 3.11 can be considered a constrained model, the unconstrained model being given in Table 3.10. The constrained model is such that $\beta_1 = 0$, meaning that the Japanese returns are not significant. It is then possible to perform a Fisher test on this hypothesis:

$$F = \frac{(RSS_c - RSS_{nc})/q}{RSS_{nc}/(T-k-1)} \sim F(q, T-k-1) \tag{3.183}$$

with $RSS_{nc} = 0.348373$, $RSS_c = 0.356678$ and $q = 1$. So we have:

$$F = \frac{(0.356678 - 0.348373)/1}{0.348373/(449-2-1)} = 10.6324 \tag{3.184}$$

This statistic follows a Fisher distribution with (1446) degrees of freedom. At the 5% significance level, the Fisher table gives us $F(1446) = 3.842$. Since $10.6324 > 3.842$, we reject the null hypothesis that the coefficient associated with the Japanese variable is not significant. The Japanese returns contribute to the explanation of the UK returns, which is of course consistent with the result derived from the t-test on the coefficient associated with this same variable.

Conclusion

This chapter has provided a detailed presentation of the multiple regression model. It should be recalled that the model is based on a number of hypotheses relating to the explanatory variables and the error term. These include the fundamental assumptions of no autocorrelation and homoskedasticity of errors. Since, in practice, one or both of these assumptions are often not met, Chap. 4 presents the procedure to be followed when autocorrelation and/or heteroskedasticity of errors occur.

The Gist of the Chapter

Multiple regression model	
	$\underset{(T,1)}{Y} = \underset{(T,k+1)}{X}\underset{(k+1,1)}{\beta} + \underset{(T,1)}{\varepsilon}$
Matrices	Vector of explained variable values: $\boldsymbol{Y}$
	Matrix of k explanatory variables: $\boldsymbol{X}$
	Vector of error term values: $\boldsymbol{\varepsilon}$
Assumptions	$\boldsymbol{X}$ is nonrandom: $E(\varepsilon_t \mid \boldsymbol{X}) = 0$
	$\boldsymbol{X}$ is of full rank: $Rank(\boldsymbol{X}) = k + 1$
	Zero mean error: $E(\boldsymbol{\varepsilon}) = 0$
	Non-autocorrelation and homoskedasticity:
	$E\left(\boldsymbol{\varepsilon}\boldsymbol{\varepsilon}'\right) = \sigma_\varepsilon^2 \boldsymbol{I}$
	Normality: $\boldsymbol{\varepsilon} \sim N\left(0, \sigma_\varepsilon^2 \boldsymbol{I}\right)$
Residuals	$\boldsymbol{e} = \boldsymbol{Y} - \boldsymbol{X}\hat{\boldsymbol{\beta}}$
OLS estimators	$\hat{\boldsymbol{\beta}} = \left(\boldsymbol{X}'\boldsymbol{X}\right)^{-1} \boldsymbol{X}'\boldsymbol{Y}$
	$\hat{\sigma}_\varepsilon^2 = \frac{e'e}{T-k-1} \equiv \frac{1}{T-k-1} \sum_{t=1}^{T} e_t^2$
Adjusted coefficient of determination	$\bar{R}^2 = 1 - \frac{T-1}{T-k-1}\left(1 - R^2\right)$
Information criteria	Akaike: $AIC = \log \hat{\sigma}_\varepsilon^2 + \frac{2(k+1)}{T}$
	Schwarz: $SIC = \log \hat{\sigma}_\varepsilon^2 + \frac{(k+1)}{T} \log T$
	Hannan-Quinn (with $c = 1$): $HQ = \log \hat{\sigma}_\varepsilon^2 + 2\frac{(k+1)\log(\log T)}{T}$

Further Reading

As in the previous chapter, developments on the multiple regression model can be found in all econometrics textbooks (see the references cited in the reference section at the end of the book). Worth mentioning are Judge et al. (1985, 1988), Johnston and Dinardo (1996), Davidson and MacKinnon (1993), and Greene (2020). For nonlinear systems of equations, interested readers may consult Gallant (1987).

Appendix 3.1: Elements of Matrix Algebra

This appendix presents the main matrix algebra concepts used in this chapter.

General

A **matrix** is an ordered array of elements (or entries):

$$\boldsymbol{A} = (a_{ij}) = \begin{pmatrix} a_{11} & a_{12} & \cdots & a_{1p} \\ a_{21} & a_{22} & \cdots & a_{2p} \\ & & & \\ a_{n1} & a_{n2} & & a_{np} \end{pmatrix} \tag{3.185}$$

a_{ij} is the element corresponding to the ith row and the jth column of the matrix $\boldsymbol{A}$. The matrix $\boldsymbol{A}$ has n rows and p columns. The **size** (or the **dimension**) of the matrix is said to be $n \times p$ (which is also noted as (n, p)).

A **row vector** is a matrix containing only one row. A **column vector** is a matrix with only one column. A matrix can therefore be thought of as a set of row vectors or column vectors.

When the number of rows is equal to the number of columns, i.e., $n = p$, we say that $\boldsymbol{A}$ is a **square matrix**. Frequently used square matrices include:

- **Symmetric matrix**: it is such that $(a_{ij}) = (a_{ji})$ for all i and j.
- **Diagonal matrix**: this is a matrix whose elements off the diagonal are zero:

$$\boldsymbol{A} = \begin{pmatrix} \alpha_1 & 0 & 0 & \cdots & 0 \\ 0 & \alpha_2 & 0 & \cdots & 0 \\ \cdots & & \ddots & & \cdots \\ \cdots & & & \ddots & 0 \\ 0 & 0 & \cdots & 0 & \alpha_p \end{pmatrix} \tag{3.186}$$

- **Scalar matrix**: this is a diagonal matrix whose elements on the diagonal are all identical:

$$\boldsymbol{A} = \begin{pmatrix} \alpha & 0 & 0 & \cdots & 0 \\ 0 & \alpha & 0 & \cdots & 0 \\ \cdots & & \ddots & & \cdots \\ \cdots & & & \ddots & 0 \\ 0 & 0 & \cdots & 0 & \alpha \end{pmatrix} \tag{3.187}$$

– **Identity matrix**: this is a scalar matrix, noted $\boldsymbol{I}$, whose elements on the diagonal are all equal to 1:

$$\boldsymbol{I} = \begin{pmatrix} 1 & 0 & 0 & \cdots & 0 \\ 0 & 1 & 0 & \cdots & 0 \\ \cdots & & \ddots & & \cdots \\ \cdots & & & \ddots & 0 \\ 0 & 0 & \cdots & 0 & 1 \end{pmatrix} \tag{3.188}$$

– **Triangular matrix** (**inferior** or **superior**): this is a matrix with only null elements above or below the diagonal.

Main Matrix Operations

Let $\boldsymbol{B}$ be a matrix such that:

$$\boldsymbol{B} = (b_{ij}) = \begin{pmatrix} b_{11} & b_{12} & \cdots & b_{1p} \\ b_{21} & b_{22} & \cdots & b_{2p} \\ & & & \\ b_{n1} & b_{n2} & & b_{np} \end{pmatrix} \tag{3.189}$$

Equality

The matrices $\boldsymbol{A}$ and $\boldsymbol{B}$ are equal if they are of the same size and if $a_{ij} = b_{ij}$ for all i and j.

Transposition

The **transpose** $\boldsymbol{A}'$ of the matrix $\boldsymbol{A}$ is the matrix whose jth row corresponds to the jth column of the matrix $\boldsymbol{A}$. Since the size of matrix $\boldsymbol{A}$ is $n \times p$, the size of matrix $\boldsymbol{A}'$ is $p \times n$. Thus, we have:

$$\boldsymbol{A}' = \begin{pmatrix} a_{11} & a_{21} & \cdots & a_{n1} \\ a_{12} & a_{22} & \cdots & a_{n2} \\ & & & \\ a_{1p} & a_{2p} & & a_{np} \end{pmatrix} \tag{3.190}$$

A symmetric matrix is therefore, by definition, a matrix such that:

$$\boldsymbol{A} = \boldsymbol{A}' \tag{3.191}$$

The transpose of the transpose is equal to the original matrix, i.e.:

$$(\boldsymbol{A}')' = \boldsymbol{A} \tag{3.192}$$

Addition and Subtraction

Two matrices $\boldsymbol{A}$ and $\boldsymbol{B}$ can be added only if they are of the same dimensions, the matrix $\boldsymbol{C}$ resulting from this sum also having this dimension:

$$\boldsymbol{C} = \boldsymbol{A} + \boldsymbol{B} = \left(a_{ij} + b_{ij}\right) \tag{3.193}$$

Similarly, we have:

$$\boldsymbol{D} = \boldsymbol{A} - \boldsymbol{B} = \left(a_{ij} - b_{ij}\right) \tag{3.194}$$

Note that:

$$(\boldsymbol{A} + \boldsymbol{B})' = \boldsymbol{A}' + \boldsymbol{B}' \tag{3.195}$$

i.e., the transpose of a sum is equal to the sum of the transpose.

Matrix Multiplication and Scalar Product

The **scalar product** of a line vector $\boldsymbol{a}$ with n elements and a column vector $\boldsymbol{b}$ with n elements is a **scalar**:

$$\boldsymbol{a}'\boldsymbol{b} = a_1 b_1 + a_2 b_2 + \ldots + a_n b_n \tag{3.196}$$

that is:

$$\boldsymbol{a}'\boldsymbol{b} = \boldsymbol{b}'\boldsymbol{a} = \sum_{i=1}^{n} a_i b_i \tag{3.197}$$

Now consider two matrices $\boldsymbol{A}$ and $\boldsymbol{B}$ and assume that $\boldsymbol{A}$ is of size $n \times p$ and $\boldsymbol{B}$ is of size $p \times q$. The matrix $\boldsymbol{C}$ resulting from the product of these two matrices is a matrix of size $n \times q$, that is:

$$\underset{(n\times q)}{\boldsymbol{C}} = \underset{(n\times p)}{\boldsymbol{A}}\underset{(p\times q)}{\boldsymbol{B}} \tag{3.198}$$

By noting $\boldsymbol{a}_i$ $(i = 1, 2, \ldots, n)$ the rows of $\boldsymbol{A}$ and $\boldsymbol{b}_j$ $(j = 1, 2, \ldots, q)$ the columns of $\boldsymbol{B}$, each element of $\boldsymbol{C}$ is the scalar product of a row vector of $\boldsymbol{A}$ and a column vector of $\boldsymbol{B}$. Noting c_{ij} the ijth element of the matrix $\boldsymbol{C}$, we thus have:

$$c_{ij} = \boldsymbol{a}_i'\boldsymbol{b}_j \tag{3.199}$$

Matrix multiplication is only possible if the number of columns in the first matrix (matrix $\boldsymbol{A}$) is equal to the number of rows in the second matrix (matrix $\boldsymbol{B}$). In this case, we speak about matrices that are **compatible for multiplication**.

The **scalar multiplication** of a matrix is the multiplication of each element of that matrix by a given scalar. Thus, for a scalar c and a matrix $\boldsymbol{A}$, we have:

$$c\boldsymbol{A} = \left(ca_{ij}\right) \tag{3.200}$$

We also note the following results:

- Multiplication by the identity matrix:

$$\boldsymbol{AI} = \boldsymbol{IA} = \boldsymbol{A} \tag{3.201}$$

- Transpose of a product of two matrices:

$$(\boldsymbol{AB})' = \boldsymbol{B}'\boldsymbol{A}' \tag{3.202}$$

- Transpose of a product of more than two matrices:

$$(\boldsymbol{ABC})' = \boldsymbol{C}'\boldsymbol{B}'\boldsymbol{A}' \tag{3.203}$$

- Multiplication of matrices is associative:

$$(\boldsymbol{AB})\,\boldsymbol{C} = \boldsymbol{A}\,(\boldsymbol{BC}) \tag{3.204}$$

- The sum and multiplication of matrices are distributive:

$$\boldsymbol{A}\,(\boldsymbol{B} + \boldsymbol{C}) = \boldsymbol{AB} + \boldsymbol{AC} \tag{3.205}$$

Idempotent Matrix

An **idempotent matrix** $\boldsymbol{A}$ is a matrix verifying: $\boldsymbol{AA} = \boldsymbol{A}$. In other words, an idempotent matrix is equal to its square. Furthermore, if $\boldsymbol{A}$ is a symmetric idempotent matrix, then $\boldsymbol{A}'\boldsymbol{A} = \boldsymbol{A}$.

Rank, Trace, Determinant, and Inverse Matrix

Rank of a Matrix

Consider a matrix $\boldsymbol{A}$ of size $n \times p$. The rows of $\boldsymbol{A}$ constitute n vectors, while the columns of $\boldsymbol{A}$ represent p vectors. Let r denote the maximum number of linearly independent rows and s the maximum number of linearly independent columns. We can demonstrate that, for any matrix $\boldsymbol{A}$ of size $n \times p$, we have:

$$r = s \tag{3.206}$$

This maximum number of linearly independent rows or columns is called the **rank** of the matrix $\boldsymbol{A}$. The maximum number of linearly independent rows is called the row rank and the maximum number of linearly independent columns is called the column rank. The row rank and the column rank of a matrix are therefore equal.

A matrix whose rank is equal to the number of its columns is called a **full rank matrix**.

The rank of a matrix is therefore necessarily less than or equal to the number of its rows or columns, i.e.:

$$Rank\,(\boldsymbol{A}) \leq \min\,(n, p) \tag{3.207}$$

Furthermore, we have the following properties:

$$Rank\,(\boldsymbol{A}) = Rank\,\left(\boldsymbol{A}'\right) \tag{3.208}$$

$$Rank\,(\boldsymbol{A}) = Rank\,\left(\boldsymbol{A}'\boldsymbol{A}\right) = Rank\,\left(\boldsymbol{A}\boldsymbol{A}'\right) \tag{3.209}$$

$$Rank\,(\boldsymbol{A}\boldsymbol{B}) \leq \min\,(Rank\,(\boldsymbol{A})\,,\,Rank(\boldsymbol{B})) \tag{3.210}$$

We can deduce from this last equation that if $\boldsymbol{A}$ is a matrix of size $n \times p$ and $\boldsymbol{B}$ a square matrix of size $n \times n$, then:

$$Rank\,(\boldsymbol{A}\boldsymbol{B}) = Rank(\boldsymbol{A}) \tag{3.211}$$

If $\boldsymbol{B}$ is a square matrix of size n and rank n, it is said to be **nonsingular**. In this case, it admits a unique **inverse matrix** (see below) noted $\boldsymbol{B}^{-1}$ such that:

$$\boldsymbol{B}\boldsymbol{B}^{-1} = \boldsymbol{B}^{-1}\boldsymbol{B} = \boldsymbol{I} \tag{3.212}$$

When the rank of the matrix $\boldsymbol{B}$ is less than n, the matrix $\boldsymbol{B}$ is said to be **singular** and has no inverse.

Trace of a Matrix

The **trace** of a square matrix $\boldsymbol{A}$ of size $n \times n$, denoted $Tr(\boldsymbol{A})$, is the sum of its diagonal elements:

$$Tr(\boldsymbol{A}) = \sum_{i=1}^{n} a_{ii} \tag{3.213}$$

Furthermore:

$$Tr(\boldsymbol{A}) = Tr(\boldsymbol{A}') \tag{3.214}$$

$$Tr(\boldsymbol{A} + \boldsymbol{B}) = Tr(\boldsymbol{A}) + Tr(\boldsymbol{B}) \tag{3.215}$$

and:

$$Tr(\boldsymbol{AB}) = Tr(\boldsymbol{BA}) \tag{3.216}$$

Determinant of a Matrix

The **determinant** of a matrix is defined for square matrices only.

As an introductory example, consider a matrix $\boldsymbol{A}$ of size 2×2:

$$\boldsymbol{A} = \begin{pmatrix} a & c \\ b & d \end{pmatrix} \tag{3.217}$$

The determinant of the matrix $\boldsymbol{A}$, denoted $det(\boldsymbol{A})$ or $|\boldsymbol{A}|$, is given by:

$$det(\boldsymbol{A}) = |\boldsymbol{A}| = \begin{vmatrix} a & c \\ b & d \end{vmatrix} = ad - bc \tag{3.218}$$

More generally, for matrices of size $n \times n$, we use the **cofactor expansion**:

$$det(\boldsymbol{A}) = |\boldsymbol{A}| = \sum_{j=1}^{n} a_{ij} (-1)^{i+j} \left|\boldsymbol{A}_{ij}\right| \tag{3.219}$$

where $\boldsymbol{A}_{ij}$ is the matrix obtained from matrix $\boldsymbol{A}$ by deleting row i and column j. $\left|\boldsymbol{A}_{ij}\right|$ is called a **minor** and the term:

$$C_{ij} = (-1)^{i+j} \left|\boldsymbol{A}_{ij}\right| \tag{3.220}$$

is called a **cofactor**.

We have the following property:

$$|\boldsymbol{A}| = \left|\boldsymbol{A}'\right| \tag{3.221}$$

If $\boldsymbol{A}$ and $\boldsymbol{B}$ are two square matrices, we have:

$$|\boldsymbol{AB}| = |\boldsymbol{A}| \times |\boldsymbol{B}| \tag{3.222}$$

Moreover, the determinant of a matrix is nonzero if and only if that matrix is of full rank. This last property thus provides a way to determine whether or not a matrix is of full rank (this is only operational if the matrix is not too large).

Inverse Matrix

For a matrix to be invertible, it must be nonsingular. Conversely, a matrix is nonsingular if and only if its inverse exists.

Consider a matrix $\boldsymbol{A}$ of size $n \times n$. We can write the determinant of this matrix as a function of the cofactors as follows:

$$|\boldsymbol{A}| = a_{i1}C_{i1} + a_{i2}C_{i2} + \ldots + a_{in}C_{in} \tag{3.223}$$

or, equivalently:

$$|\boldsymbol{A}| = a_{1j}C_{1j} + a_{2j}C_{2j} + \ldots + a_{nj}C_{nj} \tag{3.224}$$

for $i, j = 1, \ldots, n$.

The inverse of the matrix $\boldsymbol{A}$, denoted $\boldsymbol{A}^{-1}$, is defined by:

$$\boldsymbol{A}^{-1} = \frac{1}{|\boldsymbol{A}|}\begin{pmatrix} C_{11} & C_{21} & \cdots & C_{n1} \\ C_{12} & C_{22} & \cdots & C_{n2} \\ & & & \\ C_{1n} & C_{2n} & & C_{nn} \end{pmatrix} \tag{3.225}$$

Let us mention the following properties of inverse matrices:

$$\left|\boldsymbol{A}^{-1}\right| = \frac{1}{|\boldsymbol{A}|} \tag{3.226}$$

$$\left(\boldsymbol{A}^{-1}\right)^{-1} = \boldsymbol{A} \tag{3.227}$$

$$\left(\boldsymbol{A}^{-1}\right)' = \left(\boldsymbol{A}'\right)^{-1} \tag{3.228}$$

$$(\boldsymbol{AB})^{-1} = \boldsymbol{B}^{-1}\boldsymbol{A}^{-1} \tag{3.229}$$

$$(\boldsymbol{ABC})^{-1} = \boldsymbol{C}^{-1}\boldsymbol{B}^{-1}\boldsymbol{A}^{-1} \tag{3.230}$$

Appendix 3.2: Demonstrations

Appendix 3.2.1: Demonstration of the Minimum Variance Property of OLS Estimators

In order to show that $\hat{\boldsymbol{\beta}}$ is a minimum variance estimator, suppose there exists another linear estimator $\breve{\boldsymbol{\beta}}$ of $\boldsymbol{\beta}$:

$$\underset{(k+1,1)}{\breve{\boldsymbol{\beta}}} = \underset{(k+1,T)}{\boldsymbol{M}}\underset{(T,1)}{\boldsymbol{Y}} \tag{3.231}$$

We can then write:

$$\breve{\boldsymbol{\beta}} = \boldsymbol{M}\left(\boldsymbol{X}\boldsymbol{\beta} + \boldsymbol{\varepsilon}\right) = \boldsymbol{M}\boldsymbol{X}\boldsymbol{\beta} + \boldsymbol{M}\boldsymbol{\varepsilon} \tag{3.232}$$

If $\breve{\boldsymbol{\beta}}$ is the desired estimator, by virtue of (3.29), it is necessary that in the expression of $\boldsymbol{M}$:

$$\boldsymbol{M} = \left(\boldsymbol{X}'\boldsymbol{X}\right)^{-1}\boldsymbol{X}' + \boldsymbol{N} \tag{3.233}$$

$\boldsymbol{N} = \boldsymbol{0}$, where $\boldsymbol{N}$ is of dimension $(k+1, T)$. Let us show that $\boldsymbol{N} = \boldsymbol{0}$.

Knowing that $\breve{\boldsymbol{\beta}}$ must be an unbiased estimator of $\boldsymbol{\beta}$, we have:

$$E\left(\breve{\boldsymbol{\beta}}\right) = \boldsymbol{\beta} \tag{3.234}$$

Furthermore:

$$E\left(\breve{\boldsymbol{\beta}}\right) = E\left(\boldsymbol{M}\boldsymbol{X}\boldsymbol{\beta} + \boldsymbol{M}\boldsymbol{\varepsilon}\right) = \boldsymbol{M}\boldsymbol{X}\boldsymbol{\beta} \tag{3.235}$$

because $E(\boldsymbol{\varepsilon}) = 0$. We deduce that $E\left(\breve{\boldsymbol{\beta}}\right) = \boldsymbol{\beta}$ if:

$$\boldsymbol{M}\boldsymbol{X} = \boldsymbol{I} \tag{3.236}$$

Replacing $\boldsymbol{M}$ with $\left(\boldsymbol{X}'\boldsymbol{X}\right)^{-1}\boldsymbol{X}' + \boldsymbol{N}$, we have:

$$\left[\left(\boldsymbol{X}'\boldsymbol{X}\right)^{-1}\boldsymbol{X}' + \boldsymbol{N}\right]\boldsymbol{X} = \boldsymbol{I} \tag{3.237}$$

Knowing that $\left(\boldsymbol{X}'\boldsymbol{X}\right)^{-1}\boldsymbol{X}'\boldsymbol{X} = \boldsymbol{I}$, we get:

$$\boldsymbol{I} + \boldsymbol{N}\boldsymbol{X} = \boldsymbol{I} \tag{3.238}$$

Hence:

$$\boldsymbol{N}\boldsymbol{X} = 0 \tag{3.239}$$

Replacing $\boldsymbol{M}\boldsymbol{X}$ with $\boldsymbol{I}$ in (3.232), we have:

$$\breve{\boldsymbol{\beta}} = \boldsymbol{\beta} + \boldsymbol{M}\boldsymbol{\varepsilon} = E\left(\breve{\boldsymbol{\beta}}\right) + \boldsymbol{M}\boldsymbol{\varepsilon} \tag{3.240}$$

Hence:

$$\breve{\boldsymbol{\beta}} - \boldsymbol{\beta} = \breve{\boldsymbol{\beta}} - E\left(\breve{\boldsymbol{\beta}}\right) = \boldsymbol{M}\boldsymbol{\varepsilon} \tag{3.241}$$

Let us now determine the variance-covariance matrix $\boldsymbol{\Omega}_{\breve{\boldsymbol{\beta}}}$ of $\breve{\boldsymbol{\beta}}$:

$$\boldsymbol{\Omega}_{\breve{\boldsymbol{\beta}}} = E\left[\left(\breve{\boldsymbol{\beta}} - \boldsymbol{\beta}\right)\left(\breve{\boldsymbol{\beta}} - \boldsymbol{\beta}\right)'\right] \tag{3.242}$$

$$= E\left[(\boldsymbol{M\varepsilon})(\boldsymbol{M\varepsilon})'\right]$$

$$= E\left[\boldsymbol{M\varepsilon\varepsilon}'\boldsymbol{M}'\right]$$

Hence:

$$\boldsymbol{\Omega}_{\breve{\boldsymbol{\beta}}} = \sigma_{\varepsilon}^{2}\boldsymbol{MM}' \tag{3.243}$$

Let us determine the matrix product $\boldsymbol{MM}'$:

$$\boldsymbol{MM}' = \left[\left(\boldsymbol{X}'\boldsymbol{X}\right)^{-1}\boldsymbol{X}' + \boldsymbol{N}\right]\left[\left(\boldsymbol{X}'\boldsymbol{X}\right)^{-1}\boldsymbol{X}' + \boldsymbol{N}\right]' \tag{3.244}$$

$$\boldsymbol{MM}' = \left[\left(\boldsymbol{X}'\boldsymbol{X}\right)^{-1}\boldsymbol{X}' + \boldsymbol{N}\right]\left[\boldsymbol{X}\left(\boldsymbol{X}'\boldsymbol{X}\right)^{-1} + \boldsymbol{N}'\right]$$

$$\boldsymbol{MM}' = \left(\boldsymbol{X}'\boldsymbol{X}\right)^{-1}\boldsymbol{X}'\boldsymbol{X}\left(\boldsymbol{X}'\boldsymbol{X}\right)^{-1} + \left(\boldsymbol{X}'\boldsymbol{X}\right)^{-1}\boldsymbol{X}'\boldsymbol{N}' + \boldsymbol{NX}\left(\boldsymbol{X}'\boldsymbol{X}\right)^{-1} + \boldsymbol{NN}'$$

According to (3.239), $\boldsymbol{NX} = \left(\boldsymbol{X}'\boldsymbol{N}'\right)' = 0$, hence:

$$\boldsymbol{MM}' = \left(\boldsymbol{X}'\boldsymbol{X}\right)^{-1} + \boldsymbol{NN}' \tag{3.245}$$

So we have:

$$\boldsymbol{\Omega}_{\breve{\boldsymbol{\beta}}} = \sigma_{\varepsilon}^{2}\left(\left(\boldsymbol{X}'\boldsymbol{X}\right)^{-1} + \boldsymbol{NN}'\right) \tag{3.246}$$

For $\breve{\boldsymbol{\beta}}$ to have minimal variance and knowing that the variances lie on the diagonal of $\boldsymbol{\Omega}_{\breve{\boldsymbol{\beta}}}$, we need to minimize the diagonal elements of $\boldsymbol{\Omega}_{\breve{\boldsymbol{\beta}}}$. Since the diagonal elements of $\left(\boldsymbol{X}'\boldsymbol{X}\right)^{-1}$ are constants, the diagonal elements of the matrix $\boldsymbol{NN}'$ must be minimized. If we denote n_{ij} the elements of the matrix $\boldsymbol{N}$, where i stands for the row and j for the column, the diagonal elements of the matrix $\boldsymbol{NN}'$ are given by: $\sum_{j} n_{ij}^{2}$. These elements are minimal if $\sum_{j} n_{ij}^{2} = 0$, or $n_{ij} = 0\ \forall i, \forall j$. We deduce:

$$\boldsymbol{N} = 0 \tag{3.247}$$

Therefore:

$$\breve{\boldsymbol{\beta}} = \hat{\boldsymbol{\beta}} \tag{3.248}$$

It follows that the OLS estimator $\hat{\boldsymbol{\beta}}$ is of minimum variance.

Appendix 3.2.2: Calculation of the Error Variance

In order to estimate the variance σ_ε^2 of the errors, we need to use the residuals $\boldsymbol{e}$:

$$\boldsymbol{e} = \boldsymbol{Y} - \boldsymbol{X}\hat{\boldsymbol{\beta}} \tag{3.249}$$

We have:

$$\begin{aligned} \boldsymbol{e} &= \boldsymbol{X\beta} + \boldsymbol{\varepsilon} - \boldsymbol{X}\left(\boldsymbol{X}'\boldsymbol{X}\right)^{-1}\boldsymbol{X}'\boldsymbol{Y} \\ &= \boldsymbol{X\beta} + \boldsymbol{\varepsilon} - \boldsymbol{X}\left(\boldsymbol{X}'\boldsymbol{X}\right)^{-1}\boldsymbol{X}'\left(\boldsymbol{X\beta} + \boldsymbol{\varepsilon}\right) \end{aligned} \tag{3.250}$$

Hence:

$$\boldsymbol{e} = \boldsymbol{\varepsilon} - \boldsymbol{X}\left(\boldsymbol{X}'\boldsymbol{X}\right)^{-1}\boldsymbol{X}'\boldsymbol{\varepsilon} \tag{3.251}$$

Noting $\boldsymbol{P} = \boldsymbol{I} - \boldsymbol{X}\left(\boldsymbol{X}'\boldsymbol{X}\right)^{-1}\boldsymbol{X}'$, we can write:

$$\boldsymbol{e} = \boldsymbol{P\varepsilon} \tag{3.252}$$

Let us study the properties of the matrix $\boldsymbol{P}$.

- $\boldsymbol{P}' = \left[\boldsymbol{I} - \boldsymbol{X}\left(\boldsymbol{X}'\boldsymbol{X}\right)^{-1}\boldsymbol{X}'\right]' = \boldsymbol{I} - \boldsymbol{X}\left(\boldsymbol{X}'\boldsymbol{X}\right)^{-1}\boldsymbol{X}' = \boldsymbol{P}$. $\boldsymbol{P}$ is therefore a symmetric matrix.
- $\boldsymbol{P}^2 = \left[\boldsymbol{I} - \boldsymbol{X}\left(\boldsymbol{X}'\boldsymbol{X}\right)^{-1}\boldsymbol{X}'\right]\left[\boldsymbol{I} - \boldsymbol{X}\left(\boldsymbol{X}'\boldsymbol{X}\right)^{-1}\boldsymbol{X}'\right] = \boldsymbol{P}$. $\boldsymbol{P}$ is therefore an idempotent matrix.

We have:

$$\boldsymbol{e}'\boldsymbol{e} = \boldsymbol{\varepsilon}'\boldsymbol{P}'\boldsymbol{P\varepsilon} = \boldsymbol{\varepsilon}'\boldsymbol{P\varepsilon} \tag{3.253}$$

by virtue of the idempotency and symmetry properties of the matrix $\boldsymbol{P}$.

Let us now determine the mathematical expectation of $\boldsymbol{e}'\boldsymbol{e}$:

$$E\left(\boldsymbol{e}'\boldsymbol{e}\right) = E\left(\boldsymbol{\varepsilon}'\boldsymbol{P\varepsilon}\right) \tag{3.254}$$

Since $\boldsymbol{\varepsilon}'\boldsymbol{P\varepsilon}$ is a scalar and noting Tr the trace, we have:

$$E\left(\boldsymbol{e}'\boldsymbol{e}\right) = E\left[Tr\left(\boldsymbol{\varepsilon}'\boldsymbol{P\varepsilon}\right)\right] \tag{3.255}$$

or:

$$E\left(\boldsymbol{e}'\boldsymbol{e}\right) = E\left[Tr\left(\boldsymbol{P\varepsilon\varepsilon}'\right)\right] \tag{3.256}$$

using the fact that $Tr(\boldsymbol{AB}) = Tr(\boldsymbol{BA})$ with $\boldsymbol{A} = \boldsymbol{\varepsilon}'$ and $\boldsymbol{B} = \boldsymbol{P\varepsilon}$. Hence:

$$E\left(\boldsymbol{e}'\boldsymbol{e}\right) = \sigma_\varepsilon^2 Tr\left(\boldsymbol{P}\right) \tag{3.257}$$

It remains for us to determine the trace of the matrix $\boldsymbol{P}$:

$$\begin{aligned} Tr\left(\boldsymbol{P}\right) &= Tr\left[\boldsymbol{I} - \boldsymbol{X}\left(\boldsymbol{X}'\boldsymbol{X}\right)^{-1}\boldsymbol{X}'\right] \\ &= Tr\boldsymbol{I} - Tr\left[\boldsymbol{X}\left(\boldsymbol{X}'\boldsymbol{X}\right)^{-1}\boldsymbol{X}'\right] \\ &= Tr\boldsymbol{I} - Tr\left[\left(\boldsymbol{X}'\boldsymbol{X}\right)^{-1}\boldsymbol{X}'\boldsymbol{X}\right] \end{aligned} \tag{3.258}$$

$Tr\boldsymbol{I} = T$ and $Tr\left[\left(\boldsymbol{X}'\boldsymbol{X}\right)^{-1}\boldsymbol{X}'\boldsymbol{X}\right] = k + 1$ since the matrix $\boldsymbol{X}'\boldsymbol{X}$ is of size $(k + 1, k + 1)$. We deduce:

$$Tr\left(\boldsymbol{P}\right) = T - k - 1 \tag{3.259}$$

Finally:

$$E\left(\boldsymbol{e}'\boldsymbol{e}\right) = \sigma_\varepsilon^2\left(T - k - 1\right) \tag{3.260}$$

It follows that the estimator $\hat{\sigma}_\varepsilon^2$ of the error variance is therefore written as:

$$\hat{\sigma}_\varepsilon^2 = \frac{\boldsymbol{e}'\boldsymbol{e}}{T - k - 1} \equiv \frac{1}{T - k - 1}\sum_{t=1}^{T} e_t^2 \tag{3.261}$$

This is an unbiased estimator of σ_ε^2.

Appendix 3.2.3: Significance Tests of Several Coefficients

In order to derive the various significance tests, we need to determine the distribution followed by $\boldsymbol{R\beta}$. $\boldsymbol{\beta}$ being unknown, let us replace it by its estimator:

$$\boldsymbol{R}\hat{\boldsymbol{\beta}} = \boldsymbol{r} \tag{3.262}$$

and determine the distribution followed by $\boldsymbol{R}\hat{\boldsymbol{\beta}}$. Knowing that $\hat{\boldsymbol{\beta}}$ is an unbiased estimator of $\boldsymbol{\beta}$, we can write:

$$E\left(\boldsymbol{R}\hat{\boldsymbol{\beta}}\right) = \boldsymbol{R\beta} \tag{3.263}$$

Furthermore:

$$V\left(\boldsymbol{R}\hat{\boldsymbol{\beta}}\right)=E\left[\boldsymbol{R}\left(\hat{\boldsymbol{\beta}}-\boldsymbol{\beta}\right)\left(\hat{\boldsymbol{\beta}}-\boldsymbol{\beta}\right)'\boldsymbol{R}'\right] \tag{3.264}$$

Hence:

$$V\left(\boldsymbol{R}\hat{\boldsymbol{\beta}}\right)=\sigma_\varepsilon^2\boldsymbol{R}\left(\boldsymbol{X}'\boldsymbol{X}\right)^{-1}\boldsymbol{R}' \tag{3.265}$$

We know that $\hat{\boldsymbol{\beta}}$ follows a normal distribution with $(k+1)$ dimensions, therefore:

$$\boldsymbol{R}\hat{\boldsymbol{\beta}}\sim N\left(\boldsymbol{R}\hat{\boldsymbol{\beta}},\sigma_\varepsilon^2\boldsymbol{R}\left(\boldsymbol{X}'\boldsymbol{X}\right)^{-1}\boldsymbol{R}'\right) \tag{3.266}$$

and:

$$\boldsymbol{R}\left(\hat{\boldsymbol{\beta}}-\boldsymbol{\beta}\right)\sim N\left(0,\sigma_\varepsilon^2\boldsymbol{R}\left(\boldsymbol{X}'\boldsymbol{X}\right)^{-1}\boldsymbol{R}'\right) \tag{3.267}$$

Under the null hypothesis $\boldsymbol{R}\boldsymbol{\beta}=\boldsymbol{r}$, we therefore have:

$$\boldsymbol{R}\hat{\boldsymbol{\beta}}-\boldsymbol{r}\sim N\left(0,\sigma_\varepsilon^2\boldsymbol{R}\left(\boldsymbol{X}'\boldsymbol{X}\right)^{-1}\boldsymbol{R}'\right) \tag{3.268}$$

Using the result that if $\boldsymbol{w}\sim N\left(0,\boldsymbol{\Sigma}\right)$ where $\boldsymbol{\Sigma}$ is of size (K,K), we have $\boldsymbol{w}'\boldsymbol{\Sigma}^{-1}\boldsymbol{w}\sim\chi_K^2$; then:

$$\left(\boldsymbol{R}\hat{\boldsymbol{\beta}}-\boldsymbol{r}\right)'\left[\sigma_\varepsilon^2\boldsymbol{R}\left(\boldsymbol{X}'\boldsymbol{X}\right)^{-1}\boldsymbol{R}'\right]^{-1}\left(\boldsymbol{R}\hat{\boldsymbol{\beta}}-\boldsymbol{r}\right)\sim\chi_q^2 \tag{3.269}$$

Knowing that:

$$\frac{\boldsymbol{e}'\boldsymbol{e}}{\sigma_\varepsilon^2}\sim\chi_{T-k-1}^2 \tag{3.270}$$

and using the result (see Box 2.2 in Chap. 2) that if $w\sim\chi_s^2$ and $v\sim\chi_r^2$, the statistic $F=\frac{w/s}{v/r}$ follows a Fisher distribution with (s,r) degrees of freedom, we deduce:

$$F=\frac{\left(\boldsymbol{R}\hat{\boldsymbol{\beta}}-\boldsymbol{r}\right)'\left[\boldsymbol{R}\left(\boldsymbol{X}'\boldsymbol{X}\right)^{-1}\boldsymbol{R}'\right]^{-1}\left(\boldsymbol{R}\hat{\boldsymbol{\beta}}-\boldsymbol{r}\right)/q}{\boldsymbol{e}'\boldsymbol{e}/\left(T-k-1\right)}\sim F\left(q,T-k-1\right) \tag{3.271}$$

We then have the following decision rule:

- If $F \leq F(q, T-k-1)$, the null hypothesis is not rejected, i.e.: $\boldsymbol{R\beta} = \boldsymbol{r}$.
- If $F > F(q, T-k-1)$, the null hypothesis is rejected.

Let us now return to the three special cases studied—test on a single coefficient, test on all coefficients, and test on a subset of coefficients—to specify the expression of the test in each of these cases.

- *Test on a particular regression coefficient* β_i. This case corresponds to the null hypothesis $\beta_i = 0$, i.e.:

$$\boldsymbol{R} = [0 \cdots 0\ 1\ 0 \cdots 0] \text{ and } \boldsymbol{r} = 0 \tag{3.272}$$

We then have $\boldsymbol{R}\hat{\boldsymbol{\beta}} - \boldsymbol{r} = \beta_i$ and the quadratic form $\boldsymbol{R}\left(\boldsymbol{X}'\boldsymbol{X}\right)^{-1}\boldsymbol{R}'$ is equal to the $(i+1)$th element of the diagonal of the matrix $\left(\boldsymbol{X}'\boldsymbol{X}\right)^{-1}$, i.e., $a_{i+1,i+1}$.
The test statistic given in (3.271) becomes:

$$F = \frac{\beta_i^2\left[a_{i+1,i+1}\right]^{-1}/1}{\hat{\sigma}_\varepsilon^2} \sim F(1, T-k-1) \tag{3.273}$$

that is finally:

$$F = \frac{\beta_i^2}{\hat{\sigma}_\varepsilon^2 a_{i+1,i+1}} \sim F(1, T-k-1) \tag{3.274}$$

This test is equivalent to a Student's significance test on a single coefficient since $[t(T-k-1)]^2 = F(1, T-k-1)$. The same reasoning applies to the test of the null hypothesis $\beta_i = \beta_0$.
- *Test of significance of all coefficients*. This case corresponds to the null hypothesis $\beta_1 = \beta_2 = \cdots = \beta_k = 0$, i.e.:

$$\boldsymbol{R} = \begin{pmatrix} 0 & 1 & 0 & \cdots & 0 \\ 0 & 0 & 1 & \cdots & 0 \\ \vdots & \vdots & \vdots & \ddots & \vdots \\ 0 & \cdots & \cdots & 0 & 1 \end{pmatrix} \text{ and } \boldsymbol{r} = \begin{pmatrix} 0 \\ 0 \\ \vdots \\ 0 \end{pmatrix} \tag{3.275}$$

We then have $\boldsymbol{R}\hat{\boldsymbol{\beta}} - \boldsymbol{r} = \bar{\boldsymbol{\beta}}$ with $\bar{\boldsymbol{\beta}} = \left(\hat{\beta}_1\ \hat{\beta}_2 \cdots \hat{\beta}_k\right)'$, i.e., $\bar{\boldsymbol{\beta}}$ is the vector of OLS coefficients without the constant term. Furthermore, the matrix $\boldsymbol{R}\left(\boldsymbol{X}'\boldsymbol{X}\right)^{-1}\boldsymbol{R}'$ involved in calculating the test statistic (Eq. (3.271)) is equal to the submatrix of size (k, k) obtained by deleting the first row and column of the matrix $\left(\boldsymbol{X}'\boldsymbol{X}\right)^{-1}$. To clarify the expression of this submatrix, let us decompose the matrix $\boldsymbol{X}$ into two blocks:

$$\boldsymbol{X} = \left(\bar{\boldsymbol{x}}\ \bar{\boldsymbol{X}}\right) \tag{3.276}$$

where $\bar{\boldsymbol{x}}$ denotes a column vector composed of 1 and $\bar{\boldsymbol{X}}$ is the matrix of size (T, k) comprising the values of the k explanatory variables. We then have:

$$\boldsymbol{X}'\boldsymbol{X} = \begin{pmatrix} T & \bar{\boldsymbol{x}}'\bar{\boldsymbol{X}} \\ \bar{\boldsymbol{X}}'\bar{\boldsymbol{x}} & \bar{\boldsymbol{X}}'\bar{\boldsymbol{X}} \end{pmatrix} \tag{3.277}$$

The calculation of $\left(\boldsymbol{X}'\boldsymbol{X}\right)^{-1}$ shows us that the submatrix of size (k, k) that interests us here is written as:

$$\left(\bar{\boldsymbol{X}}'\bar{\boldsymbol{X}} - \bar{\boldsymbol{X}}'\bar{\boldsymbol{x}}T^{-1}\bar{\boldsymbol{x}}'\bar{\boldsymbol{X}}\right)^{-1} = \left(\bar{\boldsymbol{X}}'\boldsymbol{Z}\bar{\boldsymbol{X}}\right)^{-1} \tag{3.278}$$

where $\boldsymbol{Z}$ is the transformation matrix given by:

$$\boldsymbol{Z} = \boldsymbol{I} - T^{-1}\bar{\boldsymbol{x}}\bar{\boldsymbol{x}}' \tag{3.279}$$

Relationship (3.271) then becomes:

$$F = \frac{\bar{\boldsymbol{\beta}}'\left(\bar{\boldsymbol{X}}'\boldsymbol{Z}\bar{\boldsymbol{X}}\right)\bar{\boldsymbol{\beta}}/q}{\boldsymbol{e}'\boldsymbol{e}/\left(T - k - 1\right)} \tag{3.280}$$

or, knowing that $q = k$:

$$F = \frac{\bar{\boldsymbol{\beta}}'\left(\bar{\boldsymbol{X}}'\boldsymbol{Z}\bar{\boldsymbol{X}}\right)\bar{\boldsymbol{\beta}}/k}{\boldsymbol{e}'\boldsymbol{e}/\left(T - k - 1\right)} \tag{3.281}$$

The decision rule is given by:
- If $F \leq F\left(q, T - k - 1\right)$, the null hypothesis that all explanatory variables are not significant is not rejected.
- If $F > F\left(q, T - k - 1\right)$, the null hypothesis is rejected.

As we have seen in the chapter, this test can also be apprehended through the analysis-of-variance equation.

- *Test of significance of a subset of coefficients*. This case corresponds to the null hypothesis: $\beta_{k-s+2} = \beta_{k-s+3} = \cdots = \beta_k = 0$, i.e.:

$$\boldsymbol{R} = \left[\boldsymbol{0}\ \boldsymbol{I}_s\right] \text{ and } \boldsymbol{r} = \boldsymbol{0} \tag{3.282}$$

Let us decompose the matrix $\boldsymbol{X}$ and the vector $\boldsymbol{\beta}$ into blocks so that:

$$\begin{aligned} \boldsymbol{Y} &= \left(\boldsymbol{X}_r \ \boldsymbol{X}_s\right) \begin{pmatrix} \hat{\boldsymbol{\beta}}_r \\ \hat{\boldsymbol{\beta}}_s \end{pmatrix} + \boldsymbol{e} \\ &= \boldsymbol{X}_r \hat{\boldsymbol{\beta}}_r + \boldsymbol{X}_s \hat{\boldsymbol{\beta}}_s + \boldsymbol{e} \end{aligned} \tag{3.283}$$

where the matrix $\boldsymbol{X}_r$ is formed by the $(k+1-s)$ first columns of $\boldsymbol{X}$ and $\boldsymbol{X}_s$ is formed by the s remaining columns of the matrix $\boldsymbol{X}$. We then have: $\boldsymbol{R}\hat{\boldsymbol{\beta}} - \boldsymbol{r} = \hat{\boldsymbol{\beta}}_s$. Furthermore, the matrix $\boldsymbol{R}\left(\boldsymbol{X}'\boldsymbol{X}\right)^{-1}\boldsymbol{R}'$ involved in calculating the test statistic (Eq. (3.271)) is equal to the submatrix of order s obtained by deleting the $(k+1-s)$ first rows and columns of the matrix $\left(\boldsymbol{X}'\boldsymbol{X}\right)^{-1}$. Let us explain the form of this submatrix. We have:

$$\boldsymbol{X}'\boldsymbol{X} = \begin{pmatrix} \boldsymbol{X}_r'\boldsymbol{X}_r & \boldsymbol{X}_r'\boldsymbol{X}_s \\ \boldsymbol{X}_s'\boldsymbol{X}_r & \boldsymbol{X}_s'\boldsymbol{X}_s \end{pmatrix} \tag{3.284}$$

The calculation of $\left(\boldsymbol{X}'\boldsymbol{X}\right)^{-1}$ shows us that the submatrix we are interested in here is written as:

$$\begin{aligned} \left(\boldsymbol{X}_s'\boldsymbol{X}_s - \boldsymbol{X}_s'\boldsymbol{X}_r\left(\boldsymbol{X}_r'\boldsymbol{X}_r\right)^{-1}\boldsymbol{X}_r'\boldsymbol{X}_s\right)^{-1} &= \left[\boldsymbol{X}_s'\left(\boldsymbol{I} - \boldsymbol{X}_r\left(\boldsymbol{X}_r'\boldsymbol{X}_r\right)^{-1}\boldsymbol{X}_r'\right)\boldsymbol{X}_s\right]^{-1} \\ &= \left[\boldsymbol{X}_s'\boldsymbol{Z}_r\boldsymbol{X}_s\right]^{-1} \end{aligned} \tag{3.285}$$

where $\boldsymbol{Z}$ is the transformation matrix given by:

$$\boldsymbol{Z}_r = \boldsymbol{I} - \boldsymbol{X}_r\left(\boldsymbol{X}_r'\boldsymbol{X}_r\right)^{-1}\boldsymbol{X}_r' \tag{3.286}$$

Relationship (3.271) then becomes:

$$F = \frac{\hat{\boldsymbol{\beta}}_s'\left(\boldsymbol{X}_s'\boldsymbol{Z}_r\boldsymbol{X}_s\right)\hat{\boldsymbol{\beta}}_s/s}{\boldsymbol{e}'\boldsymbol{e}/\left(T-k-1\right)} \tag{3.287}$$

Let us now explain the expression of the numerator. To do this, consider the regression of $\boldsymbol{Y}$ on the explanatory variables listed in $\boldsymbol{X}_r$ and note $\boldsymbol{e}_r$ the residuals resulting from this regression:

$$\begin{aligned} \boldsymbol{e}_r &= \boldsymbol{Y} - \boldsymbol{X}_r\hat{\boldsymbol{\beta}}_r \\ &= \boldsymbol{Y} - \boldsymbol{X}_r\left(\left(\boldsymbol{X}_r'\boldsymbol{X}_r\right)^{-1}\boldsymbol{X}_r'\boldsymbol{Y}\right) \\ &= \boldsymbol{Z}_r\boldsymbol{Y} \end{aligned} \tag{3.288}$$

Let us multiply each member of (3.283) by the matrix $\boldsymbol{Z_r}$:

$$\boldsymbol{Z_r Y} = \boldsymbol{Z_r X_r \hat{\beta}_r} + \boldsymbol{Z_r X_s \hat{\beta}_s} + \boldsymbol{Z_r e} \tag{3.289}$$

We have:

- $\boldsymbol{Z_r X_r} = \boldsymbol{X_r} - \boldsymbol{X_r}\left(\boldsymbol{X_r' X_r}\right)^{-1}\boldsymbol{X_r' X_r} = \boldsymbol{0}$
- $\boldsymbol{Z_r} = \boldsymbol{Z_r^2} = \boldsymbol{Z_r'}$ (idempotent and symmetric matrix)
- $\boldsymbol{Z_r e} = \boldsymbol{e}$ because:

$$\begin{aligned}\boldsymbol{Z_r e} &= \left(\boldsymbol{I} - \boldsymbol{X_r}\left(\boldsymbol{X_r' X_r}\right)^{-1}\boldsymbol{X_r'}\right)\left(\boldsymbol{Y} - \boldsymbol{X\hat{\beta}}\right) \\ &= \boldsymbol{Y} - \boldsymbol{X\hat{\beta}} - \boldsymbol{X_r}\left(\boldsymbol{X_r' X_r}\right)^{-1}\boldsymbol{X_r' Y} + \boldsymbol{X_r}\left(\boldsymbol{X_r' X_r}\right)^{-1}\boldsymbol{X_r' X\hat{\beta}} \\ &= \boldsymbol{Y} - \boldsymbol{X\hat{\beta}} - \boldsymbol{X_r\hat{\beta}_r} + \boldsymbol{X_r}\left(\boldsymbol{X_r' X_r}\right)^{-1}\boldsymbol{X_r'}\left(\boldsymbol{Y} - \boldsymbol{e}\right)\end{aligned} \tag{3.290}$$

Let us find the value of $\boldsymbol{X_r' e}$. We know that:

$$\begin{aligned}\left(\boldsymbol{X'X}\right)\boldsymbol{\hat{\beta}} &= \boldsymbol{X'Y} \\ &= \boldsymbol{X'}\left(\boldsymbol{X\hat{\beta}} + \boldsymbol{e}\right)\end{aligned} \tag{3.291}$$

We therefore deduce that:

$$\boldsymbol{X'}\ \boldsymbol{e} = \begin{pmatrix}\boldsymbol{X_r' e} \\ \boldsymbol{X_r' e}\end{pmatrix} = \boldsymbol{0} \tag{3.292}$$

Finally, we have:

$$\boldsymbol{Z_r e} = \boldsymbol{Y} - \boldsymbol{X\hat{\beta}} - \boldsymbol{X_r\hat{\beta}_r} + \boldsymbol{X_r\hat{\beta}_r} = \boldsymbol{e}$$

Hence:

$$\boldsymbol{Z_r Y} = \boldsymbol{Z_r X_s \hat{\beta}_s} + \boldsymbol{e} \tag{3.293}$$

Let us multiply each member of this relation by its transpose:

$$\boldsymbol{Y' Z_r' Z_r Y} = \boldsymbol{\hat{\beta}_s' X_s' Z_r' Z_r X_s \hat{\beta}_s} + \boldsymbol{e'e} \tag{3.294}$$

Knowing that $\boldsymbol{Z_r} = \boldsymbol{Z_r'}$, we get:

$$\boldsymbol{Y' Z_r Y} = \boldsymbol{\hat{\beta}_s' X_s' Z_r X_s \hat{\beta}_s} + \boldsymbol{e'e} \tag{3.295}$$

Furthermore, since $\boldsymbol{e_r} = \boldsymbol{Z_r Y}$, we have: $\boldsymbol{e_r' e_r} = \boldsymbol{Y' Z_r' Z_r Y} = \boldsymbol{Y' Z_r Y}$. We then obtain the following result:

$$\hat{\boldsymbol{\beta}}_s' \boldsymbol{X}_s' \boldsymbol{Z_r} \boldsymbol{X}_s \hat{\boldsymbol{\beta}}_s = \boldsymbol{e_r' e_r} - \boldsymbol{e' e} \tag{3.296}$$

Relationship (3.271) is finally written as:

$$F = \frac{\left(\boldsymbol{e_r' e_r} - \boldsymbol{e' e}\right)/s}{\boldsymbol{e' e}/\left(T-k-1\right)} \sim F\left(s, T-k-1\right) \tag{3.297}$$

This test, which is very frequently employed, can be used to test the significance of a subset of explanatory variables $\boldsymbol{X}_s$. In practice, it consists in running two regressions:

- A regression of $\boldsymbol{Y}$ on the set of explanatory variables, $\boldsymbol{e'e}$ being the corresponding sum of squared residuals
- A regression of $\boldsymbol{Y}$ on the subset of explanatory variables $\boldsymbol{X_r}$ (i.e., on variables other than $\boldsymbol{X}_s$), $\boldsymbol{e_r' e_r}$ being the corresponding sum of squared residuals

The decision rule is as follows:

- If $F \leq F\left(s, T-k-1\right)$, the null hypothesis that the variables $\boldsymbol{X}_s$ are not significant is not rejected.
- If $F > F\left(s, T-k-1\right)$, the null hypothesis is rejected.

Heteroskedasticity and Autocorrelation of Errors

4

The regression models studied in previous chapters were based on a number of assumptions:

- The nonrandom nature of the matrix of explanatory variables $\boldsymbol{X}$.
- $Rank(\boldsymbol{X}) = k + 1$, k denoting the number of explanatory variables.
- $E(\boldsymbol{\varepsilon}) = 0$.
- $E(\boldsymbol{\varepsilon}\boldsymbol{\varepsilon}') = \sigma_\varepsilon^2 \boldsymbol{I}$ where $\boldsymbol{I}$ denotes the identity matrix and σ_ε^2 the variance of the error term.
- $\boldsymbol{\varepsilon} \sim N(0, \sigma_\varepsilon^2 \boldsymbol{I})$: this normality assumption is not necessary to establish the results of the multiple regression model, but it does allow statistical results to be derived and test statistics to be constructed.

We are interested here in the **sphericity of errors** hypothesis:

$$E(\boldsymbol{\varepsilon}\boldsymbol{\varepsilon}') = \sigma_\varepsilon^2 \boldsymbol{I} \tag{4.1}$$

This hypothesis combines two characteristics:

- The absence of autocorrelation of errors.
- The homoskedasticity of errors, i.e., the fact that the variance of the errors is constant. When such an assumption is violated, we speak of **heteroskedasticity**: the variance of the errors is no longer constant. It varies over time or with the observations of one or more explanatory variables.

This chapter focuses on the case where the hypothesis of sphericity of errors is not verified. We concentrate on the problems of **autocorrelation** and **heterroskedasticity** of errors by seeking to answer the following questions:

V. Mignon, *Principles of Econometrics*, Classroom Companion: Economics,
https://doi.org/10.1007/978-3-031-52535-3_4

- How can we estimate the parameters of a regression model in the presence of autocorrelation and/or heteroskedasticity of errors?
- What are the sources of autocorrelation and heteroskedasticity?
- How can we detect autocorrelation and heteroskedasticity?
- What are the solutions to autocorrelation and heteroskedasticity?

4.1 The Generalized Least Squares (GLS) Estimators

The variance-covariance matrix of the error term is written:

$$E\left(\boldsymbol{\varepsilon}\boldsymbol{\varepsilon}'\right) = \begin{pmatrix} V(\varepsilon_1) & Cov(\varepsilon_1, \varepsilon_2) & \cdots & Cov(\varepsilon_1, \varepsilon_T) \\ Cov(\varepsilon_2, \varepsilon_1) & V(\varepsilon_2) & \cdots & Cov(\varepsilon_2, \varepsilon_T) \\ \vdots & \vdots & \ddots & \vdots \\ Cov(\varepsilon_T, \varepsilon_1) & Cov(\varepsilon_T, \varepsilon_2) & \cdots & V(\varepsilon_T) \end{pmatrix} \tag{4.2}$$

If the errors are autocorrelated, the terms off the diagonal are not all zero. Similarly, if the errors are heteroskedastic, the terms on the diagonal are not all identical.

4.1.1 Properties of OLS Estimators in the Presence of Autocorrelation and/or Heteroskedasticity

Consider the following model:

$$\underset{(T,1)}{\boldsymbol{Y}} = \underset{(T,k+1)}{\boldsymbol{X}}\underset{(k+1,1)}{\boldsymbol{\beta}} + \underset{(T,1)}{\boldsymbol{\varepsilon}} \tag{4.3}$$

where $\boldsymbol{X}$ is a nonrandom matrix of full rank, with:

- $E(\boldsymbol{\varepsilon}) = 0$,
- $E\left(\boldsymbol{\varepsilon}\boldsymbol{\varepsilon}'\right) = \boldsymbol{\Omega}_{\varepsilon}$

where $\boldsymbol{\Omega}_{\varepsilon} \neq \sigma_{\varepsilon}^2 \boldsymbol{I}$ denotes the variance-covariance matrix of the errors. The fact that $\boldsymbol{\Omega}_{\varepsilon} \neq \sigma_{\varepsilon}^2 \boldsymbol{I}$ means there is autocorrelation and/or heteroskedasticity of the errors.

The OLS estimator $\hat{\boldsymbol{\beta}}$ of $\boldsymbol{\beta}$ is given by:

$$\hat{\boldsymbol{\beta}} = \boldsymbol{\beta} + \left(\boldsymbol{X}'\boldsymbol{X}\right)^{-1} \boldsymbol{X}'\boldsymbol{\varepsilon} \tag{4.4}$$

by virtue of Eq. (3.30) in Chap. 3. Given that $E(\boldsymbol{\varepsilon}) = 0$, we have:

$$E\left(\hat{\boldsymbol{\beta}}\right) = \boldsymbol{\beta} \tag{4.5}$$

Thus, when there is autocorrelation and/or heteroskedasticity of errors, the OLS estimator remains an unbiased estimator of $\boldsymbol{\beta}$.

Let us now examine whether the estimator is still of minimum variance.

We have:

$$\boldsymbol{\Omega}_{\hat{\boldsymbol{\beta}}} = E\left[\left(\hat{\boldsymbol{\beta}} - \boldsymbol{\beta}\right)\left(\hat{\boldsymbol{\beta}} - \boldsymbol{\beta}\right)'\right] \tag{4.6}$$

which can also be written (see Chap. 3):

$$\boldsymbol{\Omega}_{\hat{\boldsymbol{\beta}}} = \left(\boldsymbol{X}'\boldsymbol{X}\right)^{-1}\boldsymbol{X}'E\left(\boldsymbol{\varepsilon}\boldsymbol{\varepsilon}'\right)\boldsymbol{X}\left(\boldsymbol{X}'\boldsymbol{X}\right)^{-1} \tag{4.7}$$

that is:

$$\boldsymbol{\Omega}_{\hat{\boldsymbol{\beta}}} = \left(\boldsymbol{X}'\boldsymbol{X}\right)^{-1}\boldsymbol{X}'\boldsymbol{\Omega}_{\varepsilon}\boldsymbol{X}\left(\boldsymbol{X}'\boldsymbol{X}\right)^{-1} \tag{4.8}$$

This expression is different from that obtained when there is neither autocorrelation nor heteroskedasticity, i.e., $\sigma_{\varepsilon}^{2}\left(\boldsymbol{X}'\boldsymbol{X}\right)^{-1}$. It follows that autocorrelation and/or heteroskedasticity implies that the OLS estimators are no longer of minimum variance. It is therefore necessary to define other estimators: the generalized least squares estimators.

4.1.2 The Generalized Least Squares (GLS) Method

Consider the multiple regression model:

$$\boldsymbol{Y} = \boldsymbol{X}\boldsymbol{\beta} + \boldsymbol{\varepsilon} \tag{4.9}$$

and multiply each term by a nonsingular transformation matrix $\boldsymbol{M}$ of size (T, T), i.e.:

$$\boldsymbol{M}\boldsymbol{Y} = \boldsymbol{M}\boldsymbol{X}\boldsymbol{\beta} + \boldsymbol{M}\boldsymbol{\varepsilon} \tag{4.10}$$

Let us determine the variance-covariance matrix of the error term $(\boldsymbol{M}\boldsymbol{\varepsilon})$:

$$E\left(\boldsymbol{M}\boldsymbol{\varepsilon}\boldsymbol{\varepsilon}'\boldsymbol{M}'\right) = \boldsymbol{M}\boldsymbol{\Omega}_{\boldsymbol{\varepsilon}}\boldsymbol{M}' = \sigma_{\varepsilon}^{2}\boldsymbol{M}\boldsymbol{\Gamma}_{\boldsymbol{\varepsilon}}\boldsymbol{M}' \tag{4.11}$$

with $\boldsymbol{\Gamma}_{\boldsymbol{\varepsilon}} = \left(\sigma_{\varepsilon}^{2}\right)^{-1}\boldsymbol{\Omega}_{\boldsymbol{\varepsilon}}$. If there were a matrix $\boldsymbol{M}$ such as:

$$\boldsymbol{M}\boldsymbol{\Gamma}_{\boldsymbol{\varepsilon}}\boldsymbol{M}' = \boldsymbol{I} \tag{4.12}$$

we could apply OLS to model (4.10). The resulting estimators would then have the same properties as in the usual case.

We saw in Chap. 3 that if $\boldsymbol{\Gamma}_{\varepsilon}$ is a positive definite symmetric matrix, there exists a nonsingular—and therefore invertible—matrix $\boldsymbol{P}$ such that:

$$\boldsymbol{\Gamma}_{\varepsilon} = \boldsymbol{P}\boldsymbol{P}' \tag{4.13}$$

We can then write:

$$\boldsymbol{P}^{-1}\boldsymbol{\Gamma}_{\varepsilon}\left(\boldsymbol{P}'\right)^{-1} = \boldsymbol{P}^{-1}\boldsymbol{P}\boldsymbol{P}'\left(\boldsymbol{P}'\right)^{-1} = \boldsymbol{I} \tag{4.14}$$

Comparing (4.14) with (4.12), we get:

$$\boldsymbol{M} = \boldsymbol{P}^{-1} \tag{4.15}$$

Furthermore:

$$\boldsymbol{\Gamma}_{\varepsilon}^{-1} = \left(\boldsymbol{P}'\right)^{-1}\boldsymbol{P}^{-1} = \boldsymbol{M}'\boldsymbol{M} \tag{4.16}$$

If we apply OLS to model (4.10), we obtain the following estimator $\tilde{\boldsymbol{\beta}}$:

$$\tilde{\boldsymbol{\beta}} = \left(\boldsymbol{X}'\boldsymbol{M}'\boldsymbol{M}\boldsymbol{X}\right)^{-1}\boldsymbol{X}'\boldsymbol{M}'\boldsymbol{M}\boldsymbol{Y} \tag{4.17}$$

thus:

$$\tilde{\boldsymbol{\beta}} = \left(\boldsymbol{X}'\boldsymbol{\Gamma}_{\varepsilon}^{-1}\boldsymbol{X}\right)^{-1}\boldsymbol{X}'\boldsymbol{\Gamma}_{\varepsilon}^{-1}\boldsymbol{Y} \tag{4.18}$$

or:

$$\tilde{\boldsymbol{\beta}} = \left(\boldsymbol{X}'\boldsymbol{\Omega}_{\varepsilon}^{-1}\boldsymbol{X}\right)^{-1}\boldsymbol{X}'\boldsymbol{\Omega}_{\varepsilon}^{-1}\boldsymbol{Y} \tag{4.19}$$

The estimator $\tilde{\boldsymbol{\beta}}$ given by Eq. (4.19) is called the **generalized least squares (GLS) estimator** (or **Aitken** estimator). Since it was obtained from expressions (4.11) and (4.12) and since the model (4.10) satisfies the assumptions required for the OLS, the GLS estimator $\tilde{\boldsymbol{\beta}}$ is the best linear unbiased estimator of $\boldsymbol{\beta}$ in the model $\boldsymbol{Y} = \boldsymbol{X}\boldsymbol{\beta} + \boldsymbol{\varepsilon}$ with $E\left(\boldsymbol{\varepsilon}\boldsymbol{\varepsilon}'\right) = \boldsymbol{\Omega}_{\varepsilon}$. $\tilde{\boldsymbol{\beta}}$ is therefore a **BLUE estimator**.

Let us determine the variance-covariance matrix $\boldsymbol{\Omega}_{\tilde{\beta}}$ of $\tilde{\boldsymbol{\beta}}$:

$$\begin{aligned}\boldsymbol{\Omega}_{\tilde{\beta}} &= E\left[\left(\tilde{\boldsymbol{\beta}} - \boldsymbol{\beta}\right)\left(\tilde{\boldsymbol{\beta}} - \boldsymbol{\beta}\right)'\right] \\ &= E\left[\boldsymbol{M}\boldsymbol{\varepsilon}\boldsymbol{\varepsilon}'\boldsymbol{M}'\right] \\ &= \boldsymbol{M}\boldsymbol{\Omega}_{\varepsilon}\boldsymbol{M}'\end{aligned} \tag{4.20}$$

So:

$$\boldsymbol{\Omega}_{\tilde{\boldsymbol{\beta}}} = \left(\boldsymbol{X}'\boldsymbol{\Omega}_{\boldsymbol{\varepsilon}}^{-1}\boldsymbol{X}\right)^{-1}\boldsymbol{X}'\boldsymbol{\Omega}_{\boldsymbol{\varepsilon}}\boldsymbol{X}\left(\boldsymbol{X}'\boldsymbol{\Omega}_{\boldsymbol{\varepsilon}}^{-1}\boldsymbol{X}\right)^{-1} \tag{4.21}$$

Hence, the expression for the variance-covariance matrix of $\tilde{\boldsymbol{\beta}}$:

$$\boldsymbol{\Omega}_{\tilde{\boldsymbol{\beta}}} = \left(\boldsymbol{X}'\boldsymbol{\Omega}_{\boldsymbol{\varepsilon}}^{-1}\boldsymbol{X}\right)^{-1} \tag{4.22}$$

4.1.3 Estimation of the Variance of the Errors

We saw in Chap. 3 that the OLS estimator of the error variance was given by (see Eq. (3.261)):

$$\hat{\sigma}_{\varepsilon}^{2} = \frac{\boldsymbol{e}'\boldsymbol{e}}{T-k-1} \tag{4.23}$$

Applying OLS to model (4.10) we get:

$$\tilde{\sigma}_{\varepsilon}^{2} = \frac{\boldsymbol{e}'\boldsymbol{e}}{T-k-1} \tag{4.24}$$

where:

$$\boldsymbol{e} = \boldsymbol{MY} - \boldsymbol{MX}\tilde{\boldsymbol{\beta}} \tag{4.25}$$

Equation (4.24) then becomes:

$$\tilde{\sigma}_{\varepsilon}^{2} = \frac{\left(\boldsymbol{MY} - \boldsymbol{MX}\tilde{\boldsymbol{\beta}}\right)'\left(\boldsymbol{MY} - \boldsymbol{MX}\tilde{\boldsymbol{\beta}}\right)}{T-k-1} \tag{4.26}$$

that is:

$$\tilde{\sigma}_{\varepsilon}^{2} = \frac{\left(\boldsymbol{Y} - \boldsymbol{X}\tilde{\boldsymbol{\beta}}\right)'\boldsymbol{M}'\boldsymbol{M}\left(\boldsymbol{Y} - \boldsymbol{X}\tilde{\boldsymbol{\beta}}\right)}{T-k-1} \tag{4.27}$$

Given that $\boldsymbol{M}'\boldsymbol{M} = \boldsymbol{\Gamma}_{\boldsymbol{\varepsilon}}^{-1}$, we have:

$$\tilde{\sigma}_{\varepsilon}^{2} = \frac{\left(\boldsymbol{Y} - \boldsymbol{X}\tilde{\boldsymbol{\beta}}\right)'\boldsymbol{\Gamma}_{\boldsymbol{\varepsilon}}^{-1}\left(\boldsymbol{Y} - \boldsymbol{X}\tilde{\boldsymbol{\beta}}\right)}{T-k-1} \tag{4.28}$$

or:

$$\tilde{\sigma}_{\varepsilon}^{2} = \frac{\boldsymbol{Y}'\boldsymbol{\Gamma}_{\boldsymbol{\varepsilon}}^{-1}\boldsymbol{Y} - \tilde{\boldsymbol{\beta}}'\boldsymbol{X}'\boldsymbol{\Gamma}_{\boldsymbol{\varepsilon}}^{-1}\boldsymbol{Y}}{T-k-1} \tag{4.29}$$

Therefore, assuming that the error term is normally distributed, all the tests developed in the previous chapters can be applied here.

For the various formulas given above to be operational, the matrix $\boldsymbol{\Omega}_{\boldsymbol{\varepsilon}}$ must be known. In practice, this is not the case. Thus, to determine the matrix $\boldsymbol{\Omega}_{\boldsymbol{\varepsilon}}$, we need to specify the analytical form of autocorrelation of errors and/or heteroskedasticity. Given that the errors are unknown, it is from the residuals that we will look for such analytic forms. We start by dealing with the problem of heteroskedasticity before turning to that of autocorrelation.

4.2 Heteroskedasticity of Errors

4.2.1 The Sources of Heteroskedasticity

Recall that heteroskedasticity is present when the terms on the diagonal of the error variance-covariance matrix are not identical:[1]

$$E\left(\boldsymbol{\varepsilon}\boldsymbol{\varepsilon}'\right) = \begin{pmatrix} \sigma_{\varepsilon_1}^2 & 0 & \cdots & 0 \\ 0 & \sigma_{\varepsilon_2}^2 & \cdots & 0 \\ \vdots & \vdots & \ddots & \vdots \\ 0 & 0 & \cdots & \sigma_{\varepsilon_T}^2 \end{pmatrix} \tag{4.30}$$

We then have $E\left(\boldsymbol{\varepsilon}\boldsymbol{\varepsilon}'\right) = \boldsymbol{\Omega}_{\varepsilon} \neq \sigma_{\varepsilon}^2 \boldsymbol{I}$, which can be written as:

$$E\left(\varepsilon_t^2\right) = \sigma_{\varepsilon_t}^2 \tag{4.31}$$

for $t = 1, \ldots, T$. Note that σ_{ε}^2 has been indexed by t, meaning that the variance varies with t.

Heteroskedasticity can have several sources, including:

- The heterogeneity of the sample under consideration. This is the case, for example, if the sample studied comprises a large number of countries, bringing together developed countries and emerging or developing countries.
- The omission of an explanatory variable from the model.

[1] It is assumed here that there is no autocorrelation of errors.

- Asymmetry in the distribution of certain explanatory variables. As an example, the distribution of a variable such as income is unequal, in the sense that the richest individuals hold the lion's share of income.
- A poor variable transformation and/or a poor functional form. Heteroskedasticity may arise from the fact that the variables have not been correctly transformed (e.g., transformation of raw series into first difference or growth rates) or from the fact that the model is functionally misspecified (e.g., linear specification when the correct specification would be log linear).
- The nature of the data. If the data studied are averages of observations from samples of different sizes, this can produce heteroskedasticity.

Remark 4.1 Heteroskedasticity is a problem frequently encountered when working with cross-sectional data. Regarding time series, the most frequent cases of heteroskedasticity concern financial series, which are generally characterized by a variance varying over time.

4.2.2 Estimation When There Is Heteroskedasticity

As previously mentioned, the appropriate estimation method when there is heteroskedasticity is the generalized least squares (GLS) procedure. Let us take an illustrative example. Consider the following model with k explanatory variables:

$$Y_t = \alpha + \beta_1 X_{1t} + \beta_2 X_{2t} + \ldots + \beta_k X_{kt} + \varepsilon_t \tag{4.32}$$

with $V(\varepsilon_t) = \sigma^2_{\varepsilon_t}$. Assume that the values of $\sigma^2_{\varepsilon_t}$ are known, for $t = 1, \ldots, T$. We can then transform model (4.32) as follows:

$$\frac{Y_t}{\sigma_{\varepsilon_t}} = \frac{\alpha}{\sigma_{\varepsilon_t}} + \beta_1 \frac{X_{1t}}{\sigma_{\varepsilon_t}} + \beta_2 \frac{X_{2t}}{\sigma_{\varepsilon_t}} + \ldots + \beta_k \frac{X_{kt}}{\sigma_{\varepsilon_t}} + \frac{\varepsilon_t}{\sigma_{\varepsilon_t}} \tag{4.33}$$

or, noting $\frac{X_{it}}{\sigma_{\varepsilon_t}} = \tilde{X}_{it}$, for $i = 1, \ldots, k$, $\tilde{Y}_t = \frac{Y_t}{\sigma_{\varepsilon_t}}$ and $\tilde{\varepsilon}_t = \frac{\varepsilon_t}{\sigma_{\varepsilon_t}}$:

$$\tilde{Y}_t = \frac{\alpha}{\sigma_{\varepsilon_t}} + \beta_1 \tilde{X}_{1t} + \beta_2 \tilde{X}_{2t} + \ldots + \beta_k \tilde{X}_{kt} + \tilde{\varepsilon}_t \tag{4.34}$$

The point in transforming the original model is that the variance of the error term is now constant. Indeed:

$$\begin{aligned} V(\tilde{\varepsilon}_t) &= V\left(\frac{\varepsilon_t}{\sigma_{\varepsilon_t}}\right) \\ &= \frac{1}{\sigma^2_{\varepsilon_t}} V(\varepsilon_t) \end{aligned} \tag{4.35}$$

$$= \frac{\sigma_{\varepsilon_t}^2}{\sigma_{\varepsilon_t}^2}$$

$$= 1$$

The error term of the transformed model is therefore homoskedastic, and it is then possible to apply the OLS technique to model (4.33). The GLS method thus consists in applying the OLS method to the transformed model. Note that this technique amounts to minimizing the residual sum of squares of the transformed model, i.e.:

$$Min \sum_t \tilde{e}_t^2 = Min \sum_t \left(\frac{e_t}{\sigma_{\varepsilon_t}} \right)^2 = Min \sum_t \omega_t e_t^2 \tag{4.36}$$

with $\tilde{e}_t = \tilde{Y}_t - \tilde{\alpha} - \tilde{\beta}_1 \tilde{X}_{1t} - \tilde{\beta}_2 \tilde{X}_{2t} - \ldots - \tilde{\beta}_k \tilde{X}_{kt}$ and $\omega_t = \frac{1}{\sigma_{\varepsilon_t}^2}$. The factors ω_t play the role of weights, and the GLS method involves minimizing a weighted residual sum of squares. For this reason, this technique is also called the **weighted least squares method** (WLS), WLS being only a special case of GLS in which the transformation matrix $\boldsymbol{M}$ is given by:

$$\boldsymbol{M} = \begin{pmatrix} \frac{1}{\sigma_{\varepsilon 1}} & 0 & \cdots & 0 \\ 0 & \frac{1}{\sigma_{\varepsilon 2}} & \cdots & 0 \\ \vdots & \vdots & \ddots & \vdots \\ 0 & 0 & \cdots & \frac{1}{\sigma_{\varepsilon T}} \end{pmatrix} \tag{4.37}$$

In this example, we have assumed that $V(\varepsilon_t)$ is known, which is generally not the case in practice. We will see later what to do when the variance of the error is unknown.

4.2.3 Detecting Heteroskedasticity

Various tests can be used to address the issue of heteroskedasticity. Before presenting them, let us mention that a first intuition can be provided graphically. The technique consists first in estimating the model considered by OLS as if there were no heteroskedasticity. We then graphically represent the estimated values $\hat{Y}_t$ of Y_t (on the x-axis) as a function of the series of squared residuals e_t^2 (on the y-axis). Some examples are given in Figs. 4.1, 4.2, 4.3, 4.4 and 4.5.

These graphs allow us to detect whether the estimated mean value of the dependent variable is systematically related to the squared residuals. If this is the case, there is a presumption of heteroskedasticity. Figure 4.1 illustrates the absence of heteroskedasticity in the sense that no particular relationship appears between the squared residuals and the estimated variable. On the contrary, Figs. 4.2, 4.3, 4.4 and

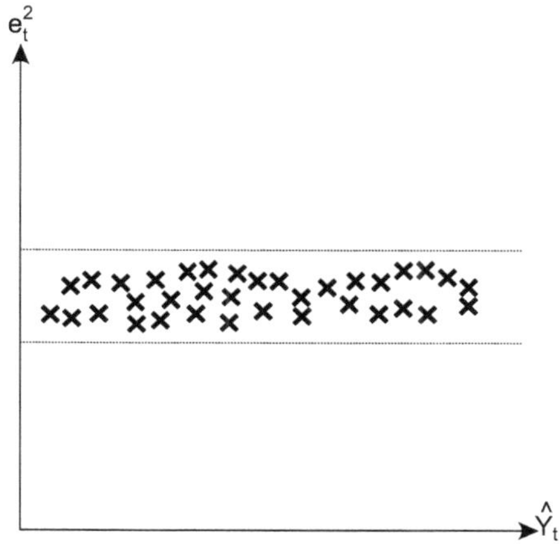

Fig. 4.1 Absence of heteroskedasticity

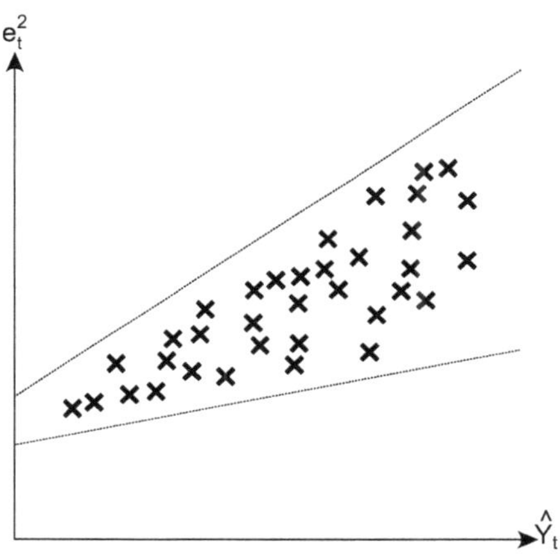

Fig. 4.2 Heteroskedasticity

4.5 highlight the existence of a relationship between the two variables, suggesting there is heteroskedasticity: a linear relationship for Fig. 4.3, a quadratic relationship according to Fig. 4.4, and a positive nonlinear relationship for Fig. 4.5.

It is also possible to produce graphs with the values of one of the explanatory variables instead of the estimated values of the dependent variable on the x-axis, with the squared residuals still shown on the y-axis. If the explanatory variable under consideration and the squared residuals appear to be related, this is an indication in favor of heteroskedasticity.

In addition to these graphical methods, there are a number of tests that we now present.

The Goldfeld and Quandt Test (1965)

The **Goldfeld and Quandt test** applies in the case where one of the explanatory variables is the cause of heteroskedasticity. It is thus assumed that the variance of

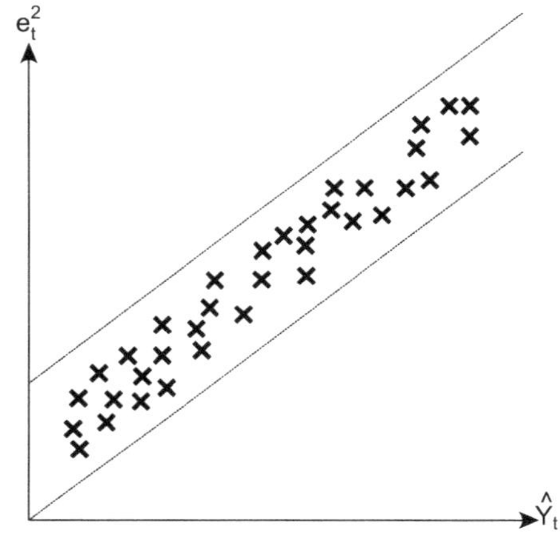

Fig. 4.3 Heteroskedasticity (linear relationship)

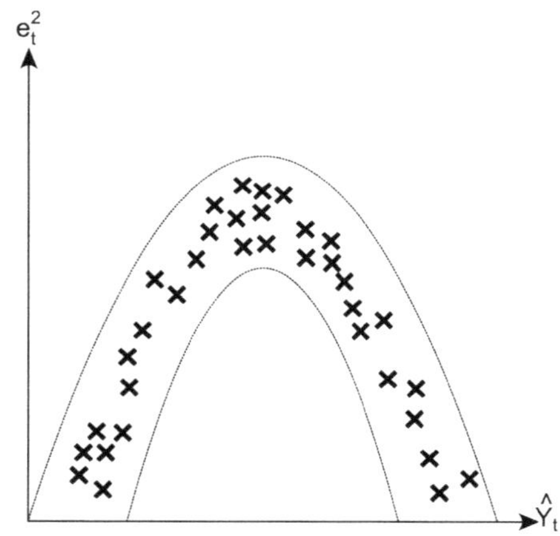

Fig. 4.4 Heteroskedasticity (quadratic relationship)

the error increases with one of the explanatory variables, for example, X_j. We then have a relationship of the type:

$$\sigma_{\varepsilon_t}^2 = aX_{jt}^2 \tag{4.38}$$

where a is a positive constant. Such a relationship means that the greater the values of X_j, the greater $\sigma_{\varepsilon_t}^2$. If this is the case, it is an indication that there is heteroskedasticity. More generally, the test is based on the idea that if we divide the sample into two subsamples, then, under the assumption of homoskedasticity, the

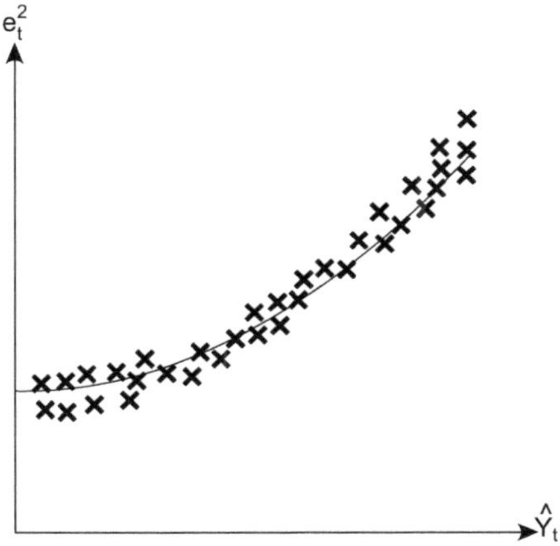

Fig. 4.5 Heteroskedasticity (positive nonlinear relationship)

error variances should be identical in both groups. Under the alternative assumption of heteroskedasticity, they are different. To capture this, Goldfeld and Quandt suggest a five-step test:

- *Step 1*. Observations of the different variables are ranked according to increasing values of X_j.
- *Step 2*. The m central values of the resulting sample are disregarded. We obtain two subsamples, each comprising $(T - m)/2$ observations: one corresponding to the low values of X_j, the other to the high values of this same variable.
- *Step 3*. The model is estimated by OLS on each of the two subsamples, both composed of $(T - m)/2$ observations.
- *Step 4*. The sum of squared residuals corresponding to each of the two regressions is calculated. We denote RSS_1 and RSS_2 the sums of squared residuals with $RSS_2 > RSS_1$, i.e., RSS_2 is associated with the regression for which the sum of squared residuals is higher.
- *Step 5*. The test statistic given by the ratio:

$$GQ = \frac{RSS_2}{RSS_1} \tag{4.39}$$

is calculated. Under the null hypothesis of homoskedasticity:

$$GQ \sim F\left(\frac{T - m - 2(k + 1)}{2}, \frac{T - m - 2(k + 1)}{2}\right) \tag{4.40}$$

The decision rule is as follows:

- If $GQ < F\left(\frac{T-m-2(k+1)}{2}, \frac{T-m-2(k+1)}{2}\right)$, the null hypothesis of homoskedasticity is not rejected.
- If $GQ > F\left(\frac{T-m-2(k+1)}{2}, \frac{T-m-2(k+1)}{2}\right)$, there is heteroskedasticity.

The power of the Goldfeld and Quandt test depends on the choice of m. Harvey and Phillips (1973) suggest choosing a value of m close to $T/3$.

The Glejser Test (1969)

The **Glejser test** is intended not only to detect possible heteroskedasticity, but also to specify its form. As before, this test is based on the assumption that only one variable is the cause of heteroskedasticity. Let us assume that it is the variable X_j. The test can be presented in three steps:

- *Step 1*. The regression of the dependent variable on the explanatory variables is estimated by OLS. The residual series e_t is deduced.
- *Step 2*. The absolute value of the residuals e_t is regressed on the variable X_j or on transformations of this variable. Glejser proposes various functional forms for this regression:
  - $|e_t| = a_0 + a_1 X_{jt} + u_t$
  - $|e_t| = a_0 + a_1 \sqrt{X_{jt}} + u_t$
  - $|e_t| = a_0 + a_1 \frac{1}{X_{jt}} + u_t$
  - $|e_t| = a_0 + a_1 \frac{1}{\sqrt{X_{jt}}} + u_t$
  - $|e_t| = \sqrt{a_0 + a_1 X_{jt}} + u_t$
  - $|e_t| = \sqrt{a_0 + a_1 X_{jt}^2} + u_t$

  where u_t is the error term.
- *Step 3*. The null hypothesis that $a_1 = 0$ is tested by means of a usual significance test. If this hypothesis is not rejected, we conclude in favor of homoskedasticity. On the other hand, if the null hypothesis is rejected, the hypothesis of heteroskedasticity is adopted.

Remark 4.2 The Glejser test was criticized by Goldfeld and Quandt (1972) who pointed out that the error term u_t in the various regressions did not have the right statistical properties, which implies that the conditions necessary for implementating the usual t-test of significance are not required.

The Breusch-Pagan Test (1979)

This is a general test in the sense that it covers a large number of cases of heteroskedasticity. Since it is an asymptotic test, it is only valid for sufficiently large sample sizes. In addition, it overcomes the difficulties of the Goldfeld and Quandt test residing in the choice of the parameter m and in the identification of the variable

X_j at the source of heteroskedasticity. An additional advantage lies in its simplicity, as it is based on the residuals from the OLS estimation of the regression model.

In the multiple regression model:

$$Y_t = \alpha + \beta_1 X_{1t} + \beta_2 X_{2t} + \ldots + \beta_k X_{kt} + \varepsilon_t \tag{4.41}$$

the error term is assumed to have a normal distribution with variance:

$$\sigma_{\varepsilon_t}^2 = f\left(a_0 + a_1 Z_{1t} + \ldots + a_p Z_{pt}\right) \tag{4.42}$$

where f is any function, the coefficients a_i, $i = 1, \ldots, p$, are not related to the coefficients of the regression model (4.41), and $Z_{1t}, \ldots, Z_{pt}$ are variables likely to be the source of heteroskedasticity. Some or all of these variables may be explanatory variables in the regression model (4.41).

Testing the null hypothesis of homoskedasticity is equivalent to testing:

$$H_0 : a_1 = a_2 = \ldots = a_p = 0 \tag{4.43}$$

since, in this case, we have:

$$\sigma_{\varepsilon_t}^2 = f\left(a_0\right) \tag{4.44}$$

which is constant, whatever t. The test can be implemented in five steps:

- *Step 1*. Regression (4.41) is estimated by OLS, and the residual series $e_t, t = 1, \ldots, T$ is deduced.
- *Step 2*. The following quantity:

$$\hat{\sigma}_{ML}^2 = \frac{1}{T} \sum_{t=1}^{T} e_t^2 \tag{4.45}$$

  is calculated, which is the maximum likelihood estimator of the variance of the error term (see Appendix "The Maximum Likelihood Method" in Chap. 2).
- *Step 3*. The quantity:

$$h_t = \frac{e_t^2}{\hat{\sigma}_{ML}^2} \tag{4.46}$$

  is computed for $t = 1, \ldots, T$.
- *Step 4*. After specifying the variables $Z_{1t}, \ldots, Z_{pt}$, we regress h_t on these variables:

$$h_t = a_0 + a_1 Z_{1t} + \ldots + a_p Z_{pt} + u_t \tag{4.47}$$

where u_t is an error term. The explained sum of squares (ESS) of this regression is calculated.

– *Step 5*. The quantity:

$$BP = \frac{1}{2} ESS \tag{4.48}$$

is computed, which, under the null hypothesis of homoskedasticity, has a Chi-squared distribution with p degrees of freedom, i.e.:

$$BP \sim \chi_p^2 \tag{4.49}$$

The decision rule is as follows:

- If $BP < \chi_p^2$, the null hypothesis of homoskedasticity is not rejected.
- If $BP > \chi_p^2$, the null hypothesis of homoskedasticity is rejected.

The White Test (1980)

This is a very general test which has the advantage of not relying on the assumption of normality of the error term. It consists in testing the null hypothesis of homoskedasticity against the alternative hypothesis of heteroskedasticity; the form of heteroskedasticity not being specified. The test procedure can be described in three steps:

– *Step 1*. The multiple regression model is estimated:

$$Y_t = \alpha + \beta_1 X_{1t} + \beta_2 X_{2t} + \ldots + \beta_k X_{kt} + \varepsilon_t \tag{4.50}$$

and we deduce the series of residuals e_t, $t = 1, \ldots, T$.

– *Step 2*. The following auxiliary regression is estimated, reflecting the existence of a relationship between the squared residuals from the previous regression and one or more explanatory variables:

$$e_t^2 = a_0 + a_1 X_{1t} + b_1 X_{1t}^2 + a_2 X_{2t} + b_2 X_{2t}^2 + \ldots + a_k X_{kt} + b_k X_{kt}^2 + u_t \tag{4.51}$$

where u_t is an error term. It is also possible to add interaction terms such as $X_{1t} X_{2t}$ to this regression. We calculate the coefficient of determination R^2 associated with this auxiliary regression.

– *Step 3*. We test the null hypothesis of homoskedasticity:

$$H_0 : a_1 = b_1 = a_2 = b_2 = \ldots = a_k = b_k = 0 \tag{4.52}$$

If at least one coefficient of the auxiliary regression is significantly different from zero, the null hypothesis is rejected. To perform this test, we calculate the White test statistic given by the product between the number of observations T and the coefficient of determination of the auxiliary regression,[2] or TR^2. Under the null hypothesis, this statistic has a Chi-squared distribution whose number of degrees of freedom is equal to the number of estimated parameters (excluding the constant) in the auxiliary regression. Thus, in the case of model (4.51), we have:

$$TR^2 \sim \chi^2_{2k} \tag{4.53}$$

The decision rule is then:

- If $TR^2 < \chi^2_{2k}$, the null hypothesis of homoskedasticity is not rejected.
- If $TR^2 > \chi^2_{2k}$, the null hypothesis of homoskedasticity is rejected.

Remark 4.3 The White test can also be used as a model misspecification test. Under the null hypothesis, the White test assumes that the errors are not only homoskedastic, but also uncorrelated with the regressors and that the linear specification of the model is correct. If one of these conditions is violated, the test statistic is above the critical value. On the contrary, if the value of the test statistic is below the critical value, this indicates that none of these three conditions is violated.

ARCH Test

ARCH (autoregressive conditionally heteroskedastic) processes were introduced by Engle (1982) and are used to model series whose variance—also called **volatility**—in t depends on its past values. This is therefore a particular form of heteroskedasticity, called **conditional heteroskedasticity**. The test procedure can be outlined in four steps:

– *Step 1*. The multiple regression model:

$$Y_t = \alpha + \beta_1 X_{1t} + \beta_2 X_{2t} + \ldots + \beta_k X_{kt} + \varepsilon_t \tag{4.54}$$

is estimated, and the series of residuals e_t, $t = 1, \ldots, T$, is calculated.

– *Step 2*. The squared residual series, i.e., e_t^2, $t = 1, \ldots, T$.

– *Step 3*. The following regression is estimated, consisting of regressing the squared residual series on its ℓ past values and a constant:

$$e_t^2 = a_0 + \sum_{i=1}^{\ell} a_i e_{t-i}^2 \tag{4.55}$$

[2] Such a statistic is called a **Lagrange multiplier statistic**.

The coefficient of determination R^2 associated with this regression is calculated.

– *Step 4*. The null hypothesis of homoskedasticity:

$$H_0 : a_1 = a_2 = \ldots = a_\ell = 0 \tag{4.56}$$

is tested against the alternative hypothesis of conditional heteroskedasticity stipulating that at least one of the coefficients a_i, $i = 1, \ldots, \ell$, is significantly different from zero. In order to implement this test, we calculate the test statistic—which is a Lagrange multiplier statistic—consisting in computing the product between the number of observations T and the coefficient of determination of the regression estimated in Step 3. Under the null hypothesis of homoskedasticity, we have:

$$TR^2 \sim \chi^2_\ell \tag{4.57}$$

The decision rule is as follows:

- If $TR^2 < \chi^2_\ell$, the null hypothesis of homoskedasticity is not rejected.
- If $TR^2 > \chi^2_\ell$, the null hypothesis of homoskedasticity is rejected in favor of the alternative hypothesis of conditional heteroskedasticity.

4.2.4 Estimation Procedures When There Is Heteroskedasticity

The presence of heteroskedasticity results in OLS estimators remaining unbiased, but no longer of minimum variance. This poses a problem, notably because the precision of the tests is affected. It is therefore relevant to seek to correct for heteroskedasticity.

We must distinguish between cases where the variance of the error term is known and those where it is unknown. When the variance of the error term is known, the GLS method should be applied when there is heteroskedasticity. The example given in Sect. 4.2.2 relating to the weighted least squares (WLS) method was based on the knowledge of the variance of the error term. In practice, this is not the case, and it is impossible to obtain efficient estimators of the parameters by WLS when the form of heteroskedasticity is unknown.

In these cases where the variance of the error term is unknown, however, it is possible to obtain estimators of the variances and covariances of the WLS estimators corrected for heteroskedasticity. For this purpose, we can use the corrections suggested by White (1980) and Newey and West (1987). These two techniques do not modify the OLS-estimated values of the coefficients, but only change the estimated standard deviations of these coefficients (and therefore their t-statistics). We can also use the results of certain tests, such as the Glejser test, to correct for heteroskedasticity.

The White Estimator of the Variance-Covariance Matrix

Consider the multiple regression model:

$$\boldsymbol{Y} = \boldsymbol{X}\boldsymbol{\beta} + \boldsymbol{\varepsilon} \tag{4.58}$$

which can still be written for an observation t:

$$Y_t = \boldsymbol{X}_t'\boldsymbol{\beta} + \boldsymbol{\varepsilon}_t \tag{4.59}$$

where $\boldsymbol{X}_t'$ denotes a column vector equal to the transpose of the tth row of the matrix $\boldsymbol{X}$.

The $\hat{\boldsymbol{\Omega}}_\varepsilon^W$ estimator of the variance-covariance matrix proposed by White (1980) is written as:

$$\hat{\boldsymbol{\Omega}}_\varepsilon^W = \frac{T}{T-k-1}\left(\boldsymbol{X}'\boldsymbol{X}\right)^{-1}\left(\sum_{t=1}^{T} \boldsymbol{e}_t^2 \boldsymbol{X}_t \boldsymbol{X}_t'\right)\left(\boldsymbol{X}'\boldsymbol{X}\right)^{-1} \tag{4.60}$$

where T is the number of observations, k is the number of explanatory variables, and $\boldsymbol{e}_t$ are the residuals from the OLS estimation of the multiple regression model.

The Newey and West Estimator of the Variance-Covariance Matrix

The variance-covariance matrix suggested by White is based on the assumption that the residuals of the estimated model are not autocorrelated. Newey and West (1987) propose a more general estimator of the variance-covariance matrix that is valid when there are both autocorrelation and heteroskedasticity of unknown form. The estimator $\hat{\boldsymbol{\Omega}}_\varepsilon^{NW}$ of the variance-covariance matrix proposed by Newey and West is written as:

$$\hat{\boldsymbol{\Omega}}_\varepsilon^{NW} = \frac{T}{T-k-1}\left(\boldsymbol{X}'\boldsymbol{X}\right)^{-1}\hat{\boldsymbol{\Sigma}}^{NW}\left(\boldsymbol{X}'\boldsymbol{X}\right)^{-1} \tag{4.61}$$

with:

$$\begin{aligned}\hat{\boldsymbol{\Sigma}}^{NW} &= \frac{T}{T-k-1} \\ &\times \left[\sum_{t=1}^{T} \boldsymbol{e}_t^2 \boldsymbol{X}_t \boldsymbol{X}_t' + \sum_{j=1}^{q}\left(\left(1-\frac{j}{q+1}\right)\sum_{t=j+1}^{T}\right.\right. \\ &\left.\left.\times\left(\boldsymbol{X}_t \boldsymbol{e}_t \boldsymbol{e}_{t-j} \boldsymbol{X}_{t-j}' + \boldsymbol{X}_{t-j} \boldsymbol{e}_{t-j} \boldsymbol{e}_t \boldsymbol{X}_t'\right)\right)\right]\end{aligned} \tag{4.62}$$

In this expression, q designates the truncation parameter. It represents the number of autocorrelations taken into account in the dynamics of the residuals $\boldsymbol{e}_t$. Newey and West have suggested the following value for this parameter:

$$q = int\left[4\left(\frac{T}{100}\right)^{2/9}\right] \tag{4.63}$$

where int denotes the integer part.

Hypotheses About the Form of Heteroskedasticity

As previously mentioned, it is also possible to use the results of the Glejser test regarding the form of heteroskedasticity or to make certain assumptions about this form.

Consider the multiple regression model:

$$Y_t = \alpha + \beta_1 X_{1t} + \beta_2 X_{2t} + \ldots + \beta_j X_{jt} + \ldots + \beta_k X_{kt} + \varepsilon_t \tag{4.64}$$

Assume that heteroskedasticity is such that the variance of the error term is proportional to X_{jt}^2 where X_{jt} is one of the explanatory variables of the regression model (4.64), i.e.:

$$\sigma_{\varepsilon_t}^2 = aX_{jt}^2 \tag{4.65}$$

where a is a constant. It is possible to transform model (4.64) by dividing each of the terms by X_j:

$$\frac{Y_t}{X_{jt}} = \alpha\frac{1}{X_{jt}} + \beta_1\frac{X_{1t}}{X_{jt}} + \beta_2\frac{X_{2t}}{X_{jt}} + \ldots + \beta_j + \ldots + \beta_k\frac{X_{kt}}{X_{jt}} + \frac{\varepsilon_t}{X_{jt}} \tag{4.66}$$

Denote by $u_t = \frac{\varepsilon_t}{X_{jt}}$ the transformed error term. Let us determine its variance:

$$E\left(u_i^2\right) = E\left(\frac{\varepsilon_t}{X_{jt}}\right)^2 = \frac{1}{X_{jt}^2}E\left(\varepsilon_t^2\right) \tag{4.67}$$

According to (4.65), $E\left(\varepsilon_t^2\right) = aX_{jt}^2$, hence:

$$E\left(u_i^2\right) = a \tag{4.68}$$

The variance of the transformed error term u_t is constant and it is, therefore, possible to apply OLS to the transformed model (4.66).

The same type of reasoning can, of course, be applied to other forms of heteroskedasticity. As an example, suppose the form of heteroskedasticity is such that:

$$\sigma_{\varepsilon_t}^2 = aX_{jt} \tag{4.69}$$

Thus, the variance of the error term is now assumed to be proportional to X_{jt}, with a being a constant. The transformed model is then written as:

$$\frac{Y_t}{\sqrt{X_{jt}}} = \alpha \frac{1}{\sqrt{X_{jt}}} + \beta_1 \frac{X_{1t}}{\sqrt{X_{jt}}} + \beta_2 \frac{X_{2t}}{\sqrt{X_{jt}}} + \ldots + \beta_j \sqrt{X_{jt}} + \ldots + \beta_k \frac{X_{kt}}{\sqrt{X_{jt}}} + \frac{\varepsilon_t}{\sqrt{X_{jt}}} \tag{4.70}$$

The variance of the transformed error term $u_t = \frac{\varepsilon_t}{\sqrt{X_{jt}}}$ is given by:

$$E\left(u_i^2\right) = E\left(\frac{\varepsilon_t}{\sqrt{X_{jt}}}\right)^2 = \frac{1}{X_{jt}} E\left(\varepsilon_t^2\right) = a \tag{4.71}$$

which is indeed constant. It is therefore possible to apply OLS to the transformed model (4.70).

Note on the Logarithmic Transformation

The logarithmic transformation, giving rise to a log-linear model (see Chap. 2), can also be used to reduce heteroskedasticity. The log-linear multiple regression model is written:

$$\log Y_t = \alpha + \beta_1 \log X_{1t} + \beta_2 \log X_{2t} + \ldots + \beta_k \log X_{kt} + \varepsilon_t \tag{4.72}$$

The reduction in heteroskedasticity comes from the fact that the logarithmic transformation "compresses" the scales on which the variables are measured.

4.2.5 Empirical Application

Let us use the series of stock market returns studied in the previous chapter and consider the following model at monthly frequency over the period from February 1984 to June 2021 (449 observations):

$$RFTSE_t = \alpha + \beta RDJIND_t + \varepsilon_t \tag{4.73}$$

where $RFTSE$ denotes the returns of the $FTSE$ 100 index of the London Stock Exchange and $RDJIND$ the returns of the Dow Jones Industrial Average index of

Table 4.1 FTSE and Dow Jones industrial average returns

	$RFTSE$	$RDJIND$
1984.02	−0.0216	−0.0555
1984.03	0.0671	0.0088
1984.04	0.0229	0.0050
…	…	…
2021.04	0.0374	0.0267
2021.05	0.0075	0.0191
2021.06	0.0021	−0.0008

Data source: Macrobond

Table 4.2 OLS estimation results

Dependent variable: RFTSE				
Variable	Coefficient	Std. error	t-Statistic	Prob.
C	−0.001614	0.001352	−1.193763	0.2332
RDJIND	0.782505	0.030388	25.75060	0.0000
R-Squared	0.597331	Mean dependent var	0.004210	
Adjusted R-squared	0.596430	S.D. dependent var	0.044466	
S.E. of regression	0.028248	Akaike info criterion	−4.291158	
Sum squared resid	0.356678	Schwarz criterion	−4.272864	
Log likelihood	965.3650	Hannan-Quinn criterion	−4.283947	
F-statistic	663.0935	Durbin-Watson stat	2.224700	
Prob(F-statistic)	0.0000			

Table 4.3 Reordered returns for the FTSE and Dow Jones industrial average indexes

	$RFTSE$	$RDJIND$
1987.10	−0.3017	−0.2642
1998.08	−0.1061	−0.1641
2008.10	−0.1133	−0.1515
…	…	…
2020.04	0.0396	0.1051
2020.11	0.1165	0.1119
1987.01	0.0742	0.1295

the New York Stock Exchange (see Table 4.1). The series are extracted from the Macrobond database.

The OLS estimation of model (4.73) leads to the results reported in Table 4.2. Let us apply the various homoskedasticity tests previously presented.

The Goldfeld and Quandt Test

Implementing the Goldfeld and Quandt test first involves ordering the observations according to the increasing values of the explanatory variable $RDJIND$. Table 4.3 reports the observations thus reordered.

We then ignore a number m of central observations. Let us take, as suggested by Harvey and Phillips (1973), $m = T/3 = 449/3$, which leads us to omit 149

Table 4.4 Regression on the first 150 observations

Dependent variable: RFTSE				
Variable	Coefficient	Std. error	t-Statistic	Prob.
C	0.005742	0.003463	1.658159	0.0994
RDJIND	0.959400	0.065123	14.73205	0.0000
R-Squared	0.594557	Mean dependent var	−0.031056	
Adjusted R-squared	0.591818	S.D. dependent var	0.045985	
S.E. of regression	0.029379	Akaike info criterion	−4.203797	
Sum squared resid	0.127747	Schwarz criterion	−4.163655	
Log likelihood	317.2848	Hannan-Quinn criterion	−4.187489	
F-statistic	217.0332	Durbin-Watson stat	1.908283	
Prob(F-statistic)	0.000000			

Table 4.5 Regression on the last 150 observations

Dependent variable: RFTSE				
Variable	Coefficient	Std. error	t-Statistic	Prob.
C	0.001252	0.005647	0.221775	0.8248
RDJIND	0.664592	0.104489	6.360391	0.0000
R-Squared	0.214665	Mean dependent var	0.034396	
Adjusted R-squared	0.209359	S.D. dependent var	0.029963	
S.E. of regression	0.026643	Akaike info criterion	−4.399360	
Sum squared resid	0.105055	Schwarz criterion	−4.359218	
Log likelihood	331.9520	Hannan-Quinn criterion	−4.383051	
F-statistic	40.45457	Durbin-Watson stat	2.399228	
Prob(F-statistic)	0.000000			

observations. We obtain two subsamples, each composed of 150 observations. The OLS estimates of the two resulting regressions are shown in Tables 4.4 and 4.5.

The test statistic GQ is given by:

$$GQ = \frac{RSS_2}{RSS_1} \tag{4.74}$$

where RSS_1 and RSS_2 are the sums of squares of the residuals from each of the two regressions, with $RSS_2 > RSS_1$. We see that the residual sum of squares corresponding to the model estimated on the first 150 observations (Table 4.4) is greater than that relating to the model estimated on the last 150 observations (Table 4.5). We therefore have $RSS_1 = 0.105055$ and $RSS_2 = 0.127747$. So:

$$GQ = 1.2160 \tag{4.75}$$

Under the null hypothesis of homoskedasticity, the GQ statistic has a Fisher distribution with $\left(\frac{T-m-2(k+1)}{2}, \frac{T-m-2(k+1)}{2}\right) = \left(\frac{449-149-2(1+1)}{2}, \frac{449-149-2(1+1)}{2}\right)$ degrees of freedom. The Fisher table at the 5% significance level gives us: $F\,(148, 148) = 1.00$. Since $1.2160 > 1.00$, the null hypothesis of homoskedasticity is rejected.

The Glejser Test

The first step in the Glejser test is to estimate model (4.73). The results of this estimation are reported in Table 4.2. We derive the residual series:

$$e_t = RFTSE_t + 0.0016 - 0.7825 RDJIND_t \tag{4.76}$$

We then calculate the series of residuals in absolute values $|e_t|$. We regress $|e_t|$ on the explanatory variable $RDJIND$ or on various transformations of this variable.

Consider, for example, the following two models:

- $|e_t| = \hat{a}_0 + \hat{a}_1 RDJIND_t$ OLS estimation of this model leads to the following results:

$$|e_t| = \underset{(25.9251)}{0.0221} - \underset{(-1.4395)}{0.0275}\, RDJIND_t \tag{4.77}$$

- $|e_t| = \hat{a}_0 + \hat{a}_1\,(RDJIND_t)^{-1}$ OLS estimation of this regression yields:

$$|e_t| = \underset{(26.0510)}{0.0219} - \underset{(-1.1275)}{8.24.10^{-7}}\,(RDJIND_t)^{-1} \tag{4.78}$$

In both models, the values in parentheses are the t-statistics associated with the estimated coefficients.

We proceed to test the null hypothesis of homoskedasticity according to which $a_1 = 0$. To test this hypothesis, we compare the absolute values of the t-statistics of the coefficient a_1 with the value read from the Student's t table (1.96 at the 5% significance level). We deduce that such a hypothesis is not rejected by models (4.77) and (4.78).

The Breusch-Pagan Test

Consider the residuals from the regression shown in Table 4.2:

$$e_t = RFTSE_t + 0.0016 - 0.7825 RDJIND_t \tag{4.79}$$

We calculate:

$$\hat{\sigma}^2_{ML} = \frac{1}{T}\sum_{t=1}^{T} e_t^2 \tag{4.80}$$

that is:

$$\hat{\sigma}^2_{ML} = 7.9438.10^{-4} \tag{4.81}$$

This gives the series:

$$h_t = \frac{e_t^2}{\hat{\sigma}^2_{ML}} \tag{4.82}$$

for $t = 1, \ldots, 449$. We then regress h_t on $RDJIND_t$ and obtain:

$$h_t = \underset{(13.4166)}{1.0263} - \underset{(-2.0564)}{3.5352}\, RDJIND_t \tag{4.83}$$

We calculate the explained sum of squares of this regression, i.e., $ESS = 10.80$ and the Breusch-Pagan statistic:

$$BP = \frac{1}{2} ESS = 5.40 \tag{4.84}$$

Under the null hypothesis of homoskedasticity, the BP statistic follows a Chi-squared distribution with 1 degree of freedom (since only one explanatory variable is included in the regression (4.83)). At the 5% significance level, the critical value read from the Chi-squared table is equal to 3.841. Consequently, $5.40 > 3.841$, which means we reject the null hypothesis of homoskedasticity.

The White Test

For the estimation of model (4.73) giving the series of residuals e_t, we deduce the series of squared residuals in order to estimate the auxiliary regression:

$$e_t^2 = a_0 + a_1 RDJIND_t + b_1 RDJIND_t^2 + u_t \tag{4.85}$$

which gives:

$$e_t^2 = \underset{(10.3910)}{0.0007} - \underset{(-0.8232)}{0.0012}\, RDJIND_t + \underset{(3.8127)}{0.0528} RDJIND_t^2 \tag{4.86}$$

The coefficient of determination associated with this auxiliary regression is equal to $R^2 = 0.0406$. The number of observations being equal to 449, we deduce the value of the test statistic: $TR^2 = 18.2478$. Under the null hypothesis of homoskedasticity, this statistic follows a Chi-squared distribution with 2 degrees of freedom (since two variables appear in the auxiliary regression). The Chi-squared table gives us a critical value equal to 5.991 at the 5% significance level. Given that $18.2478 > 5.991$, the null hypothesis of homoskedasticity is rejected.

ARCH Test

For the estimation of model (4.73) giving the series of residuals e_t, we derive the series of squared residuals e_t^2. The e_t^2 series is then regressed on a constant and its past values. Using three lags we obtain the following results:

$$e_t^2 = \underset{(6.7282)}{0.0006} + \underset{(1.4162)}{0.0670}e_{t-1}^2 + \underset{(2.5508)}{0.11936}e_{t-2}^2 + \underset{(2.1429)}{0.1008}e_{t-3}^2 \quad (4.87)$$

The coefficient of determination associated with this regression is $R^2 = 0.0345$. We test the null hypothesis of homoskedasticity by calculating the test statistic $TR^2 = 446 \times 0.0345 = 15.3946$. The number of observations equals 446, not 449, corresponding to the number of observations used in the regression (4.87): since this regression involves three lags on the squared residual variable, three observations are "lost." Under the null hypothesis, the test statistic has a Chi-squared distribution with three degrees of freedom (since three variables appear in the regression (4.87)). The Chi-squared table gives us a critical value equal to 7.815 at the 5% significance level. Since $15.3946 > 7.815$, it follows that the null hypothesis of homoskedasticity is rejected.

With the exception of Glejser's test, all the tests conclude that the null hypothesis of homoskedasticity is rejected and that heteroskedasticity is present. Such a result is not surprising since our example concerns financial series. These series are frequently heteroskedastic since their variance changes over time. The occurrence of heteroskedasticity leads us to correct the OLS estimators of the variance of the coefficients. To this end, let us apply the methods of White and Newey and West.

Heteroskedasticity-Corrected Estimations

Tables 4.6 and 4.7 report the results from the OLS estimation of relationship (4.73), the variance-covariance matrix estimators being given by White (Table 4.6) and by Newey and West (Table 4.7). These two techniques allow for heteroskedasticity by correcting the estimators of the variances and covariances of the OLS estimators. Thus, the estimated values of the coefficients are identical to those shown in Table 4.2, but the standard deviations of the coefficients (and therefore t-statistics) are different. The coefficient associated with the $RDJIND$ variable remains significantly different from zero, both with the White correction and with the Newey-West correction.

4.3 Autocorrelation of Errors

4.3.1 Sources of Autocorrelation

There is autocorrelation when the terms off the diagonal of the variance-covariance matrix of the errors are not all zero:[3]

[3] It is assumed here that the errors are homoskedastic.

Table 4.6 White heteroskedasticity consistent estimations

Dependent variable: RFTSE				
White heteroskedasticity-consistent standard errors and Covariance				
Variable	Coefficient	Std. error	t-Statistic	Prob.
C	−0.001614	0.001401	−1.152289	0.2498
RDJIND	0.782505	0.040668	19.24132	0.0000
R-Squared	0.597331	Mean dependent var	0.004210	
Adjusted R-squared	0.596430	S.D. dependent var	0.044466	
S.E. of regression	0.028248	Akaike info criterion	−4.291158	
Sum squared resid	0.356678	Schwarz criterion	−4.272864	
Log likelihood	965.3650	Hannan-Quinn criterion	−4.283947	
F-statistic	663.0935	Durbin-Watson stat	2.224700	
Prob(F-statistic)	0.000000	Wald F-statistic	370.2283	
Prob(Wald F-statistic)	0.000000			

Table 4.7 Newey-West heteroskedasticity and autocorrelation consistent estimation

Dependent variable: RFTSE				
Newey-West HAC standard errors and covariance (bandwidth = 6)				
Variable	Coefficient	Std. error	t-Statistic	Prob.
C	−0.001614	0.001165	−1.385650	0.1665
RDJIND	0.782505	0.039089	20.01861	0.0000
R-Squared	0.597331	Mean dependent var	0.004210	
Adjusted R-squared	0.596430	S.D. dependent var	0.044466	
S.E. of regression	0.028248	Akaike info criterion	−4.291158	
Sum squared resid	0.356678	Schwarz criterion	−4.272864	
Log likelihood	965.3650	Hannan-Quinn criterion	−4.283947	
F-statistic	663.0935	Durbin-Watson stat	2.224700	
Prob(F-statistic)	0.000000	Wald F-statistic	400.7447	
Prob(Wald F-statistic)	0.000000			

$$E\left(\boldsymbol{\varepsilon}\boldsymbol{\varepsilon}'\right) = \begin{pmatrix} \sigma_\varepsilon^2 & Cov(\varepsilon_1, \varepsilon_2) & \cdots & Cov(\varepsilon_1, \varepsilon_T) \\ Cov(\varepsilon_2, \varepsilon_1) & \sigma_\varepsilon^2 & \cdots & Cov(\varepsilon_2, \varepsilon_T) \\ \vdots & \vdots & \ddots & \vdots \\ Cov(\varepsilon_T, \varepsilon_1) & Cov(\varepsilon_T, \varepsilon_2) & \cdots & \sigma_\varepsilon^2 \end{pmatrix} \tag{4.88}$$

Thus, in the presence of autocorrelation of errors, the regression model assumption that $E\ (\varepsilon_t \varepsilon_{t'}) = 0$ for $t \neq t'$ is violated. This results in the error term in t being related to the error term in t'.

The **autocovariance** of order h, denoted γ_h, is defined as:

$$\gamma_h = E\left(\varepsilon_t \varepsilon_{t-h}\right) = cov\left(\varepsilon_t, \varepsilon_{t-h}\right) \tag{4.89}$$

for $h = 0, \pm 1, \pm 2, \ldots$.

When $h = 0$, the autocovariance is equal to the variance, i.e.:

$$\gamma_0 = E\left(\varepsilon_t^2\right) = \sigma_\varepsilon^2 \tag{4.90}$$

The variance-covariance matrix can be written as:

$$E\left(\boldsymbol{\varepsilon}\boldsymbol{\varepsilon}'\right) = \begin{pmatrix} \gamma_0 & \gamma_1 & \cdots & \gamma_{h-1} \\ \gamma_1 & \gamma_0 & \cdots & \gamma_{h-2} \\ \vdots & \vdots & \ddots & \vdots \\ \gamma_{h-1} & \gamma_{h-2} & \cdots & \gamma_0 \end{pmatrix} \tag{4.91}$$

Let ρ_h be the **autocorrelation coefficient** of order h:

$$\rho_h = \frac{\gamma_h}{\gamma_0} = \frac{\gamma_h}{\sigma_\varepsilon^2} \tag{4.92}$$

for $h = 0, \pm 1, \pm 2, \ldots$. The variance-covariance matrix of the errors is written as:

$$E\left(\boldsymbol{\varepsilon}\boldsymbol{\varepsilon}'\right) = \sigma_\varepsilon^2 \begin{pmatrix} 1 & \rho_1 & \cdots & \rho_{h-1} \\ \rho_1 & 1 & \cdots & \rho_{h-2} \\ \vdots & \vdots & \ddots & \vdots \\ \rho_{h-1} & \rho_{h-2} & \cdots & 1 \end{pmatrix} \tag{4.93}$$

We then have $E\left(\boldsymbol{\varepsilon}\boldsymbol{\varepsilon}'\right) = \boldsymbol{\Omega}_\varepsilon \neq \sigma_\varepsilon^2 \boldsymbol{I}$.

Remark 4.4 When working with time series, the autocorrelation in question is of a temporal type: the error term at a given date depends on the same error term at another date. When working with cross-sectional data, we speak of spatial autocorrelation; the correlation being in space rather than time.

Autocorrelation can have several sources, including:

- the omission of one or more important explanatory variables from the model. It should be remembered that the error term can be interpreted as a set of explanatory variables not included in the model. Consequently, omitting one or more explanatory variables may result in autocorrelation of the error term, particularly if the omitted explanatory variables are themselves autocorrelated.

- Misspecification of the model. The functional form chosen is incorrect; for example, the specification chosen is linear when the model is in fact nonlinear.
- The measurement error of the dependent variable. For example, some data, such as macroeconomic data, are revised regularly, leading to the publication of values that are slightly different from those previously published. These measurement errors can produce autocorrelation in the error term.
- The use of nonseasonally adjusted time series. The use of nonseasonally adjusted series can produce cycles that generate autocorrelation of errors;
- Data transformations:
  - A first example is provided by averages or moving averages. For example, if we want to work at a quarterly frequency and have monthly series, a commonly used method is to average the values over three months. In this way, a certain regularity is introduced into the data by smoothing them: the change to quarterly data attenuates the fluctuations observed at monthly frequency. Such smoothing can lead to autocorrelation of errors.
  - A second example is given by data interpolation or extrapolation techniques. Suppose that data are available every ten years and that we wish to obtain values for all years in between. The standard practice is to interpolate the series on the basis of some ad hoc assumptions. This interpolation can impose a systematic pattern in the errors, giving rise to autocorrelation.
  - A third example is given by the use—which is very frequent in practice—of the first difference operator. Consider, for example, the following model, in which the error term is non-autocorrelated:

$$Y_t = \alpha + \beta X_t + \varepsilon_t \tag{4.94}$$

We can write this model at date $t-1$:

$$Y_{t-1} = \alpha + \beta X_{t-1} + \varepsilon_{t-1} \tag{4.95}$$

Taking the difference between (4.94) and (4.95), we obtain:

$$\Delta Y_t = \beta \Delta X_t + \Delta \varepsilon_t \tag{4.96}$$

where Δ is the first difference operator, such that: $\Delta Z_t = Z_t - Z_{t-1}$. Such a first-difference transformation can produce autocorrelation of errors in that, if the error term of (4.94) is non-autocorrelated, the error term $\Delta\varepsilon_t$ of (4.96) exhibits autocorrelation.
- The nonstationarity of the series considered. If the regression model selected involves nonstationary series and the error term is itself nonstationary, the latter will be characterized by the presence of autocorrelation (see, for example, Lardic and Mignon, 2002 for details).

As shown, the sources of autocorrelation can be very diverse. Finally, autocorrelation can be positive or negative. We speak of positive autocorrelation when the error term moves either upwards or downwards over a fairly long period. Conversely, negative autocorrelation occurs when a positive value of the error term is followed by a negative value, then a positive value, and so on.

4.3.2 Estimation When There Is Autocorrelation

As previously mentioned, in the presence of error autocorrelation, the OLS estimators remain unbiased but are no longer of minimum variance. In other words, the estimators are not efficient and the usual testing procedures no longer hold. To study this property in more detail, let us consider the following example:

$$Y_t = \alpha + \beta X_t + \varepsilon_t \tag{4.97}$$

with:

$$\varepsilon_t = \rho \varepsilon_{t-1} + u_t \tag{4.98}$$

where $|\rho| < 1$, $E(\boldsymbol{\varepsilon}) = 0$, and $E\left(\boldsymbol{\varepsilon}\boldsymbol{\varepsilon}'\right) \neq \sigma_{\varepsilon}^2 \boldsymbol{I}$. The term u_t is assumed to be white noise.

The process (4.98) is called a **first-order autoregressive process**, denoted $AR(1)$: the error term at t is a function of itself at $t-1$. In other words, this process illustrates a first-order autocorrelation of the error term. The coefficient ρ is the **first-order autocorrelation coefficient**.

The application of OLS to the model formed by Eqs. (4.97) and (4.98) leads to the following result (see Eq. (4.8)):

$$\boldsymbol{\Omega}_{\hat{\boldsymbol{\beta}}} = \left(\boldsymbol{X}'\boldsymbol{X}\right)^{-1} \boldsymbol{X}'\boldsymbol{\Omega}_{\varepsilon}\boldsymbol{X}\left(\boldsymbol{X}'\boldsymbol{X}\right)^{-1} \tag{4.99}$$

Let us explain $\boldsymbol{\Omega}_{\varepsilon}$. To this end, we iterate Equation (4.98), which gives:

$$\varepsilon_t = \rho\left(\rho\varepsilon_{t-2} + u_{t-1}\right) + u_t = \rho^2\varepsilon_{t-2} + \rho u_{t-1} + u_t \tag{4.100}$$

Continuing, we have:

$$\varepsilon_t = \rho^2\left(\rho\varepsilon_{t-3} + u_{t-2}\right) + \rho u_{t-1} + u_t = \rho^3\varepsilon_{t-3} + \rho^2 u_{t-2} + \rho u_{t-1} + u_t \tag{4.101}$$

that is finally:

$$\varepsilon_t = u_t + \rho u_{t-1} + \rho^2 u_{t-2} + \ldots \tag{4.102}$$

We seek to write the variance-covariance matrix of the error term:

$$\boldsymbol{\Omega}_{\boldsymbol{\varepsilon}} = E\left(\boldsymbol{\varepsilon}\boldsymbol{\varepsilon}'\right) = \begin{pmatrix} V(\varepsilon_1) & Cov(\varepsilon_1, \varepsilon_2) & \cdots & Cov(\varepsilon_1, \varepsilon_T) \\ Cov(\varepsilon_2, \varepsilon_1) & V(\varepsilon_2) & \cdots & Cov(\varepsilon_2, \varepsilon_T) \\ \vdots & \vdots & \ddots & \vdots \\ Cov(\varepsilon_T, \varepsilon_1) & Cov(\varepsilon_T, \varepsilon_2) & \cdots & V(\varepsilon_T) \end{pmatrix} \tag{4.103}$$

To do this, let us square the two sides of Eq. (4.102) and consider the expectation:

$$E\left(\varepsilon_t^2\right) = E\left(u_t + \rho u_{t-1} + \rho^2 u_{t-2} + \ldots\right)^2 \tag{4.104}$$

that is:

$$E\left(\varepsilon_t^2\right) = \sigma_\varepsilon^2 = \frac{\sigma_u^2}{1-\rho^2} \tag{4.105}$$

Furthermore, we have:

$$E\left(\varepsilon_t \varepsilon_{t-1}\right) = E\left(\left(\rho\varepsilon_{t-1} + u_t\right)\varepsilon_{t-1}\right) = \rho\sigma_\varepsilon^2 \tag{4.106}$$

since $E\left(\varepsilon_{t-1}\varepsilon_{t-1}\right) = \sigma_\varepsilon^2$ and $E\left(u_t\varepsilon_{t-1}\right) = 0$. In addition:

$$E\left(\varepsilon_t \varepsilon_{t-2}\right) = E\left(\left(\rho\varepsilon_{t-1} + u_t\right)\varepsilon_{t-2}\right) = \rho E\left(\varepsilon_{t-1}\varepsilon_{t-2}\right) \tag{4.107}$$

and:

$$E\left(\varepsilon_{t-1} \varepsilon_{t-2}\right) = E\left(\left(\rho\varepsilon_{t-2} + u_{t-1}\right)\varepsilon_{t-2}\right) = \rho\sigma_\varepsilon^2 \tag{4.108}$$

hence:

$$E\left(\varepsilon_t \varepsilon_{t-2}\right) = \rho^2\sigma_\varepsilon^2 \tag{4.109}$$

More generally, we have:

$$E\left(\varepsilon_t \varepsilon_{t-i}\right) = \rho^i\sigma_\varepsilon^2 \tag{4.110}$$

It is then possible to write the variance-covariance matrix:

$$\boldsymbol{\Omega}_{\boldsymbol{\varepsilon}} = E\left(\boldsymbol{\varepsilon}\boldsymbol{\varepsilon}'\right) = \begin{pmatrix} \sigma_\varepsilon^2 & \rho\sigma_\varepsilon^2 & \cdots & \rho^{T-1}\sigma_\varepsilon^2 \\ \rho\sigma_\varepsilon^2 & \sigma_\varepsilon^2 & \cdots & \rho^{T-2}\sigma_\varepsilon^2 \\ \vdots & \vdots & \ddots & \vdots \\ \rho^{T-1}\sigma_\varepsilon^2 & \rho^{T-2}\sigma_\varepsilon^2 & \cdots & \sigma_\varepsilon^2 \end{pmatrix} \tag{4.111}$$

that is:

$$\boldsymbol{\Omega}_\varepsilon = \sigma_\varepsilon^2 \begin{pmatrix} 1 & \rho & \rho^2 \cdots & \rho^{T-1} \\ \rho & 1 & \cdots & \rho^{T-2} \\ \rho^2 & & & \\ \vdots & \vdots & \ddots & \vdots \\ \rho^{T-1} & \rho^{T-2} & \cdots & 1 \end{pmatrix} \tag{4.112}$$

with $\sigma_\varepsilon^2 = \frac{\sigma_u^2}{1-\rho^2}$.

Transferring this expression to (4.99), we can show that the variance of the OLS estimator is written as:

$$V\left(\hat{\beta}_{OLS}\right) = \frac{\sigma_\varepsilon^2}{\sum\limits_{t=1}^{T} x_t^2}\left[1 + 2\rho\frac{\sum\limits_{t=2}^{T} x_t x_{t-1}}{\sum\limits_{t=1}^{T} x_t^2} + 2\rho^2\frac{\sum\limits_{t=3}^{T} x_t x_{t-2}}{\sum\limits_{t=1}^{T} x_t^2} + \ldots + 2\rho^{T-1}\frac{x_1 x_T}{\sum\limits_{t=1}^{T} x_t^2}\right] \tag{4.113}$$

where $x_t = X_t - \bar{X}$.

We have seen that when there is autocorrelation, it is appropriate to use the generalized least squares (GLS) estimator given by:

$$\tilde{\boldsymbol{\beta}} = \left(\boldsymbol{X}'\boldsymbol{\Omega}_\varepsilon^{-1}\boldsymbol{X}\right)^{-1}\boldsymbol{X}'\boldsymbol{\Omega}_\varepsilon^{-1}\boldsymbol{Y} \tag{4.114}$$

In this case, we know that the variance of the estimator is written as:

$$\Omega_{\tilde{\boldsymbol{\beta}}} = \left(\boldsymbol{X}'\boldsymbol{\Omega}_\varepsilon^{-1}\boldsymbol{X}\right)^{-1} \tag{4.115}$$

If we replace the matrix $\boldsymbol{\Omega}_\varepsilon$ by its expression (4.112) in (4.115), we obtain:

$$V\left(\tilde{\beta}_{GLS}\right) = \frac{\sigma_\varepsilon^2}{\sum\limits_{t=1}^{T} x_t^2}\left[\frac{1-\rho^2}{1 + \rho^2 - 2\rho\frac{\sum\limits_{t=2}^{T} x_t x_{t-1}}{\sum\limits_{t=1}^{T} x_t^2}}\right] \tag{4.116}$$

where $x_t = X_t - \bar{X}$.

To assess the effectiveness of the OLS estimator, we compare the bracketed terms in relationships (4.113) and (4.116). We see that these terms depend on the variable

x_t and the first-order autocorrelation coefficient ρ. Obviously, in the absence of autocorrelation of errors, i.e., if $\rho = 0$, the two expressions are identical. When there is autocorrelation, the efficiency of the OLS estimator depends on the values taken by the coefficient ρ and the nature of x_t. Calculations by Johnston and Dinardo (1996) show that, if we assume that x_t also follows an autoregressive process of order 1, the efficiency of the OLS estimator decreases from 90 to 10% when the first-order autocorrelation coefficient ρ increases from 0.2 to 0.9. These calculations are only illustrative, but they show that the effectiveness of the OLS estimator is undermined if OLS is applied to a model with an autocorrelated error term.

Furthermore, we know that the variance of the OLS estimator when the error term satisfies the good statistical properties is given by:

$$V\left(\hat{\beta}\right) = \frac{\sigma_{\varepsilon}^{2}}{\sum_{t=1}^{T} x_t^2} \tag{4.117}$$

As we can see from expression (4.113), this is no longer the true variance when there is error autocorrelation. Consequently, using the usual OLS formulas when there is autocorrelation leads to an erroneous evaluation of the variance of the estimators and, consequently, of their t-statistics. The results of the significance tests can then be significantly affected, leading to incorrect interpretations.

All in all, it should be recalled that when there is autocorrelation of errors, the GLS method should be used. However, as in the case of heteroskedasticity, GLS can only be used if the variance-covariance matrix of the errors $\mathbf{\Omega}_{\boldsymbol{\varepsilon}}$ is known. In practice, this is generally not the case, which is why we will present operational estimation procedures in the following.

4.3.3 Detecting Autocorrelation

Since the error term is unobservable, autocorrelation is detected by studying the residuals that are deduced from estimating the regression model. There are various tests for detecting autocorrelation. Before presenting them, let us mention that a first intuition can be provided graphically. Figures 4.6, 4.7, 4.8 and 4.9 reproduce various cases of autocorrelation of residuals, with Fig. 4.10 representing the case of absence of autocorrelation. Figure 4.6 illustrates positive autocorrelation, Fig. 4.7 an increasing trend in the residuals, Fig. 4.8 a decreasing trend, and Fig. 4.9 a case of negative autocorrelation of the residuals. In Fig. 4.10, no particular pattern stands out, illustrating the case of no autocorrelation in the residuals.

The Geary Test (1970)

This is a nonparametric test in the sense that no hypothesis is made about the probability distribution from which the residuals are derived. This test is based on the study of the sign of the residuals or, more precisely, the change in sign of the

Fig. 4.6 Positive autocorrelation of residuals

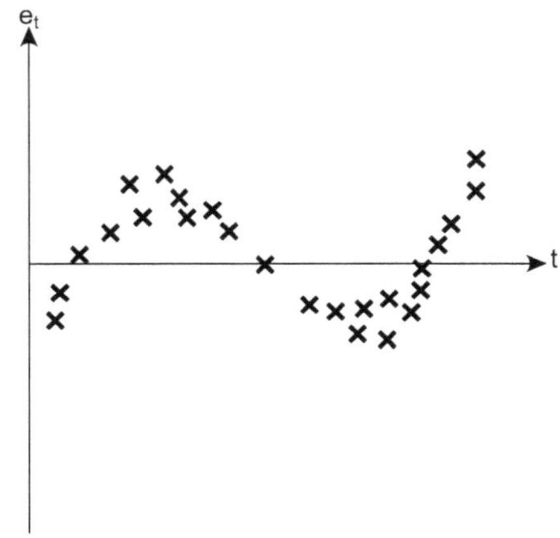

Fig. 4.7 Increasing trend in residuals

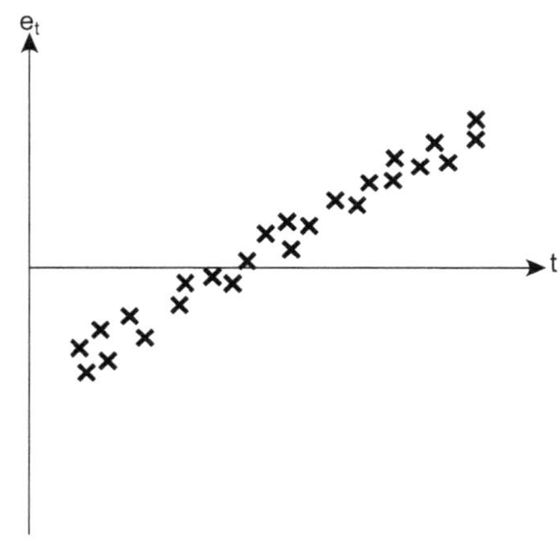

Fig. 4.8 Decreasing trend in residuals

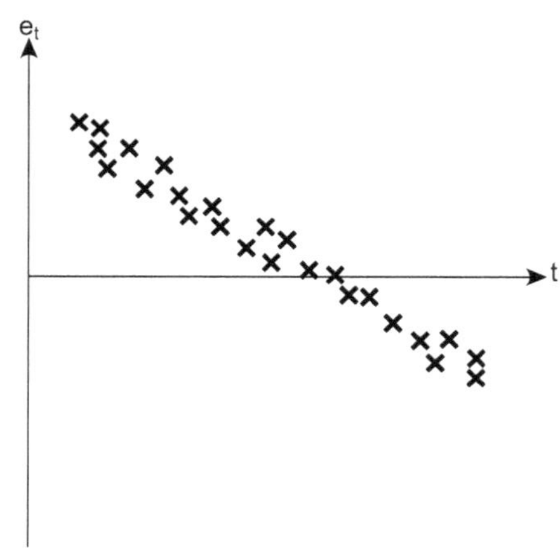

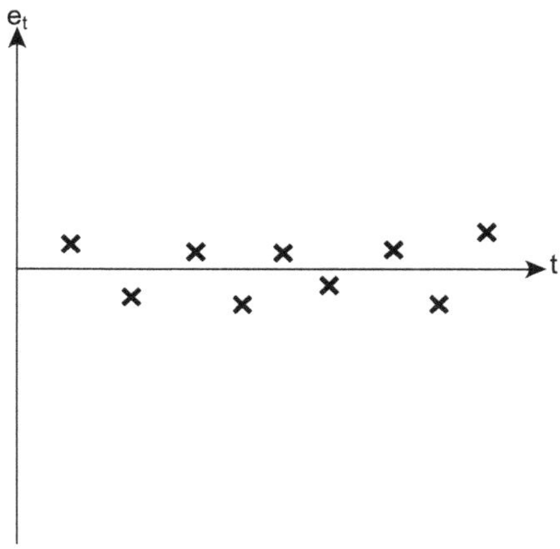

Fig. 4.9 Negative autocorrelation of residuals

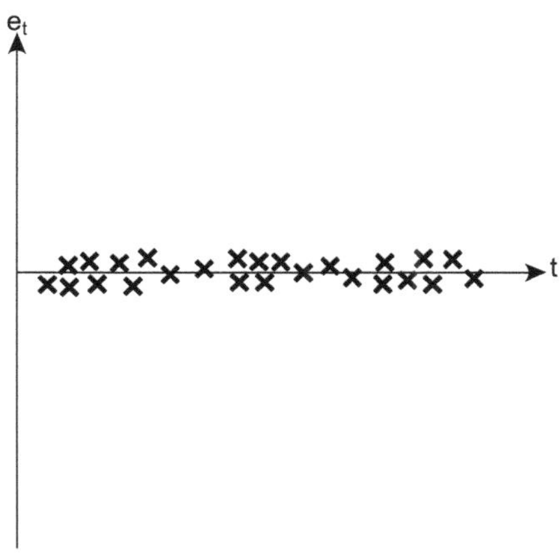

Fig. 4.10 Absence of autocorrelation of residuals

residuals. Let us define a run (or sequence) of length i as a sequence of i consecutive values of the same sign. For example, let us denote positive residuals by $+$ and negative residuals by $-$. Suppose that the estimation of a regression model on 30 observations leads to the following results for the residuals:

$$- - - - - - - + + + + + + + + + + + + + - - - - - - - - - -$$

Here we have three runs: a negative run of length 7, a positive run of length 13, and then a negative run of length 10. We wonder whether these three runs are from a purely random series of 30 observations. Intuitively, we might think that if the number of runs is very large, the residuals frequently change sign, which is an indication in favor of a negative autocorrelation of the residuals. Similarly, if the number of runs is very small, indicating that residuals rarely change sign, this may indicate a positive autocorrelation of residuals.

To test these assumptions, let us note T the total number of observations, T_1 the number of positive residuals, and T_2 the number of negative residuals ($T = T_1 + T_2$). Let R be the number of runs. Under the null hypothesis of independence of observations (here of residuals) and assuming that $T_1 > 10$ and $T_2 > 10$, the number of runs follows a normal distribution with mean:

$$E(R) = 1 + \frac{2T_1T_2}{T} \tag{4.118}$$

and variance:

$$\sigma_R^2 = \frac{2T_1T_2(2T_1T_2 - 1)}{(T-1)T^2} \tag{4.119}$$

Under the null hypothesis of independence, we can construct a 95% confidence interval:

$$E(R) \pm 1.96\sigma_R \tag{4.120}$$

The decision rule is as follows:

- If the number of runs R lies within the confidence interval, the null hypothesis is not rejected.
- If the number of runs R lies outside the confidence interval, the null hypothesis is rejected.

Returning to our example, we have: $T = 30$, $T_1 = 13$, $T_2 = 17$, and $R = 3$. We deduce: $E(R) = 15.7333$ and $\sigma_R = 2.7328$. The 95% confidence interval is then given by: [10.3770; 21.0896]. The number of runs, 3, does not lie within this interval. The null hypothesis is rejected, reflecting the presence of autocorrelation at the 5% significance level.

The Durbin and Watson Test (1950, 1951)

Unlike the previous test, the **Durbin-Watson** test is a parametric test in the sense that it relies on an assumption about the probability distribution of residuals. It is the best-known test for detecting the presence of autocorrelation, is included in almost all econometric software, and is widely used.

The Durbin-Watson test is based on the calculation of the following statistic:

$$DW = \frac{\sum_{t=2}^{T}(e_t - e_{t-1})^2}{\sum_{t=1}^{T} e_t^2} \tag{4.121}$$

where e_t are the residuals resulting from the estimation of the regression model (simple or multiple). It allows us to test the null hypothesis of no **first-order** autocorrelation of the residuals against the alternative hypothesis of first-order autocorrelation of the residuals. If we assume that the error term follows a first-order autoregressive process:

$$\varepsilon_t = \rho\varepsilon_{t-1} + u_t \tag{4.122}$$

the Durbin-Watson test consists in testing the null hypothesis:

$$H_0 : \rho = 0 \tag{4.123}$$

against the alternative hypothesis:

$$H_1 : \rho \neq 0 \tag{4.124}$$

The statistic DW is thus used to measure the magnitude of the first-order autocorrelation. To show some of the characteristics of this test, let us expand the numerator of (4.121):

$$DW = \frac{\sum_{t=2}^{T} e_t^2 + \sum_{t=2}^{T} e_{t-1}^2 - 2\sum_{t=2}^{T} e_t e_{t-1}}{\sum_{t=1}^{T} e_t^2} \tag{4.125}$$

For T sufficiently large, we can write:

$$DW \simeq \frac{2\sum_{t=2}^{T} e_t^2 - 2\sum_{t=2}^{T} e_t e_{t-1}}{\sum_{t=1}^{T} e_t^2} \tag{4.126}$$

that is:

$$DW \simeq 2\left(1 - \frac{\sum_{t=2}^{T} e_t e_{t-1}}{\sum_{t=1}^{T} e_t^2}\right) \tag{4.127}$$

Given that $E(e_t) = 0$, the term $\frac{\sum_{t=2}^{T} e_t e_{t-1}}{\sum_{t=1}^{T} e_t^2}$ represents the estimate of the first-order autocorrelation coefficient of the residual series. In other words, this is the estimate of the coefficient ρ in the regression of e_t on e_{t-1}. Let us denote this estimated coefficient $\hat{\rho}$. We can write:

$$DW \simeq 2\left(1 - \hat{\rho}\right) \tag{4.128}$$

The expression (4.128) shows that there is a relationship between the statistic DW and the first-order autocorrelation coefficient of the residuals. Furthermore, this relationship allows us to highlight various characteristics of the statistic DW:

- Given that a coefficient of autocorrelation varies between -1 and 1, the statistic DW varies between 0 and 4. It is 0 when there is perfect positive autocorrelation ($\hat{\rho} = 1$) and 4 when there is perfect negative autocorrelation ($\hat{\rho} = -1$).
- When $DW \simeq 2$, the residuals are not autocorrelated ($\hat{\rho} = 0$).
- When $DW > 2$, the autocorrelation of the residuals is negative.
- When $DW < 2$, the autocorrelation of the residuals is positive.

To carry out the test, Durbin and Watson tabulated a lower bound and an upper bound for the critical values of the statistic DW as a function of the number of observations and the number of explanatory variables k included in the model under consideration. The table thus gives two values d_1 (lower bound) and d_2 (upper bound) allowing us to perform the test according to the table below.

0 $\quad d_1 \quad d_2 \quad 2 \quad 4-d_2 \quad 4-d_1 \quad 4$

$\rho > 0$ Rejection of H_0	?	$\rho = 0$ Non-rejection of H_0	?	$\rho < 0$ Rejection of H_0

We can see that there are two regions of "doubt" or indecision. In practice, if we find ourselves in one of these regions, we tend to reject the null hypothesis of no autocorrelation. This is because the consequences of not rejecting the null hypothesis of no autocorrelation even though it is false are considered more "severe" than the consequences of wrongly assuming the absence of autocorrelation. Thus, in practice, when in a region of doubt, we use the upper bound d_2 as if it were a usual critical value: we reject the null hypothesis of no autocorrelation if $DW < d_2$. The region of doubt decreases as the sample size increases.

The Durbin-Watson test is very frequently used. However, it is important to specify certain conditions of use:

- The regression model must include a constant term. The critical values given in the tables of Durbin and Watson have indeed been tabulated assuming the presence of a constant in the regression model.[4]

[4] Let us mention, however, that Farebrother (1980) tabulated the critical values of the statistic DW in the absence of a constant term.

- The explanatory variables must be nonrandom. Consequently, the Durbin-Watson test cannot be used when the model includes the lagged endogenous variable among the explanatory variables. In this case, the Durbin test or the Breusch-Godfrey test can be used (see below).
- The error term is assumed to follow a normal distribution (see Mittelhammer et al., 2000 for a study of the consequences of not observing the normality assumption).
- The Durbin-Watson test only detects the first-order autocorrelation of residuals. In other words, it is not appropriate if the residuals exhibit higher-order autocorrelation. It is then possible to use the Wallis test or more general tests such as the Breusch-Godfrey, Box-Pierce, or Ljung-Box tests.

Remark 4.5 (The Wallis Test (1972)) As mentioned by Wallis (1972), it is common, when working with quarterly frequency series, to expect a fourth-order error autocorrelation. Wallis proposes a modification of the Durbin-Watson statistic:

$$DW_4 = \frac{\sum_{t=5}^{T} (e_t - e_{t-4})^2}{\sum_{t=1}^{T} e_t^2} \tag{4.129}$$

allowing us to test the null hypothesis of absence of autocorrelation at order 4 of the residuals, i.e., $\rho_4 = 0$ in the equation:

$$\varepsilon_t = \rho_4 \varepsilon_{t-4} + u_t \tag{4.130}$$

Wallis (1972) derived tables of critical values including an upper and lower bound for DW_4 (see also Giles and King, 1978).

The Durbin Test (1970)

As we have previously pointed out, the Durbin-Watson test cannot be applied if the regression model under consideration includes the lagged endogenous variable among the explanatory variables. The test proposed by Durbin (1970) can be implemented in such a case.

Consider the following regression model:

$$Y_t = \alpha + \phi_1 Y_{t-1} + \ldots + \phi_p Y_{t-p} + \beta_1 X_{1t} + \ldots + \beta_k X_{kt} + \varepsilon_t \tag{4.131}$$

with:

$$\varepsilon_t = \rho \varepsilon_{t-1} + u_t \tag{4.132}$$

The **Durbin test** allows us to detect the presence of first-order autocorrelation of the residuals in a model with the lagged endogenous variable among the explanatory variables.

The test statistic proposed by Durbin is written as:

$$h = \hat{\rho}\sqrt{\frac{T}{1 - TV\left(\hat{\phi}_1\right)}} \tag{4.133}$$

where $V\left(\hat{\phi}_1\right)$ denotes the estimated variance of the coefficient associated with Y_{t-1} in regression (4.131) and $\hat{\rho}$ is the estimator of the first-order autoregressive coefficient of the residuals, obtained from the regression of e_t on e_{t-1}:

$$\hat{\rho} = \frac{\sum_{t=2}^{T} e_t e_{t-1}}{\sum_{t=1}^{T} e_t^2} \tag{4.134}$$

e_t denoting residuals from the OLS estimation of regression (4.131).

Under the null hypothesis of no autocorrelation at order 1 ($\rho = 0$), the Durbin statistic has a standard normal distribution, i.e.:

$$h \sim N(0, 1) \tag{4.135}$$

At the 5% significance level, the decision rule is:

- If $|h| < 1.96$, the null hypothesis of no autocorrelation at order 1 is not rejected.
- If $|h| > 1.96$, the null hypothesis of no autocorrelation at order 1 is rejected.

Remark 4.6 If $TV\left(\hat{\phi}_1\right) \geq 1$, the formula (4.133) cannot be applied. In such a case, Durbin suggests proceeding as follows:

- Estimate model (4.131) by OLS and derive the residual series e_t.
- Estimate by OLS the regression of e_t on $e_{t-1}, Y_{t-1}, \ldots, Y_{t-p}, X_{1t}, \ldots, X_{kt}$.
- Perform a test of significance (t-test) on the coefficient associated with e_{t-1}. If this coefficient is significantly different from zero, the residuals are autocorrelated to order 1.

The Breusch-Godfrey Test

Breusch (1978) and Godfrey (1978) proposed—independently—a test to detect the presence of autocorrelation of order greater than 1 that remains valid when the model includes the lagged endogenous variable among the explanatory variables.

The **Breusch-Godfrey test** is a Lagrange multiplier test based on the search for a relationship between the errors $\varepsilon_t, t = 1, \ldots, T$.

Suppose that the error term ε_t of the multiple regression model:

$$Y_t = \alpha + \beta_1 X_{1t} + \beta_2 X_{2t} + \ldots + \beta_j X_{jt} + \ldots + \beta_k X_{kt} + \varepsilon_t \tag{4.136}$$

follows an autoregressive process of order p, which we note $AR(p)$, or:

$$\varepsilon_t = \phi_1 \varepsilon_{t-1} + \phi_2 \varepsilon_{t-2} + \ldots + \phi_p \varepsilon_{t-p} + u_t \tag{4.137}$$

where u_t is white noise.

The Breusch-Godfrey test consists of testing the null hypothesis of no autocorrelation of errors, i.e.:

$$H_0 : \phi_1 = \phi_2 = \ldots = \phi_p = 0 \tag{4.138}$$

The test procedure can be described in three steps:

- The multiple regression model (4.136) is estimated and the residuals $e_t, t = 1, \ldots, T$, are deduced. The test concerns the residuals since the errors are obviously unknown.
- We regress the residuals e_t on their p past values as well as on the k explanatory variables, i.e.:

$$\begin{aligned} e_t = &\alpha + \beta_1 X_{1t} + \beta_2 X_{2t} + \ldots + \beta_j X_{jt} + \ldots + \beta_k X_{kt} \\ &+ \phi_1 e_{t-1} + \phi_2 e_{t-2} + \ldots + \phi_p e_{t-p} + u_t \end{aligned} \tag{4.139}$$

  and we calculate the coefficient of determination R^2 associated with this regression.
- We calculate the test statistic:

$$BG = (T - p) R^2 \tag{4.140}$$

  Under the null hypothesis, $BG \sim \chi_p^2$.
  The decision rule is then:
  - If $BG < \chi_p^2$, the null hypothesis of no autocorrelation is not rejected.
  - If $BG > \chi_p^2$, the null hypothesis of no autocorrelation is rejected: at least one of the coefficients $\phi_i, i = 1, \ldots, p$, is significantly different from zero.

Remark 4.7 In the previous developments, it has been assumed that the error term ε_t of the multiple regression model follows an autoregressive process of order p (Eq. (4.137)). The Breusch-Godfrey test can also be applied in the case where the error process follows a moving average process of order p, which is noted $MA(p)$, i.e.:

$$\varepsilon_t = \theta_1 u_{t-1} + \theta_2 u_{t-2} + \ldots + \theta_p u_{t-p} \tag{4.141}$$

where u_t is white noise.

The test procedure is exactly the same as described above.

The Box-Pierce (1970) and Ljung-Box (1978) Tests

The **Box-Pierce test**, also known as the "**portmanteau**" test, is designed to test the non-autocorrelated nature of residuals. Noting $\rho_{h(e_t)}$ the coefficient of autocorrelation of order h of the residuals, $h = 1, \ldots, H$, the test consists in testing the null hypothesis:

$$\rho_{1(e_t)} = \rho_{2(e_t)} = \cdots = \rho_{h(e_t)} = \cdots = \rho_{H(e_t)} = 0 \tag{4.142}$$

against the alternative hypothesis that there is at least one coefficient $\rho_{h(e_t)}$ significantly different from zero.

The test statistic is written as:

$$BP(H) = T \sum_{h=1}^{H} \hat{\rho}^2_{h(e_t)} \tag{4.143}$$

where $\hat{\rho}_{h(e_t)}$ is the estimator of the autocorrelation coefficient of order h of the residuals:

$$\hat{\rho}_h = \frac{\sum\limits_{t=h+1}^{T} e_t e_{t-h}}{\sum\limits_{t=1}^{T} e_t^2} \tag{4.144}$$

and H is the maximum number of lags.

Under the null hypothesis of no autocorrelation:

$$\rho_{1(e_t)} = \rho_{2(e_t)} = \cdots = \rho_{H(e_t)} = 0 \tag{4.145}$$

the statistic $BP(H)$ follows a Chi-squared distribution with H degrees of freedom.[5]

Ljung and Box (1978) suggest an improvement to the Box-Pierce test when the sample size is small. The distribution of the **Ljung-Box test** statistic is indeed closer

[5] It is assumed here that the lagged dependent variable is not among the explanatory variables. We will come back to the Box-Pierce test in Chap. 7.

to that of the Chi-squared in small samples than is that of the Box-Pierce test. The test statistic is written as:

$$LB(H) = T(T+2)\sum_{h=1}^{H}\frac{\hat{\rho}^2_{h(e_t)}}{T-h} \tag{4.146}$$

Under the null hypothesis of no autocorrelation:

$$\rho_{1(e_t)} = \rho_{2(e_t)} = \cdots = \rho_{H(e_t)} = 0 \tag{4.147}$$

the statistic $LB(H)$ has a Chi-squared distribution with H degrees of freedom.

4.3.4 Estimation Procedures in the Presence of Error Autocorrelation

In the presence of error autocorrelation, the OLS estimators remain unbiased, but are no longer of minimum variance. As in the case of heteroskedasticity, this has the consequence of affecting the precision of the tests. So how can we correct for error autocorrelation?

To answer this question, we need to distinguish between cases where the variance of the error term is known and those where it is unknown. When the variance of the error term is known, we have seen (see Sect. 4.1.2) that the GLS method should be applied in the presence of autocorrelation. When the variance of the error term is unknown, various methods are available, which we describe below.

Case Where the Variance of the Error Term Is Known: General Principle of GLS

Consider the multiple regression model:

$$\boldsymbol{Y} = \boldsymbol{X\beta} + \boldsymbol{\varepsilon} \tag{4.148}$$

with $E\left(\boldsymbol{\varepsilon\varepsilon}'\right) = \boldsymbol{\Omega}_{\varepsilon}$.

As we have seen previously (see Sect. 4.1.2), the GLS method can be applied provided we find a transformation matrix $\boldsymbol{M}$ of known parameters, such as:

$$\boldsymbol{M}'\boldsymbol{M} = \boldsymbol{\Gamma}_{\varepsilon}^{-1} \tag{4.149}$$

with:

$$\boldsymbol{\Gamma}_{\varepsilon} = \left(\sigma^2_{\varepsilon}\right)^{-1}\boldsymbol{\Omega}_{\varepsilon} \tag{4.150}$$

It is then sufficient to apply OLS to the transformed variables $\boldsymbol{MY}$ and $\boldsymbol{MX}$. To get a clearer picture, let us consider the simple regression model:

$$Y_t = \alpha + \beta X_t + \varepsilon_t \tag{4.151}$$

and assume that the error term follows a first-order autoregressive process ($AR(1)$), i.e.:

$$\varepsilon_t = \rho\varepsilon_{t-1} + u_t \tag{4.152}$$

where $|\rho| < 1$ and u_t is white noise.

As previously shown, the variance-covariance matrix of the error term is given by:

$$\boldsymbol{\Omega}_{\boldsymbol{\varepsilon}} = \sigma_\varepsilon^2 \begin{pmatrix} 1 & \rho & \cdots & \rho^{T-1} \\ \rho & 1 & \cdots & \rho^{T-2} \\ \vdots & \vdots & \ddots & \vdots \\ \rho^{T-1} & \rho^{T-2} & \cdots & 1 \end{pmatrix} \tag{4.153}$$

with $\sigma_\varepsilon^2 = \frac{\sigma_u^2}{1-\rho^2}$.

If ρ is known, the GLS estimator:

$$\tilde{\boldsymbol{\beta}} = \left(\boldsymbol{X}'\boldsymbol{\Omega}_{\boldsymbol{\varepsilon}}^{-1}\boldsymbol{X}\right)^{-1}\boldsymbol{X}'\boldsymbol{\Omega}_{\boldsymbol{\varepsilon}}^{-1}\boldsymbol{Y} \tag{4.154}$$

can be obtained with:

$$\boldsymbol{\Omega}_{\boldsymbol{\varepsilon}}^{-1} = \frac{1}{\sigma_u^2}\begin{pmatrix} 1 & -\rho & 0 & 0 & \cdots & 0 \\ -\rho & 1+\rho^2 & -\rho & 0 & \cdots & 0 \\ 0 & -\rho & 1+\rho^2 & -\rho & & 0 \\ \vdots & \vdots & & & \ddots & \vdots \\ 0 & \cdots & 0 & -\rho & 1+\rho^2 & -\rho \\ 0 & \cdots & & 0 & -\rho & 1 \end{pmatrix} \tag{4.155}$$

Consider the transformation matrix $\boldsymbol{M}$ such that:

$$\boldsymbol{M} = \begin{pmatrix} -\rho & 1 & 0 & 0 & \cdots & 0 \\ 0 & -\rho & 1 & 0 & \cdots & 0 \\ 0 & 0 & -\rho & 1 & & 0 \\ \vdots & \vdots & & & \ddots & \vdots \\ 0 & 0 & 0 & \cdots & -\rho & 1 \end{pmatrix} \tag{4.156}$$

Then we have:

$$
\boldsymbol{M}'\boldsymbol{M} = \begin{pmatrix} \rho^2 & -\rho & 0 & 0 & \cdots & 0 \\ -\rho & 1+\rho^2 & -\rho & 0 & \cdots & 0 \\ 0 & -\rho & 1+\rho^2 & -\rho & & 0 \\ \vdots & \vdots & & & \ddots & \vdots \\ 0 & 0 & 0 & \cdots & 1+\rho^2 & -\rho \\ 0 & 0 & \cdots & 0 & -\rho & 1 \end{pmatrix} \tag{4.157}
$$

$\boldsymbol{M}'\boldsymbol{M}$ is identical to $\sigma_u^2 \boldsymbol{\Omega}_{\varepsilon}^{-1}$, except for the first element of the diagonal (ρ^2 instead of 1).

By applying the matrix $\boldsymbol{M}$ to model (4.151), we obtain the transformed variables:

$$
\boldsymbol{M}\boldsymbol{Y} = \begin{pmatrix} Y_2 - \rho Y_1 \\ Y_3 - \rho Y_2 \\ \vdots \\ Y_T - \rho Y_{T-1} \end{pmatrix} \tag{4.158}
$$

and

$$
\boldsymbol{M}\boldsymbol{X} = \begin{pmatrix} 1 & X_2 - \rho X_1 \\ 1 & X_3 - \rho X_2 \\ \vdots & \vdots \\ 1 & X_T - \rho X_{T-1} \end{pmatrix} \tag{4.159}
$$

The GLS method amounts to applying the OLS to the regression model formed by the $(T-1)$ transformed observations $\boldsymbol{M}\boldsymbol{Y}$ and $\boldsymbol{M}\boldsymbol{X}$:

$$
\begin{pmatrix} Y_2 - \rho Y_1 \\ Y_3 - \rho Y_2 \\ \vdots \\ Y_T - \rho Y_{T-1} \end{pmatrix} = \begin{pmatrix} 1 & X_2 - \rho X_1 \\ 1 & X_3 - \rho X_2 \\ \vdots & \vdots \\ 1 & X_T - \rho X_{T-1} \end{pmatrix} \begin{pmatrix} \alpha\,(1-\rho) \\ \beta \end{pmatrix} + \begin{pmatrix} u_2 \\ u_3 \\ \vdots \\ u_T \end{pmatrix} \tag{4.160}
$$

The variables thus transformed are said to be transformed into **first-quasi differences**.

Remark 4.8 In order not to lose the first observation, we can add a first row to the matrix M. This first row is such that all the elements are zero, except the first one which is equal to $\sqrt{1-\rho^2}$.

However, such a method is only applicable if the coefficient ρ is known, which is rarely the case in practice. It is a parameter that has to be estimated. Once this estimation has been made, the method previously described can be applied by replacing ρ by its estimator $\hat{\rho}$ in the transformed model. Various methods are available for this purpose and are discussed below.

Case Where the Variance of the Error Term Is Unknown: Pseudo GLS Methods

We can distinguish iterative methods from other techniques. These different methods are called **pseudo GLS** methods. Generally speaking, they consist of estimating the parameters of the residuals' generating model, transforming the variables of the model using these parameters, and applying OLS to the model formed by the variables thus transformed.

Non-iterative Methods

Among the non-iterative methods, it is possible to find the estimator $\hat{\rho}$ of the coefficient ρ in two different ways: by relying on the Durbin-Watson statistic or by performing regressions using residuals.

The Use of the Durbin-Watson Test

We know that (see Eq. (4.128)):

$$DW \simeq 2\left(1-\hat{\rho}\right) \tag{4.161}$$

where $\hat{\rho}$ denotes the estimate of ρ in the regression of the residuals e_t on e_{t-1}. Using this expression leads directly to the estimator:

$$\hat{\rho} \simeq 1 - \frac{DW}{2} \tag{4.162}$$

Once this estimator has been obtained, we transform the variables as follows:

$$Y_t - \hat{\rho} Y_{t-1} \text{ and } X_{it} - \hat{\rho} X_{it-1} \tag{4.163}$$

where $i = 1, \ldots, k$, k denotes the number of explanatory variables, and apply OLS to the transformed model.

Method Based on Residuals

This technique consists in regressing e_t on e_{t-1} and deducing the estimator $\hat{\rho}$ of ρ:

$$\hat{\rho} = \frac{\sum\limits_{t=2}^{T} e_t e_{t-1}}{\sum\limits_{t=1}^{T} e_t^2} \tag{4.164}$$

It then remains for us to transform the variables and apply OLS to the transformed model.[6]

Iterative Methods

Various iterative pseudo GLS techniques are available to estimate the coefficient ρ. The best known are those of Cochrane and Orcutt (1949) and Hildreth and Lu (1960).

The Cochrane-Orcutt Method

This is the most popular iterative technique. It can be described in five steps.

- *Step 1*. The regression model under consideration is estimated and the residuals e_t are deduced. An initial estimate $\hat{\rho}_0$ of ρ is obtained:

$$\hat{\rho}_0 = \frac{\sum_{t=2}^{T} e_t e_{t-1}}{\sum_{t=1}^{T} e_t^2} \tag{4.165}$$

- *Step 2*. The transformed variables $Y_t - \hat{\rho}_0 Y_{t-1}$ and $X_{it} - \hat{\rho}_0 X_{it-1}$ are constructed for $i = 1, \ldots, k$, with k denoting the number of explanatory variables.
- *Step 3*. OLS is applied to the model in quasi-differences:

$$\begin{aligned} Y_t - \hat{\rho}_0 Y_{t-1} &= \alpha\left(1 - \hat{\rho}_0\right) + \beta_1\left(X_{1t} - \hat{\rho}_0 X_{1t-1}\right) + \ldots \\ &\quad + \beta_k\left(X_{kt} - \hat{\rho}_0 X_{kt-1}\right) + u_t \end{aligned} \tag{4.166}$$

- *Step 4*. From the new estimation residuals $e_t^{(1)}$, a new estimation $\hat{\rho}_1$ of ρ is performed:

$$\hat{\rho}_1 = \frac{\sum_{t=2}^{T} e_t^{(1)} e_{t-1}^{(1)}}{\sum_{t=1}^{T} e_t^{(1)2}} \tag{4.167}$$

- *Step 5*. We construct the transformed variables $Y_t - \hat{\rho}_1 Y_{t-1}$ and $X_{it} - \hat{\rho}_1 X_{it-1}$ and apply the OLS to the model in quasi-differences:

$$\begin{aligned} Y_t - \hat{\rho}_1 Y_{t-1} &= \alpha\left(1 - \hat{\rho}_1\right) + \beta_1\left(X_{1t} - \hat{\rho}_1 X_{1t-1}\right) + \ldots \\ &\quad + \beta_k\left(X_{kt} - \hat{\rho}_1 X_{kt-1}\right) + u_t \end{aligned} \tag{4.168}$$

[6] It is unnecessary to introduce a constant term in the regression of e_t on e_{t-1} since the mean of the residuals is zero.

A new set of residuals $e_t^{(2)}$ is deduced, from which a new estimate $\hat{\rho}_2$ of ρ is obtained and so on.

These calculations are continued until the estimated regression coefficients $\hat{\beta}_1, \ldots, \hat{\beta}_k$ and $\alpha\left(1-\hat{\rho}_0\right)$ are stable.

Remark 4.9 We previously noted (see Remark 4.8) that it was possible not to omit the first observation during the variable transformation step. When this observation is not omitted, the method of Cochrane-Orcutt is slightly modified and is called the **Prais-Winsten method** (see Prais and Winsten, 1954).

The Hildreth-Lu Method

Consider the following quasi-difference model:

$$Y_t - \hat{\rho} Y_{t-1} = \alpha\left(1-\hat{\rho}\right) + \beta_1\left(X_{1t} - \hat{\rho} X_{1t-1}\right) + \ldots + \beta_k\left(X_{kt} - \hat{\rho} X_{kt-1}\right) + u_t \tag{4.169}$$

The procedure can be described in three steps.

- *Step 1*. We give ourselves a grid of possible values for $\hat{\rho}$, between -1 and 1. For example, we can set a step size of 0.1 and consider the values -0.9, -0.8, ..., 0.8, 0.9.
- *Step 2*. Relationship (4.169) is estimated for each of the previously fixed values of $\hat{\rho}$. The value of $\hat{\rho}$ that minimizes the sum of squared residuals is retained.
- *Step 3*. To refine the estimates, we repeat the previous two steps, setting a smaller step size (e.g., 0.01) and so on.

Other Methods

Two other techniques can also be implemented to account for autocorrelation.

The first technique involves applying the **maximum likelihood method** to the regression model. This method simultaneously estimates the usual parameters of the regression model as well as the value of ρ (see Beach and MacKinnon, 1978).

The second technique has already been discussed in the treatment of heteroskedasticity. This is the correction proposed by Newey and West (1987). Recall that this technique allows us to apply OLS to the regression model, despite the presence of error autocorrelation, and to correct the standard deviations of the estimated coefficients. We do not describe this technique again, since it has already been outlined (see Sect. 4.2.4).

4.3.5 Prediction in the Presence of Error Autocorrelation

Let us consider the bivariate regression model:

$$Y_t = \alpha + \beta X_t + \varepsilon_t \tag{4.170}$$

and assume that the error term follows a first-order autoregressive process, i.e.:

$$\varepsilon_t = \rho\varepsilon_{t-1} + u_t \tag{4.171}$$

where $|\rho| < 1$ and u_t is white noise.

We can then write:

$$Y_t = \alpha + \beta X_t + \rho\varepsilon_{t-1} + u_t \tag{4.172}$$

The prediction of Y for the date $T + 1$ is given by:

$$\hat{Y}_{T+1} = \hat{\alpha} + \hat{\beta} X_{T+1} + \rho\hat{\varepsilon}_T \tag{4.173}$$

Thus, compared with the usual regression model without error autocorrelation, the term $\rho\hat{\varepsilon}_T$ is added.

4.3.6 Empirical Application

Let us consider our monthly-frequency model over the period February 1984–June 2021 linking the returns of the $RFTSE$ London Stock Exchange index to the returns of the $RDJIND$ New York Stock Exchange index (see Table 4.8):

$$RFTSE_t = \alpha + \beta RDJIND_t + \varepsilon_t \tag{4.174}$$

The OLS estimation of model (4.174) leads to the results shown in Table 4.9. The residuals resulting from the estimation of this model are plotted in Fig. 4.11. In order to determine whether or not they are autocorrelated, let us apply the tests of absence of autocorrelation.

The value of the Durbin-Watson test statistic is given in Table 4.9: $DW = 2.2247$. At the 5% significance level, the reading of the Durbin-Watson table in the case where only one exogenous variable appears in the model gives $d_1 = 1.65$ and $d_2 = 1.69$. Since $d_2 < DW < 4 - d_2$, we do not reject the null hypothesis of absence of first-order autocorrelation of the residuals.

Table 4.8 $FTSE$ and Dow Jones industrial returns

	$RFTSE$	$RDJIND$
1984.02	−0.0216	−0.0555
1984.03	0.0671	0.0088
1984.04	0.0229	0.0050
...	...	...
2021.04	0.0374	0.0267
2021.05	0.0075	0.0191
2021.06	0.0021	−0.0008

Data source: Macrobond

Table 4.9 OLS estimation

Dependent variable: RFTSE				
Variable	Coefficient	Std. error	t-Statistic	Prob.
C	−0.001614	0.001352	−1.193763	0.2332
RDJIND	0.782505	0.030388	25.75060	0.0000
R-Squared	0.597331	Mean dependent var	0.004210	
Adjusted R-squared	0.596430	S.D. dependent var	0.044466	
S.E. of regression	0.028248	Akaike info criterion	−4.291158	
Sum squared resid	0.356678	Schwarz criterion	−4.272864	
Log likelihood	965.3650	Hannan-Quinn criterion	−4.283947	
F-statistic	663.0935	Durbin-Watson stat	2.224700	
Prob(F-statistic)	0.0000			

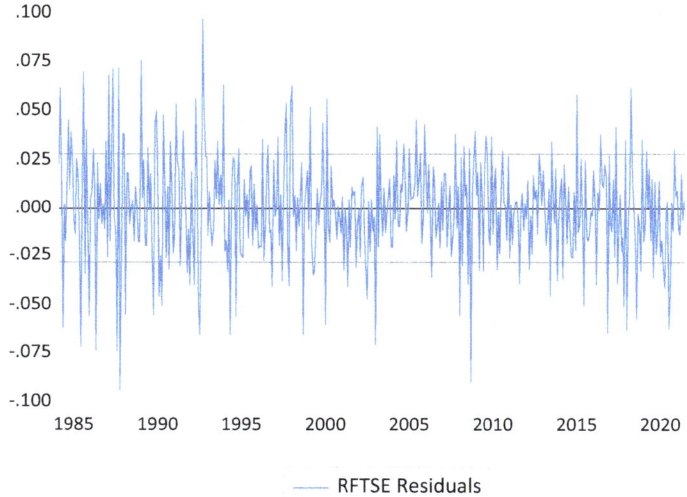

Fig. 4.11 Graphical representation of residuals

In order to apply the Breusch-Godfrey test, we regress the residuals (denoted $RESID$) obtained from the estimation of model (4.174) on the explanatory variable $RDJIND$ and on the one- and two-period lagged residuals. The results of this estimation are shown in Table 4.10.

The test statistic is given by:

$$BG = (T - p) R^2 = (449 - 2) \times 0.0208 = 9.2824 \tag{4.175}$$

Under the null hypothesis of no autocorrelation, the statistic BG follows a Chi-squared distribution with $p = 2$ degrees of freedom. At the 5% significance level, the critical value is 5.991. Given that $9.2824 > 5.991$, the null hypothesis

Table 4.10 The Breusch-Godfrey test

Dependent variable: RESID				
Variable	Coefficient	Std. error	t-Statistic	Prob.
C	−5.17E–05	0.001341	−0.038557	0.9693
RDJIND	0.006827	0.030235	0.225801	0.8215
RESID(-1)	−0.123734	0.047264	−2.617914	0.0091
RESID(-2)	−0.089929	0.047333	−1.899944	0.0581
R-Squared	0.020766	Mean dependent var	−6.99E–19	
Adjusted R-squared	0.014164	S.D. dependent var	0.028216	
S.E. of regression	0.028016	Akaike info criterion	−4.303234	
Sum squared resid	0.349272	Schwarz criterion	−4.266646	
Log likelihood	970.0760	Hannan-Quinn criterion	−4.288812	
F-statistic	3.145607	Durbin-Watson stat	2.016886	
Prob(F-statistic)	0.024998			

Table 4.11 The Ljung-Box test

H	$LB(H)$	χ^2_H	H	$LB(H)$	χ^2_H
1	5.787	3.841	7	11.81	14.067
2	8.351	5.991	8	11.811	15.507
3	10.79	7.815	12	13.055	21.026
4	10.99	9.488	18	16.307	28.869
5	10.993	11.07	24	21.142	36.415
6	11.77	12.592	30	31.654	43.773

of no autocorrelation is rejected. The residuals are therefore autocorrelated. This conclusion was to be expected from the results shown in Table 4.10 since the coefficient associated with the one-period lagged residuals is significantly different from zero and that associated with the two-period lagged residuals is significant at the 10% level.

Table 4.11 reports the calculated values of the Ljung-Box statistic and the critical value given by the Chi-squared distribution at the 5% significance level for a number of lags H ranging from 1 to 30. We find that, for values of H ranging from 1 to 4, the calculated value of the Ljung-Box statistic is higher than the critical value. The null hypothesis of no autocorrelation is therefore rejected. For higher values of H ($H \geq 5$), the null hypothesis is no longer rejected at the 5% significance level (it is rejected up to 6 lags if the 10% level is used).

While the Durbin-Watson test concludes in the absence of first-order autocorrelation of the residuals, the Breusch-Godfrey and Ljung-Box tests reject the null hypothesis of no autocorrelation—particularly for higher orders. To take account of this feature, it is possible to reestimate model (4.174) by OLS by applying the correction suggested by Newey-West. The results have already been presented in the study of heteroskedasticity and are shown in Table 4.7.

For illustrative purposes, let us also apply the Cochrane-Orcutt method. The first step is to obtain an initial estimate $\hat{\rho}_0$ of ρ. To do this, we regress the residuals

Table 4.12 The Cochrane-Orcutt procedure

Dependent variable: DRFTSE				
Variable	Coefficient	Std. error	t-Statistic	Prob.
C	−0.001929	0.001350	−1.429024	0.1537
DRDJIND	0.791900	0.030019	26.37990	0.0000
R-Squared	0.609423	Mean dependent var	0.004744	
Adjusted R-squared	0.608547	S.D. dependent var	0.044869	
S.E. of regression	0.028073	Akaike info criterion	−4.303587	
Sum squared resid	0.351481	Schwarz criterion	−4.285262	
Log likelihood	966.0034	Hannan-Quinn criterion	−4.296363	
F-statistic	695.8990	Durbin-Watson stat	2.020324	
Prob(F-statistic)	0.0000			

obtained from the estimation of model (4.174) on the first-lagged residuals. We obtain:

$$RESID_t = -0.1132 \times RESID_{t-1} \tag{4.176}$$

We thus have: $\hat{\rho}_0 = -0.1132$. We construct the variables in quasi-differences:

$$DRFTSE_t = RFTSE_t + 0.1132 \times RFTSE_{t-1} \tag{4.177}$$

and:

$$DRDJIND_t = RDJIND_t + 0.1132 \times RDJIND_{t-1} \tag{4.178}$$

We then regress $DRFTSE_t$ on a constant and $DRDJIND_t$. The results are shown in Table 4.12.

We can calculate the constant term:

$$\hat{\alpha} = -0.0019/\,(1 + 0.1132) = -0.0017 \tag{4.179}$$

The procedure can then be continued by estimating a new value of ρ based on the residuals from the estimation of the quasi-difference model (Table 4.12).

Conclusion

This chapter has focused on error-related problems, namely, autocorrelation and heteroskedasticity. The next chapter is still concerned with the violation of the assumptions of the regression model, but now focuses on problems related to the explanatory variables. It specifies the procedure to follow when the matrix of explanatory variables is no longer random, when the explanatory variables are not independent of each other (collinearity), and when there is some instability in the estimated model.

The Gist of the Chapter

Multiple regression model	$\underset{(T,1)}{\boldsymbol{Y}} = \underset{(T,k+1)}{\boldsymbol{X}}\underset{(k+1,1)}{\boldsymbol{\beta}} + \underset{(T,1)}{\boldsymbol{\varepsilon}}$
Heteroskedasticity and/or autocorrelation	$E\left(\boldsymbol{\varepsilon}\boldsymbol{\varepsilon}'\right) = \boldsymbol{\Omega}_{\varepsilon} \neq \sigma_{\varepsilon}^{2}\boldsymbol{I}$
GLS estimators	$\tilde{\boldsymbol{\beta}} = \left(\boldsymbol{X}'\boldsymbol{\Omega}_{\varepsilon}^{-1}\boldsymbol{X}\right)^{-1}\boldsymbol{X}'\boldsymbol{\Omega}_{\varepsilon}^{-1}\boldsymbol{Y}$
Tests of homoskedasticity	
	Goldfeld and Quandt (1965)
	Glejser (1969)
	Breusch and Pagan (1979)
	White (1980)
	ARCH
Tests for absence of autocorrelation	
	Geary (1970)
	Durbin and Watson (1950, 1951): $DW = \frac{\sum_{t=2}^{T}(e_t - e_{t-1})^2}{\sum_{t=1}^{T} e_t^2}$, e: residual; $DW \simeq 2$: no autocorrelation
	Durbin (1970)
	Breusch (1978) and Godfrey (1978)
	Box and Pierce (1970)
	Ljung and Box (1978)

Further Reading

This chapter includes a large number of references related to methods for detecting heteroskedasticity and autocorrelation, as well as the solutions provided. In addition to these references, most econometric textbooks contain developments on heteroskedasticity and autocorrelation problems. In particular, the books by Dhrymes (1978), Judge et al. (1985, 1988), Davidson and MacKinnon (1993), Hendry (1995), Wooldridge (2012), Gujarati et al. (2017), or Greene (2020) can be recommended.

Problems with Explanatory Variables: Random Variables, Collinearity, and Instability 5

As we saw in the third chapter, the multiple regression model is based on a number of assumptions. Here, we focus more specifically on the first two assumptions, which relate to explanatory variables:

- The matrix X of explanatory variables is non-random. This hypothesis amounts to assuming that the matrix X is independent of the error term.
- The matrix X is of full rank. In other words, the explanatory variables in the matrix X are linearly independent.

In this chapter, we look at what happens when these assumptions do not hold. If the first assumption is violated, the implication is that the explanatory variables are dependent on the error term. Under these conditions, the OLS estimators are no longer consistent and it is necessary to use another estimator called the **instrumental variables estimator**. This is the subject of the first section of the chapter.

The consequence of violating the second assumption is that the explanatory variables are not linearly independent. In other words, they are **collinear**. This issue of **multicollinearity** is addressed in the second section of the chapter.

Finally, we turn our attention to the third problem related to the explanatory variables, namely, the question of the **stability** of the estimated model.

5.1 Random Explanatory Variables and the Instrumental Variables Method

The aim of this section is to find an estimator that remains valid in the presence of correlation between the explanatory variables and the error term. We know that when the independence assumption between the matrix of explanatory variables and the error term is violated, the OLS estimator is no longer consistent: even if

V. Mignon, *Principles of Econometrics*, Classroom Companion: Economics,
https://doi.org/10.1007/978-3-031-52535-3_5

we increase the sample size, the estimator does not tend towards its true value. It is therefore necessary to find another estimator that does not suffer from this consistency problem. This is precisely the purpose of the instrumental variables method, which consists in finding a set of variables that are uncorrelated with the error term but that are correlated with the explanatory variables, in order to represent them correctly. Applying this method yields an estimator, called the **instrumental variables estimator**, which remains valid in the presence of correlation between the explanatory variables and the error term.

5.1.1 Instrumental Variables Estimator

If the explanatory variables are random and correlated with the error term, it can be shown that the OLS estimator is no longer consistent (see in particular Greene 2020). In other words, even if the sample size grows indefinitely, the OLS estimators $\hat{\boldsymbol{\beta}}$ do not approach their true values $\boldsymbol{\beta}$:

$$P \lim \hat{\boldsymbol{\beta}} \neq \boldsymbol{\beta} \tag{5.1}$$

where $P \lim$ denotes the probability limit (or convergence in probability).

The problem is then to find a consistent estimator of $\boldsymbol{\beta}$. To this end, we use the **instrumental variables method**.

Consider the following general linear model:

$$\boldsymbol{Y} = \boldsymbol{X\beta} + \boldsymbol{\varepsilon} \tag{5.2}$$

The purpose of the instrumental variables method is to find a set of k variables $Z_{1t}, Z_{2t}, \ldots, Z_{kt}$ that are uncorrelated with the error term. By noting $\boldsymbol{Z}$ the matrix composed of these k variables ($\boldsymbol{Z} = (Z_1, Z_2, \ldots, Z_k)$), we seek to obtain $\boldsymbol{Z}$ such that:

$$Cov(\boldsymbol{Z}'\boldsymbol{\varepsilon}) = 0 \tag{5.3}$$

In other words, the aim is to find a matrix $\boldsymbol{Z}$ of variables that are uncorrelated at each period with the error term, i.e.:

$$E\left(Z_{it}\varepsilon_t\right) = 0 \tag{5.4}$$

for $i = 1, \ldots, k$ and $t = 1, \ldots, T$.

Let us premultiply the model (5.2) by $\boldsymbol{Z}'$:

$$\boldsymbol{Z}'\boldsymbol{Y} = \boldsymbol{Z}'\boldsymbol{X\beta} + \boldsymbol{Z}'\boldsymbol{\varepsilon} \tag{5.5}$$

Assuming, by analogy with the OLS method, that $\boldsymbol{Z}'\boldsymbol{\varepsilon} = \boldsymbol{0}$, we can write:

$$\boldsymbol{Z}'\boldsymbol{Y} = \boldsymbol{Z}'\boldsymbol{X}\boldsymbol{\beta} \tag{5.6}$$

Under the assumption that the matrix $(\boldsymbol{Z}'\boldsymbol{X})$ is non-singular, we obtain the **instrumental variables estimator**, denoted $\hat{\boldsymbol{\beta}}_{IV}$, defined by:

$$\hat{\boldsymbol{\beta}}_{IV} = (\boldsymbol{Z}'\boldsymbol{X})^{-1}\,\boldsymbol{Z}'\boldsymbol{Y} \tag{5.7}$$

It can be shown (see in particular Johnston and Dinardo, 1996) that the estimator of instrumental variables is a consistent estimator of $\boldsymbol{\beta}$, i.e.:

$$P\lim \hat{\boldsymbol{\beta}}_{IV} = \boldsymbol{\beta} \tag{5.8}$$

The variables that appear in the matrix $\boldsymbol{Z}$ are called **instrumental variables** or **instruments**. Some of these variables may be variables that are present in the original explanatory variables matrix $\boldsymbol{X}$. The instrumental variables must be correlated with the explanatory variables, that is:

$$Cov(\boldsymbol{Z}'\boldsymbol{X}) \neq 0 \tag{5.9}$$

Otherwise, the matrix $(\boldsymbol{Z}'\boldsymbol{X})$ would indeed be zero, and the procedure could not be applied.

By positing $\hat{\boldsymbol{X}} = \boldsymbol{Z}\,(\boldsymbol{Z}'\boldsymbol{Z})^{-1}\,\boldsymbol{Z}'\boldsymbol{X}$, we can also write the instrumental variables estimator as follows:

$$\hat{\boldsymbol{\beta}}_{IV} = \left(\hat{\boldsymbol{X}}'\boldsymbol{X}\right)^{-1}\hat{\boldsymbol{X}}'\,\boldsymbol{Y} = \left(\hat{\boldsymbol{X}}'\hat{\boldsymbol{X}}\right)^{-1}\hat{\boldsymbol{X}}'\boldsymbol{Y} \tag{5.10}$$

because $\hat{\boldsymbol{X}}'\boldsymbol{X} = \hat{\boldsymbol{X}}'\hat{\boldsymbol{X}}$.

Employing a technique similar to that used in Chap. 3 for the OLS estimator, it can easily be shown that the variance-covariance matrix $\boldsymbol{\Omega}_{\hat{\beta}_{IV}}$ of the instrumental variables estimator is given by:

$$\boldsymbol{\Omega}_{\hat{\beta}_{IV}} = \sigma_{\varepsilon}^{2}\left(\hat{\boldsymbol{X}}'\hat{\boldsymbol{X}}\right)^{-1} \tag{5.11}$$

It now remains for us to find a procedure to assess whether or not the explanatory variables are correlated with the error term in order to determine which estimator to choose between the OLS estimator and the instrumental variables estimator. To this end, the Hausman (1978) specification test is used.

5.1.2 The Hausman (1978) Specification Test

When the explanatory variables are not correlated with the error term, it is preferable to use the OLS estimator rather than the instrumental variables estimator, as the OLS estimator is more accurate (for demonstrations, see in particular Greene, 2020). It is therefore important to have a test that can be used to determine whether or not there is a correlation between the explanatory variables and the error term. This is the purpose of the **Hausman test** (Hausman, 1978).

This test consists of testing the null hypothesis that the explanatory variables and the error term are uncorrelated, against the alternative hypothesis that the correlation between the two types of variables is non-zero. Under the null hypothesis, the OLS and instrumental variables estimators are consistent, but the OLS estimator is more accurate. Under the alternative hypothesis, the OLS estimator is no longer consistent, unlike the instrumental variables estimator.

The idea behind the Hausman test is to test the significance of the difference between the two estimators. If the difference is not significant, the null hypothesis is not rejected. On the other hand, if the difference is significant, the null hypothesis is rejected and the instrumental variables estimator should be used. We calculate the following statistic, known as the **Wald** statistic:

$$H = \left(\hat{\boldsymbol{\beta}}_{IV} - \hat{\boldsymbol{\beta}}\right)' \left[\hat{\sigma}_{\varepsilon}^2 \left(\left(\hat{\boldsymbol{X}}'\hat{\boldsymbol{X}}\right)^{-1} - \left(\boldsymbol{X}'\boldsymbol{X}\right)^{-1}\right)\right]^{-1} \left(\hat{\boldsymbol{\beta}}_{IV} - \hat{\boldsymbol{\beta}}\right) \tag{5.12}$$

where $\hat{\sigma}_{\varepsilon}^2$ denotes the estimator of the variance of the error term σ_{ε}^2, i.e. (see Chap. 3):

$$\hat{\sigma}_{\varepsilon}^2 = \frac{\boldsymbol{e}'\boldsymbol{e}}{T - k - 1} \tag{5.13}$$

Under the null hypothesis, the statistic H follows a Chi-squared distribution whose number of degrees of freedom depends on the context studied:

- If the matrices $\boldsymbol{X}$ and $\boldsymbol{Z}$ have no common variables, the number of degrees of freedom is $(k + 1)$.
- If the matrices $\boldsymbol{X}$ and $\boldsymbol{Z}$ have k' common variables, the number of degrees of freedom is $\left(k - k'\right)$.

5.1.3 Application Example: Measurement Error

The data used are, of course, assumed to be accurate measurements of their theoretical equivalents. So far, we have supposed that the variables (dependent and explanatory) are measured without error. In practice, however, this is not always the case. For example, survey data (obtained by sampling), aggregate data (GDP, household consumption, investment, etc.), and so on do not always represent exact measures of the theoretical variables. In this case, we speak of **measurement error** on the variables. This may arise from errors in data reporting, calculation errors, etc.

Consider the following model with centered variables:

$$y_t = \beta x_t^* + \varepsilon_t \tag{5.14}$$

and assume that the observations x_t available are not a perfect measure of x_t^*. In other words, the observed variable x_t is subject to measurement errors, i.e.:

$$x_t = x_t^* + \mu_t \tag{5.15}$$

where μ_t is an error term that follows a normal distribution of zero mean and variance σ_μ^2. It is further assumed that the two error terms ε_t and μ_t are independent. Such a model can, for example, be representative of the link between consumption and permanent income, where y_t denotes current consumption and x_t^* permanent income. Permanent income is not observable, only current income x_t being observable. μ_t thus denotes the measurement error on permanent income x_t^*.

We can rewrite the model as follows:

$$y_t = \beta x_t - \beta \mu_t + \varepsilon_t \tag{5.16}$$

To simplify the notations, let us posit:

$$\eta_t = -\beta \mu_t + \varepsilon_t \tag{5.17}$$

Then we have:

$$y_t = \beta x_t + \eta_t \tag{5.18}$$

Let us calculate the covariance between x_t and η_t:

$$Cov\,(x_t, \eta_t) = Cov\,\left(x_t^* + \mu_t, -\beta \mu_t + \varepsilon_t\right) = -\beta \sigma_\mu^2 \neq 0 \tag{5.19}$$

Since the covariance between x_t and η_t is non-zero, it follows that the OLS estimator is biased[1] and is not consistent. Thus, when there is a measurement error on the explanatory variable, the OLS estimator is no longer consistent and the instrumental variables estimator should be used.

[1] In the case where it is the explained variable that is observed with error, then the OLS estimator is still non-consistent, but is no longer biased.

5.2 Multicollinearity and Variable Selection

As we previously recalled, one of the basic assumptions of the multiple regression model is that the rank of the matrix $\boldsymbol{X}$ is equal to $k + 1$, i.e., to the number of explanatory variables plus the constant. This assumption means that the explanatory variables are linearly independent, or orthogonal. In other words, there is no **multicollinearity** between the explanatory variables.

In this section, we study what happens when such an assumption is violated. In practice, it is quite common for the explanatory variables to be more or less related to each other.

5.2.1 Presentation of the Problem

We speak of **perfect** (or exact) **collinearity** between two explanatory variables if they are perfectly dependent on each other. Thus, two variables X_{1t} and X_{2t} are perfectly collinear if:

$$X_{2t} = \lambda X_{1t} \tag{5.20}$$

where λ is a non-zero constant.

We speak of **perfect** (or exact) **multicollinearity** when an explanatory variable is the result of a linear combination of several other explanatory variables. In this case, the coefficient of determination is equal to one.

In a model comprising k explanatory variables, we speak of perfect multicollinearity if there is a linear combination:

$$\lambda_1 X_{1t} + \lambda_2 X_{2t} + \ldots + \lambda_k X_{kt} = 0 \tag{5.21}$$

where $\lambda_1, \ldots, \lambda_k$ are constants that are not all zero simultaneously.

In these cases of perfect collinearity or multicollinearity, the rank of the matrix $\boldsymbol{X}$ is less than $k + 1$, which means that the assumption of linear independence between the columns of $\boldsymbol{X}$ no longer holds. It follows that the rank of $\left(\boldsymbol{X}'\boldsymbol{X}\right)$ is also less than $k + 1$. It is therefore theoretically impossible to invert the matrix $\left(\boldsymbol{X}'\boldsymbol{X}\right)$, as the latter is singular (its determinant is zero). The regression coefficients are then indeterminate.

Cases of perfect collinearity and multicollinearity are rare. In practice, explanatory variables frequently exhibit strong, but not perfect, multicollinearity. We then speak of quasi-multicollinearity or, more simply, **multicollinearity**. There is multicollinearity if, in a model with k explanatory variables, we have the following relationship:

$$\lambda_1 X_{1t} + \lambda_2 X_{2t} + \ldots + \lambda_k X_{kt} + u_t = 0 \tag{5.22}$$

where u_t is an error term.

As we will see later, when there is multicollinearity, the regression coefficients can be estimated—they are determined—but their standard deviations are very high, making the estimation very imprecise.

5.2.2 The Effects of Multicollinearity

Multicollinearity has several effects. Firstly, the variances and covariances of the estimators tend to increase. Let us explain this point.

We demonstrated in Chap. 3 that the variance-covariance matrix $\boldsymbol{\Omega}_{\hat{\boldsymbol{\beta}}}$ of the OLS coefficients is given by:

$$\boldsymbol{\Omega}_{\hat{\boldsymbol{\beta}}} = \sigma_{\varepsilon}^{2} \left(\boldsymbol{X}'\boldsymbol{X}\right)^{-1} \tag{5.23}$$

We have also shown that the variance of the OLS coefficient $\hat{\beta}_i$ associated with the ith explanatory variable X_{it} is written as:

$$V\left(\hat{\beta}_i\right) = \sigma_{\varepsilon}^{2} a_{i+1,i+1} \tag{5.24}$$

where $a_{i+1,i+1}$ denotes the $(i+1)$ th element of the diagonal of $\left(\boldsymbol{X}'\boldsymbol{X}\right)^{-1}$. It is possible to show that:

$$a_{i+1,i+1} = \frac{1}{1-R_i^2} = VIF_i \tag{5.25}$$

where VIF_i is the **variance inflation factor** and R_i^2 is the coefficient of determination associated with the regression of the variable X_{it} on the $(k-1)$ other explanatory variables.

The statistic VIF_i indicates how the variance of an estimator increases when there is multicollinearity. In this case, R_i^2 tends to 1 and $a_{i+1,i+1}$ tends to infinity. It follows that the variance $V\left(\hat{\beta}_i\right)$ also tends to infinity. Multicollinearity therefore increases the variance of the estimators.

The second effect of multicollinearity is that the OLS estimators are highly sensitive to small changes in the data. A small change in one observation or in the number of observations can result in a large change in the estimated values of the coefficients.

Let us take an example.[2] From the data in Table 5.1, we estimate the following models by OLS:

$$Y_t = \alpha + \beta_1 X_{1t} + \beta_2 X_{2t} + \varepsilon_t \tag{5.26}$$

[2] Of course, this example is purely illustrative in the sense that only six observations are considered.

Table 5.1 Example of multicollinearity

t	Y	X_1	X_2	X_3
1	3	4	8	8
2	5	11	22	22
3	12	6	12	12
4	8	9	19	19
5	9	7	14	14
6	4	3	6	7

Table 5.2 Example of multicollinearity. Estimation results

Model (5.26)	α	β_1	β_2
Coefficient	5.09	−1.20	0.72
Standard deviation	4.67	10.41	5.07
t-statistic	1.09	−0.11	0.14
Model (5.27)	α'	β_1'	β_2'
Coefficient	5.50	2.25	−1.00
Standard deviation	4.95	7.37	3.73
t-statistic	−1.11	0.30	0.27

and

$$Y_t = \alpha' + \beta_1' X_{1t} + \beta_2' X_{3t} + \varepsilon_t' \tag{5.27}$$

The variables X_{2t} and X_{3t} differ only in the final observation (Table 5.1). Looking at the results in Table 5.2, we see that this small change in the data significantly alters the estimates. Although not significant, the values of the coefficients of the explanatory variables differ markedly between the two regressions; the same is true for their standard deviations.

This example also highlights the first mentioned effect of multicollinearity, namely, the high value of the standard deviations of the estimated coefficients.

There are also other effects of multicollinearity. These include the following consequences:

- Because of the high value of the variances of the estimators, the t-statistics associated with certain coefficients can be very low, even though the values taken by the coefficients are high.
- Despite the non-significance of one or more explanatory variables, the coefficient of determination of the regression can be very high. This is frequently considered to be one of the most visible symptoms of multicollinearity. Thus, if the coefficient of determination is very high, the Fisher test tends to reject the null hypothesis of non-significance of the regression as a whole, even though the t-statistics of several coefficients indicate that the latter are not significant.
- Some variables are sensitive to the exclusion or inclusion of other explanatory variables.

– It is difficult, if not impossible, to distinguish between the effects of different explanatory variables on the variable being explained.

5.2.3 Detecting Multicollinearity

Strictly speaking, there is no test of multicollinearity as such. However, several techniques can be used to detect it and assess how important it is.

Correlation Between Explanatory Variables

A simple method is to calculate the linear correlation coefficients between the explanatory variables. If the latter are strongly correlated, this is an indication in favor of multicollinearity. The presence of strong correlations is not, however, necessary to observe multicollinearity. Multicollinearity can occur even when the correlation coefficients are relatively low (e.g., below 0.5). Let us take an example to illustrate this.

Example 5.1 Consider the following model:

$$Y_t = \alpha + \beta_1 X_{1t} + \beta_2 X_{2t} + \beta_3 X_{3t} + \varepsilon_t \tag{5.28}$$

and suppose that X_{3t} is a linear combination of the other two explanatory variables:

$$X_{3t} = \lambda_1 X_{1t} + \lambda_2 X_{2t} \tag{5.29}$$

where λ_1 and λ_2 are simultaneously non-zero constants. Because of the existence of this linear combination, the coefficient of determination R^2 from the regression of X_{3t} on X_{1t} and X_{2t} is equal to 1. By virtue of the relationship (3.106) from Chap. 3, we can write:

$$R^2 = \frac{r_{X_3X_1}^2 + r_{X_3X_2}^2 - 2r_{X_3X_1}r_{X_3X_2}r_{X_1X_2}}{1 - r_{X_1X_2}^2} = 1 \tag{5.30}$$

The previous relationship is satisfied for $r_{X_3X_1} = r_{X_3X_2} = 0.6$ and $r_{X_1X_2} = -0.28$. It is worth mentioning that these values are not very high even though there is multicollinearity.

Consequently, in a model with more than two explanatory variables, care must be taken when interpreting the values of the correlation coefficients.

The Klein Test (1962)

This is not strictly speaking a statistical test. The method proposed by Klein (1962) consists in calculating the linear correlation coefficients between the different explanatory variables: $r_{X_iX_j}$ for $i \neq j$ and comparing these values with the coefficient of determination R^2 associated with the regression of Y_t on all the k explanatory variables. If $r^2_{X_i X_j} > R^2$, there is a presumption of multicollinearity.

The Farrar and Glauber Test (1967)

Farrar and Glauber (1967) proposed a technique for detecting multicollinearity based on the matrix of correlation coefficients between the explanatory variables:

$$\begin{pmatrix} 1 & r_{X_1X_2} & r_{X_1X_3} & \cdots & r_{X_1X_k} \\ r_{X_2X_1} & 1 & r_{X_2X_3} & \cdots & r_{X_2X_k} \\ \vdots & \cdots & \cdots & \cdots & \vdots \\ r_{X_kX_1} & r_{X_kX_2} & r_{X_kX_3} & \cdots & 1 \end{pmatrix} \tag{5.31}$$

The underlying idea is that, if the variables are perfectly correlated, the determinant of this matrix is zero. Let us take an example to visualize this property.

Example 5.2 Consider a model with two explanatory variables X_{1t} and X_{2t}. The determinant D of the correlation coefficient matrix is given by:

$$D = \begin{vmatrix} 1 & r_{X_1X_2} \\ r_{X_2X_1} & 1 \end{vmatrix} \tag{5.32}$$

If the variables are perfectly correlated, $r_{X_1X_2} = 1$. Therefore:

$$D = \begin{vmatrix} 1 & 1 \\ 1 & 1 \end{vmatrix} = 0 \tag{5.33}$$

The determinant of the correlation coefficient matrix is zero when the variables are perfectly correlated.

Conversely, when the explanatory variables are orthogonal, $r_{X_1X_2} = 0$ and the determinant of the correlation coefficient matrix is 1.

The method proposed by Farrar and Glauber (1967) consists of investigating whether or not the determinant of the correlation coefficient matrix between the explanatory variables is close to 0. If this is the case, multicollinearity is presumed. The authors suggested a Chi-squared test to test the null hypothesis that the determinant of the correlation coefficient matrix is 1, meaning that the variables are orthogonal, against the alternative hypothesis that the determinant is less than 1, indicating that the variables are dependent. The test statistic is given by:

$$FG = -\left(T - 1 - \frac{1}{6}\left[2(k+1) + 5\right]\right) \log D \tag{5.34}$$

where D denotes the determinant of the correlation coefficient matrix of the k explanatory variables. Under the null hypothesis:

$$FG \sim \chi^2_{\frac{1}{2}k(k+1)} \tag{5.35}$$

The decision rule is written:

- If $FG < \chi^2_{\frac{1}{2}k(k+1)}$, the orthogonality hypothesis is not rejected.
- If $FG > \chi^2_{\frac{1}{2}k(k+1)}$, the orthogonality hypothesis is rejected.

The Eigenvalue Method

This technique is based on the calculation of the eigenvalues of the matrix $\left(\boldsymbol{X}'\boldsymbol{X}\right)$ or, similarly, of the matrix of correlation coefficients of the explanatory variables. Knowing that the determinant of the matrix $\left(\boldsymbol{X}'\boldsymbol{X}\right)$ is equal to the product of the eigenvalues, a low value of the determinant means that one or more of these eigenvalues are low. There is then a presumption of multicollinearity.

Belsley et al. (1980) suggest calculating the following statistic, called the **state indicator**:

$$\varkappa = \frac{\sqrt{\lambda_{\max}}}{\sqrt{\lambda_{\min}}} \tag{5.36}$$

where $\lambda_{\max}$ (respectively $\lambda_{\min}$) denotes the largest (respectively smallest) eigenvalue of the matrix $\left(\boldsymbol{X}'\boldsymbol{X}\right)$. If the matrix $\boldsymbol{X}$ has been normalized, so that the length of each of its columns is 1, then the $\varkappa$ statistic is equal to 1 when the columns are orthogonal and greater than 1 when the columns exhibit multicollinearity. This technique is not a statistical test as such, but it is frequently considered that values of $\varkappa$ between 10 and 30 correspond to a situation of moderate multicollinearity, and that values above 30 are an indication in favor of strong multicollinearity.

Variance Inflation Factors

Variance inflation factors (VIF) can be used as indicators of multicollinearity. Again, it is worth mentioning that this is not a statistical test per se.

We have previously seen (see Eq. (5.25)) that the variance inflation factor associated with the ith explanatory variable is written as:

$$VIF_i = \frac{1}{1 - R_i^2} \tag{5.37}$$

where R_i^2 is the coefficient of determination relating to the regression of the variable X_{it} on the $(k-1)$ other explanatory variables. Obviously, the value of VIF_i is higher the closer R_i^2 is to 1. Consequently, the higher VIF_i is, the more collinear the variable X_{it} is.

The VIF_i statistics can be calculated for the various explanatory variables ($i = 1, \ldots, k$). When the differences between the VIF_i statistics are large, it is possible to identify the highest values and, thus, identify collinear variables. But, if the differences between the VIF_i statistics for the different explanatory variables are small, it is impossible to detect the variables responsible for multicollinearity.

In practice, if the value of the VIF_i statistic is greater than 10, which corresponds to the case where $R_i^2 > 0.9$, the variable X_{it} is considered to be strongly collinear.

Empirical Application

Consider the following model:

$$REURO_t = \alpha + \beta_1 RDJIND_t + \beta_2 RFTSE_t + \beta_3 RNIKKEI_t + \varepsilon_t \qquad (5.38)$$

where:

- $REURO$ denotes the series of returns of the European stock market index, Euro Stoxx 50.
- $RDJIND$ is the series of returns of the Dow Jones Industrial Average index.
- $RFTSE$ is the series of returns of the UK stock market index, $FTSE$ 100.
- $RNIKKEI$ is the series of returns of the $NIKKEI$ index of the Tokyo Stock Exchange.

The data, taken from the Macrobond database, are quarterly and cover the period from the second quarter of 1987 to the second quarter of 2021 ($T = 137$).

We are interested in the possible multicollinearity between the three explanatory variables under consideration. Let us start by calculating the matrix of correlation coefficients among the explanatory variables:

$$\begin{pmatrix} 1 & r_{RDJIND,RFTSE} & r_{RDJIND,RNIKKEI} \\ r_{RFTSE,RDJIND} & 1 & r_{RFTSE,RNIKKEI} \\ r_{NIKKEI,RDJIND} & r_{RNIKKEI,RFTSE} & 1 \end{pmatrix}$$
$$= \begin{pmatrix} 1 & 0.8562 & 0.6059 \\ 0.8562 & 1 & 0.5675 \\ 0.6059 & 0.5675 & 1 \end{pmatrix} \qquad (5.39)$$

It appears that the most strongly correlated explanatory variables are $RFTSE$ and $RDJIND$.

The estimation of the model (5.38) leads to the results in Table 5.3.

If we refer to the method proposed by Klein, we find that the coefficient of determination of the model (5.38), equal to 0.7959, is higher than the linear correlation coefficients between $RDJIND$ and $RNIKKEI$ (0.6059) and between $RFTSE$ and $RNIKKEI$ (0.5675), but is lower than the linear correlation coefficient between $RDJIND$ and $RFTSE$ (0.8562), suggesting there is collinearity.

Table 5.3 Estimation of the relationship between the series of stock market returns

Variable	Coefficient	Std. Error	t-Statistic	Prob.
C	−0.005134	0.004447	−1.154642	0.2503
RDJIND	0.525822	0.107257	4.902431	0.0000
RFTSE	0.633759	0.101216	6.261419	0.0000
RNIKKEI	0.084954	0.046478	1.827827	0.0698
R-squared	0.795886	Mean dependent var	0.011257	
Adjusted R-squared	0.791282	S.D. dependent var	0.107580	
S.E. of regression	0.049149	Akaike info criterion	−3.159171	
Sum squared resid	0.321274	Schwarz criterion	−3.073916	
Log likelihood	220.4032	Hannan-Quinn criterion	−3.124526	
F-statistic	172.8658	Durbin-Watson stat	1.945250	
Prob(F-statistic)	0.0000			

To investigate whether the Farrar and Glauber test leads to the same conclusion, let us calculate the determinant D of the matrix of correlations between the explanatory variables. We obtain:

$$D = 0.1666 \tag{5.40}$$

This determinant being closer to 0 than to 1, the presumption of multicollinearity remains valid. Let us calculate the test statistic:

$$FG = -\left(137 - 1 - \frac{1}{6}\left[2\left(3+1\right)+5\right]\right)\log(0.1666) = 239.8618 \tag{5.41}$$

The value read from the table of the Chi-squared distribution is equal to $\chi^2_6 = 12.592$ at the 5% significance level. As the calculated value is higher than the critical value, the null hypothesis of orthogonality between the explanatory variables is rejected and the presumption of multicollinearity is confirmed.

Let us now apply the technique based on the calculation of the variance inflation factors (VIF). To do this, we regress each of the explanatory variables on the other two and calculate the coefficient of determination associated with each regression. The results are reported in Table 5.4. The values of the VIF statistics are relatively low (less than 10), suggesting that multicollinearity, if present, is not very strong. This is consistent with the fact that the coefficients of determination R_i^2

Table 5.4 Calculation of VIF

i	R_i^2	VIF_i
$RDJIND$	0.7543	4.0698
$RFTSE$	0.7368	3.7994
$RNIKKEI$	0.3760	1.6025

associated with each of the three regressions are lower than the overall coefficient of determination ascertained by estimating the model (5.38).

5.2.4 Solutions to Multicollinearity

A frequently used technique is to increase the number of observations in order to increase the sample size. Such a procedure is useless if the data added are the same as those already in the sample. In such a case, multicollinearity will effectively be repeated. Other techniques have therefore been proposed.

Use of Preliminary Estimates

This first, sequential technique consists in using estimation results from a previous study. To do this, we decompose the matrix $\boldsymbol{X}$ of explanatory variables and the vectors of coefficients as follows:

$$\boldsymbol{X} = \left(\boldsymbol{X}_r \ \boldsymbol{X}_s\right), \boldsymbol{\beta} = \begin{pmatrix} \boldsymbol{\beta}_r \\ \boldsymbol{\beta}_s \end{pmatrix} \text{ and } \hat{\boldsymbol{\beta}} = \begin{pmatrix} \hat{\boldsymbol{\beta}}_r \\ \hat{\boldsymbol{\beta}}_s \end{pmatrix} \tag{5.42}$$

where $\boldsymbol{X}_r$ is the submatrix of size (T, r) formed by the first r columns of $\boldsymbol{X}$ and $\boldsymbol{X}_s$ is the submatrix composed of the $s = k + 1 - r$ remaining columns.

Suppose that, in a previous study, the coefficient $\hat{\boldsymbol{\beta}}_s$ was obtained and that it is an unbiased estimator of $\boldsymbol{\beta}_s$. It then remains for us to estimate $\boldsymbol{\beta}_r$. To do this, we start by calculating a new dependent variable $\tilde{\boldsymbol{Y}}$, which consists in correcting the dependent variable of the observations already used, $\boldsymbol{X}_s$:

$$\tilde{\boldsymbol{Y}} = \boldsymbol{Y} - \boldsymbol{X}_s \hat{\boldsymbol{\beta}}_s \tag{5.43}$$

We then regress $\tilde{\boldsymbol{Y}}$ on the explanatory variables appearing in $\boldsymbol{X}_r$ and obtain the following OLS estimator $\hat{\boldsymbol{\beta}}_r$:

$$\hat{\boldsymbol{\beta}}_r = \left(\boldsymbol{X}_r'\boldsymbol{X}_r\right)^{-1} \boldsymbol{X}_r'\tilde{\boldsymbol{Y}} \tag{5.44}$$

Given that:

$$\boldsymbol{Y} = \boldsymbol{X}\boldsymbol{\beta} + \boldsymbol{\varepsilon} = \boldsymbol{X}_r\boldsymbol{\beta}_r + \boldsymbol{X}_s\boldsymbol{\beta}_s + \boldsymbol{\varepsilon} \tag{5.45}$$

we can write:

$$\hat{\boldsymbol{\beta}}_r = \left(\boldsymbol{X}_r'\boldsymbol{X}_r\right)^{-1} \boldsymbol{X}_r' \left(\boldsymbol{X}_r\boldsymbol{\beta}_r + \boldsymbol{X}_s\boldsymbol{\beta}_s + \boldsymbol{\varepsilon} - \boldsymbol{X}_s\hat{\boldsymbol{\beta}}_s\right) \tag{5.46}$$

Hence:

$$\hat{\boldsymbol{\beta}}_r = \boldsymbol{\beta}_r + \left(\boldsymbol{X}_r'\boldsymbol{X}_r\right)^{-1}\boldsymbol{X}_r'\boldsymbol{X}_s\left(\boldsymbol{\beta}_s - \hat{\boldsymbol{\beta}}_s\right) + \left(\boldsymbol{X}_r'\boldsymbol{X}_r\right)^{-1}\boldsymbol{X}_r'\boldsymbol{\varepsilon} \tag{5.47}$$

Knowing that $E(\boldsymbol{\varepsilon}) = 0$ and $E\left(\hat{\boldsymbol{\beta}}_s\right) = \boldsymbol{\beta}_s$, we deduce:

$$E\left(\hat{\boldsymbol{\beta}}_r\right) = \boldsymbol{\beta}_r \tag{5.48}$$

meaning that $\hat{\boldsymbol{\beta}}_r$ is an unbiased estimator of $\boldsymbol{\beta}_r$.

Remark 5.1 A technique similar to this is to combine time series and cross-sectional data (see in particular Tobin, 1950).

The Ridge Regression

This is a technique proposed by Hoerl and Kennard (1970a,b) involving the numerical mechanical treatment of multicollinearity. The underlying idea is simple. The problem arises because multicollinearity means there is a column, in the matrix $(\boldsymbol{X}'\boldsymbol{X})$, representing a linear combination (within one error term) of other columns. Hoerl and Kennard (1970a,b) suggest destroying this linear combination by adding a constant to the diagonal elements of the matrix $(\boldsymbol{X}'\boldsymbol{X})$. The principle of the ridge regression is to define the **ridge estimator**:

$$\hat{\boldsymbol{\beta}}_R = \left(\boldsymbol{X}'\boldsymbol{X} + c\boldsymbol{I}\right)^{-1}\boldsymbol{X}'\boldsymbol{Y} \tag{5.49}$$

where $c > 0$ is an arbitrary constant.

The ridge estimator can be expressed in terms of the usual OLS estimator $\hat{\boldsymbol{\beta}}$:

$$\hat{\boldsymbol{\beta}}_R = \left[\boldsymbol{I} + c\left(\boldsymbol{X}'\boldsymbol{X}\right)^{-1}\right]^{-1}\hat{\boldsymbol{\beta}} \tag{5.50}$$

Furthermore, replacing $\boldsymbol{Y}$ by $\boldsymbol{X}\boldsymbol{\beta} + \boldsymbol{\varepsilon}$ in (5.49) and taking the expectation, we have:

$$E\left(\hat{\boldsymbol{\beta}}_R\right) = \left(\boldsymbol{X}'\boldsymbol{X} + c\boldsymbol{I}\right)^{-1}\boldsymbol{X}'\boldsymbol{X}\boldsymbol{\beta} \tag{5.51}$$

The ridge estimator $\hat{\boldsymbol{\beta}}_R$ is therefore a biased estimator of $\boldsymbol{\beta}$. However, Schmidt (1976) showed that the variances of the elements of $\hat{\boldsymbol{\beta}}_R$ are lower than those associated with the elements of the vector of OLS estimators.

The difficulty inherent in the ridge regression lies in the choice of the value of c. Hoerl and Kennard (1970a,b) suggest estimating using several values for c in order to study the stability of $\hat{\boldsymbol{\beta}}_R$. The technique, known as **ridge trace**, consists in plotting the different values of $\hat{\boldsymbol{\beta}}_R$ on the y-axis for various values of c on the x-axis. The value of c is then selected as the one for which the estimators $\hat{\boldsymbol{\beta}}_R$ are stable.

Remark 5.2 The ridge regression method can be generalized to the case where a value different from c is added to each of the elements of the diagonal of the matrix $(\boldsymbol{X}'\boldsymbol{X})$. This technique is called **generalized ridge regression**.

Other Techniques

There are other procedures for dealing with the multicollinearity problem, which we briefly mention below:

- *The method of Marquardt generalized inverses*. The underlying idea is to calculate the inverse of the matrix $(\boldsymbol{X}'\boldsymbol{X})$ without needing to compute the determinant of the same matrix. Since the determinant is equal to the product of the eigenvalues, the technique consists in calculating the inverse of the eigenvalue matrix.
- *Principal component analysis*. We place ourselves in a space where the axes represent the k variables and the points stand for time (or individuals). The technique consists in determining in this space k new axes possessing the property of orthogonality. Factor analysis can also be used.
- *The transformation of variables*. Rather than carrying out a regression on the raw series, i.e., on the series in levels, the regression is estimated on the series in first differences. Such a technique frequently reduces the extent of multicollinearity because, although series in levels may be highly correlated, there is a priori no reason for first-differenced series to be so too. However, this procedure is not without its critics, especially because of the possible occurrence of autocorrelation in the error term of the first-difference regression.
- *Elimination of explanatory variables*. The underlying idea is simple: remove the variable(s) that cause multicollinearity. However, such a procedure can lead to model misspecification. For example, if a significant variable is removed, the OLS estimators become biased and the variance of the error term can no longer be correctly estimated. If a non-significant variable is selected, the OLS estimators remain unbiased and the variance of the error term can be correctly estimated. However, selecting or including insignificant variables in a regression reduces the precision of the coefficient estimates of the significant variables. In the following section, we present various methods for selecting explanatory variables.

5.2.5 Variable Selection Methods

In addition to the model comparison criteria presented in Chap. 3, which may also be useful here, there are various methods for selecting explanatory variables. These techniques can guide us in choosing which variables to remove or add to a model.

The Method of All Possible Regressions

This technique consists in estimating all possible regressions. Thus, from k explanatory variables, $2^k - 1$ regressions are to be estimated. We then select the model that maximizes the adjusted coefficient of determination (or that minimizes the information criteria if we wish to use criteria other than the adjusted coefficient of determination). Obviously, this method is easily applicable for small values of k. For a high number of explanatory variables, it becomes difficult to use. For example, if we have 10 explanatory variables, the number of regressions to be estimated is equal to 1023.

Backward Elimination of Explanatory Variables

This technique involves first estimating the complete model, i.e., the one including the k explanatory variables. We then eliminate the explanatory variable whose estimated coefficient has the lowest t-statistic. We then re-estimate the model on the remaining $k - 1$ explanatory variables and again eliminate the variable whose coefficient has the lowest t-statistic. We reiterate the procedure.

Forward Selection of Explanatory Variables

This is the symmetrical technique to the previous one. We begin by calculating the correlation coefficients between the dependent variable and each of the explanatory variables. We select the explanatory variable most strongly correlated with the dependent variable. Let us note this variable X_i. We then calculate the partial correlation coefficients r_{YX_j,X_i}, for $i \neq j$, i.e., the correlation coefficients between the dependent variable and each of the $k - 1$ other explanatory variables X_j, the influence of the variable X_i having been removed. We select the explanatory variable for which the partial correlation coefficient is highest. We continue in this way. We stop when the t-statistics of the coefficients of the explanatory variables are below the selected critical value, or when the gain measured by the adjusted coefficient of determination is below a certain threshold that we set.

Remark 5.3 A variant of this technique is the **stagewise procedure**. As in the forward selection method, we begin by selecting the explanatory variable X_i that is most highly correlated with the dependent variable. We then determine the residual series resulting from the regression of Y on X_i. We calculate the correlation coefficient between this residual series and each of the $k - 1$ other explanatory variables. We then select the explanatory variable X_j for which the coefficient of correlation is the highest. The next step is to determine the residual series from the regression of Y on X_i and X_j. We again calculate the correlation coefficients between this residual series and each of the $k - 2$ other remaining explanatory variables. The one with the highest correlation coefficient is selected, and so on. The procedure stops when the correlation coefficients are no longer significantly different from zero.

The Stepwise Method

This is an extension of the previous method and is a technique of progressive selection of explanatory variables with the possibility of elimination. Thus, the stepwise method is based on the same principle as the forward selection method, except that each time an explanatory variable is introduced, we examine the coefficients' t-statistics of each of the previously selected variables and eliminate the one(s) whose associated coefficients are not significantly different from zero.

These various methods are not based on any economic considerations and should therefore be used with caution. The most frequently used technique is the stepwise method.

Empirical Application

Consider the previous empirical application aimed at explaining the returns $REURO$ of the European stock index (Euro Stoxx 50) by three explanatory variables:

- $RDJIND$: series of returns of the Dow Jones Industrial Average index
- $RFTSE$: series of returns of the UK stock market index $FTSE$ 100
- $RNIKKEI$: series of returns of the $NIKKEI$ index of the Tokyo Stock Exchange

Let us apply each of the methods presented above.

The technique of all possible regressions involves estimating the following $2^3 - 1 = 7$ models:

- Model (1): Regression of $REURO$ on $RDJIND$
- Model (2): Regression of $REURO$ on $RFTSE$
- Model (3): Regression of $REURO$ on $RNIKKEI$
- Model (4): Regression of $REURO$ on $RDJIND$ and $RFTSE$
- Model (5): Regression of $REURO$ on $RDJIND$ and $RNIKKEI$
- Model (6): Regression of $REURO$ on $RFTSE$ and $RNIKKEI$
- Model (7): Regression of $REURO$ on $RDJIND$, $RFTSE$, and $RNIKKEI$

The results are summarized in Table 5.5.

As shown, except for the constant in some cases, all the variables have significant coefficients in each of the 7 regressions estimated at the 5% significance level—at the 10% level in the case of the coefficient associated with the variable $RNIKKEI$ in model (7). If we choose the model that maximizes the adjusted coefficient of determination, model (7) must be selected. No explanatory variable is therefore eliminated.

Applying the backward method consists in (i) starting from model (7) and (ii) eliminating the variable whose coefficient has the lowest t-statistics. Using a 5% significance level, this technique leads to the elimination of the $RNIKKEI$ variable.

Table 5.5 Estimation of all possible regressions

Model	Constant	$RDJIND$	$RFTSE$	$RNIKKEI$	Adjusted R^2
(1)	−0.0116 (−2.3232)	1.1559 (18.8861)			0.7234
(2)	0.0008 (0.1617)		1.1416 (19.6947)		0.7399
(3)	0.0101 (1.3610)			0.5592 (8.6398)	0.3513
(4)	−0.0062 (−1.4065)	0.5811 (5.5986)	0.6557 (6.4680)		0.7876
(5)	−0.0098 (−1.9629)	1.0511 (13.8745)		0.1195 (2.2836)	0.7318
(6)	0.0015 (0.3380)		1.0219 (14.9681)	0.1492 (3.0913)	0.7554
(7)	−0.0051 (−1.1546)	0.5258 (4.9024)	0.6338 (6.2614)	0.0850 (1.8278)	0.7913

t-statistics of the estimated coefficients are in parentheses

The implementation of the forward selection method involves calculating the correlation coefficients between the dependent variable and each of the three explanatory variables: $r_{REURO,RDJIND} = 0.8517$, $r_{REURO,RFTSE} = 0.8613$, and $r_{REURO,RNIKKEI} = 0.5967$. The first variable selected is therefore $RFTSE$. We then estimate the models with two explanatory variables: ($RFTSE$ and $RDJIND$) and ($RFTSE$ and $RNIKKEI$). These are models (4) and (6), respectively. In each of these models, the new variable has a coefficient significantly different from zero. Since the coefficient associated with $RDJIND$ has a higher t-statistic than that for $RNIKKEI$, the second explanatory variable is $RDJIND$. Finally, we estimate the model with three explanatory variables, model (7), which is the model that we select, if we consider a 10% significance level, since the coefficients of the three variables are significant. If the usual 5% significance level is used, model (4) should be selected.

The application of the stepwise method is identical to the previous case, and the same model is selected, with the three explanatory variables having significant coefficients at the 10% significance level—model (4) being chosen if a 5% significance level is considered.

5.3 Structural Changes and Indicator Variables

The focus here is on studying the **stability** of the estimated model. When estimating a model over a certain period of time, it is possible that a **structural change** may appear in the relationship between the dependent variable and the explanatory variables. It is thus possible that the values of the estimated parameters do not remain identical over the entire period studied. In some cases, the introduction of **indicator variables** allows us to take account of these possible structural changes. We also present various stability tests of the estimated coefficients. Beforehand, we outline the **constrained least squares method** consisting in estimating a model under constraints.

5.3.1 The Constrained Least Squares Method

In Chap. 3, we presented various tests of the hypothesis that the parameter vector $\boldsymbol{\beta}$ is subject to the existence of q constraints:

$$H_0 : \boldsymbol{R\beta} = \boldsymbol{r} \tag{5.52}$$

where $\boldsymbol{R}$ is a given matrix of size $(q, k+1)$ and $\boldsymbol{r}$ is the vector of constraints of dimension $(q, 1)$. It is further assumed that $q \leq k+1$ and the matrix $\boldsymbol{R}$ is of full rank, meaning that the q constraints are linearly independent.

If the null hypothesis is not rejected and we wish to re-estimate the model taking the constraints into account, we should apply the **constrained least squares method** (CLS). We then obtain an estimator $\hat{\boldsymbol{\beta}}_0$ verifying the relationship:

$$\boldsymbol{R}\hat{\boldsymbol{\beta}}_0 = \boldsymbol{r} \tag{5.53}$$

This estimator, called the **constrained least squares estimator**, is given by:[3]

$$\hat{\boldsymbol{\beta}}_0 = \hat{\boldsymbol{\beta}} + \left(\boldsymbol{X}'\boldsymbol{X}\right)^{-1} \boldsymbol{R}' \left(\boldsymbol{R}\left(\boldsymbol{X}'\boldsymbol{X}\right)^{-1} \boldsymbol{R}'\right)^{-1} \left(\boldsymbol{r} - \boldsymbol{R}\hat{\boldsymbol{\beta}}\right) \tag{5.54}$$

The null hypothesis H_0: $\boldsymbol{R\beta} = \boldsymbol{r}$ can be tested using a Fisher test (see Chap. 3):

$$F = \frac{\left(RSS_c - RSS_{nc}\right)/q}{RSS_{nc}/\left(T-k-1\right)} \sim F\left(q, T-k-1\right) \tag{5.55}$$

where RSS_{nc} is the sum of the squared residuals of the unconstrained model (i.e., that associated with the vector $\hat{\boldsymbol{\beta}}$) and RSS_c denotes the sum of the squares of the residuals of the constrained model (i.e., that associated with the vector $\hat{\boldsymbol{\beta}}_0$), q being the number of constraints and k the number of explanatory variables included in the model. As we will see later in this chapter, such a test can also be used to assess the possibility of structural changes.

Example 5.3 In simple cases, CLS are reduced to OLS on a previously transformed model. Consider the following model:

$$Y_t = \alpha + \beta_1 X_{1t} + \beta_2 X_{2t} + \varepsilon_t \tag{5.56}$$

[3] The demonstration is given in the appendix to this chapter.

with $\beta_1 + \beta_2 = 1$. This is a model with two explanatory variables ($k = 2$) and one constraint ($q = 1$), so we have $q < k$. Noting that $\beta_2 = 1 - \beta_1$, we can write the model as follows:

$$Y_t = \alpha + \beta_1 X_{1t} + X_{2t} - \beta_1 X_{2t} + \varepsilon_t \tag{5.57}$$

that is:

$$Z_t = \alpha + \beta_1 W_t + \varepsilon_t \tag{5.58}$$

with $Z_t = Y_t - X_{2t}$ and $W_t = X_{1t} - X_{2t}$. It is then possible to apply the OLS method to Eq. (5.58) to obtain $\hat{\alpha}$ and $\hat{\beta}_1$. We then deduce: $\hat{\beta}_2 = 1 - \hat{\beta}_1$.

5.3.2 The Introduction of Indicator Variables

Definition

Indicator variables, also known as **dummy variables**, are **binary variables** composed of 0 and 1. They are used to reflect the presence or absence of a phenomenon or characteristic. Introducing dummy variables in regressions makes it possible to answer various questions, such as:

- Is the crime rate higher in urban or rural areas?
- Does women's labor supply depend on there being children in the household?
- Is there gender discrimination in hiring?
- For the same level of education, do men's and women's hiring salaries differ? If so, by how much?
- Does location (urban/rural) have an impact on educational attainment?
- Do terrorist attacks change tourist behavior?
- Etc.

Thus, dummy variables are introduced into a regression model when we wish to take a binary explanatory factor into account among the explanatory variables. As an example, such a factor could be:

- The phenomenon either takes place or does not; the dummy variable is then 1 if the phenomenon takes place, 0 otherwise.
- The male or female factor; the dummy variable is equal to 1 if the person is a man, 0 if it is a woman (or vice versa).
- The place of residence: urban or rural; the dummy variable is equal to 1 if the person lives in an urban zone, 0 if in a rural area (or vice versa).
- etc.

The dummy variables enable data to be classified into subgroups based on various characteristics or attributes. Such variables can be introduced into a regression model in the same way as "traditional" explanatory variables. A regression model can simultaneously contain "traditional" explanatory variables and dummy variables, but it can also contain dummy variables only. From a theoretical viewpoint, the introduction of dummy variables into a regression model does not change the estimation method, nor the tests to be implemented.

Introductory Examples

One frequent use of dummy variables is to take account of an exceptional or even aberrant phenomena. Examples include the following: German reunification in 1991, the launch of the euro in 1999, the September 11, 2001 attacks in the United States, the winter 1995 strikes in France, the December 1999 storm in France, the October 1987 stock market crash, the Covid-19 pandemic that broke out at the end of 2019, etc.

Consider, for example, the following regression model:

$$Y_t = \alpha + \beta_1 X_t + \varepsilon_t \tag{5.59}$$

for $t = 1, \ldots, T$. Suppose that at a date t_0, between 1 and T, a disturbance or a shock of any origin affects the variable X_t so that this value is considered an outlier in the regression. We can write the regression model:

$$Y_t = \alpha + \beta_1 X_t + \beta_2 D_t + \varepsilon_t \tag{5.60}$$

with:

$$D_t = \begin{cases} 1 \text{ if } t = t_0 \\ 0 \text{ otherwise} \end{cases} \tag{5.61}$$

The model (5.60) is then written:

$$Y_t = (\alpha + \beta_2) + \beta_1 X_t + \varepsilon_t \text{ if } t = t_0 \tag{5.62}$$

and:

$$Y_t = \alpha + \beta_1 X_t + \varepsilon_t \text{ if } t \neq t_0 \tag{5.63}$$

The two models differ only in the value of the intercept: a perturbation taken into account via a dummy variable affects only the intercept of the model.

There are, however, cases where the perturbation also impacts the slope of the regression model:

$$Y_t = \alpha + \beta_1 X_t + \beta_2 D_t + \beta_3 X_t D_t + \varepsilon_t \tag{5.64}$$

This model can thus be written as:

$$Y_t = (\alpha + \beta_2) + (\beta_1 + \beta_3) X_t + \varepsilon_t \text{ if } t = t_0 \tag{5.65}$$

and

$$Y_t = \alpha + \beta_1 X_t + \varepsilon_t \text{ if } t \neq t_0 \tag{5.66}$$

In this example, the intercept and the slope are simultaneously modified.

The choice between the specifications (5.60) and (5.64) can be guided by theoretical considerations. It is also possible to carry out a posteriori tests in order to make this choice. To this end, we start by estimating the model without dummy variables:

$$Y_t = \alpha + \beta_1 X_t + \varepsilon_t \tag{5.67}$$

We then estimate the two models incorporating the dummy variables:

$$Y_t = \alpha' + \beta_1' X_t + \varepsilon_t \tag{5.68}$$

with $\alpha' = (\alpha + \beta_2)$ and $\beta_1' = \beta_1$ in the case of model (5.60) and $\beta_1' = \beta_1 + \beta_3$ in the case of model (5.64).

We then perform coefficient comparison tests:

- If $\alpha' \neq \alpha$ and if $\beta_1' = \beta_1$: we are in the case of the specification (5.60).
- If $\alpha' \neq \alpha$ and if $\beta_1' \neq \beta_1$: we are in the case of the specification (5.64).

Model Containing Only Indicator Variables

Models whose regressors consist solely of dummy variables are sometimes called **analysis of variance (ANOVA)** models. They are frequently used to compare differences in means between two or more categories of individuals. Let us take an example. Consider consumer spending on any good B in a country by subregion:

$$Y_i = \alpha + \beta_1 D_{1i} + \beta_2 D_{2i} + \varepsilon_i \tag{5.69}$$

where:

- Y_i denotes the average consumption expenditure of the good B in the subregion i.
- $D_{1i} = \begin{cases} 1 \text{ if the subregion is located in the South} \\ 0 \text{ otherwise.} \end{cases}$
- $D_{2i} = \begin{cases} 1 \text{ if the subregion is located in the Southeast} \\ 0 \text{ otherwise.} \end{cases}$

D_{1i} and D_{2i} are two dummy variables representing a **qualitative** variable. The qualitative variable here is the region to which the subregion belongs, and each of the dummy variables represents one of the modalities associated with this variable. The average consumption expenditure of the good B in the North corresponds to the case where $D_{1i} = 1$ and $D_{2i} = 0$ and is given by the model:

$$Y_i = \alpha + \beta_1 \tag{5.70}$$

Similarly, the average consumption expenditure of the good B in the Southeast is such that $D_{1i} = 0$ and $D_{2i} = 1$ and is given by:

$$Y_i = \alpha + \beta_2 \tag{5.71}$$

We deduce that the average consumption expenditure of the good B in the Southwest corresponds to the case where $D_{1i} = D_{2i} = 0$ and is given by:

$$Y_i = \alpha \tag{5.72}$$

Thus, the average consumption expenditure of the good B in the Southwest is given by the value of the intercept, the slope coefficients β_1 and β_2 indicating, respectively, the magnitude of the difference between the average consumption expenditure in the North and the Southwest, and the value of the difference between the average consumption expenditure in the Southeast and the Southwest.

To illustrate this, let us assign values to the different coefficients. Suppose that the estimation of the model (5.69) has led to the following results:

$$\hat{Y}_i = \underset{(24.20)}{350} - \underset{(-1.54)}{30}\, D_{1i} - \underset{(-3.20)}{60}\, D_{2i} \tag{5.73}$$

where the values in parentheses correspond to the t-statistics of the estimated coefficients. This example shows that the average consumption expenditure on the good B in the Southwest is €350, those in the Northern subregions are €30 lower, and those in subregions located in the Southeast are €60 lower. The average consumption expenditure on the good B in the North is therefore €350 − 30 = 320 and €350 − 60 = 290 in the Southeast.

Let us now look at the significance of the coefficients. We see that the coefficient for subregions in the North is not significant, while that for subregions in the Southeast is. As a result, the average consumption expenditure of the good B is not significantly different between subregions located in the North and in the Southwest. On the other hand, there is a significant difference of €60 between the average consumption expenditure on the good B in the Southeast and in the Southwest.

This very simple example highlighted that, to distinguish the three regions (North, Southeast, and Southwest), only two dummy variables were introduced. It is very important to note that the introduction of three dummy variables, each representing a region, is impossible.[4] This would lead to a situation of **perfect collinearity**, i.e., in a case where the three variables would be perfectly dependent (see above). In other words, the sum of the three dummy variables would be equal to the unit vector, which would be, by construction, collinear with the unit vector of the constant term, making the matrix $(X'X)$ non-invertible, with X denoting the matrix of explanatory variables. Consequently, when a variable has m categories or attributes, it is appropriate to introduce $(m-1)$ dummy variables. In our example, we had three regions, $m = 3$, so only two dummy variables should be included in the model. Similarly, if we want to study the consumption expenditure on the good B by gender, only one dummy variable should be introduced taking the value 1 for men and 0 for women (or vice versa).

Remark 5.4 In the example studied here, we have considered a single qualitative variable comprising three attributes (North, Southeast, and Southwest). It is possible to introduce more than one qualitative variable into a model. This is the case, for example, with the following model:

$$Y_i = \alpha + \beta_1 D_{1i} + \beta_2 D_{2i} + \varepsilon_i \tag{5.74}$$

where Y_i denotes the consumption expenditure on the good B in the subregion i, D_{1i} denotes gender ($D_{1i} = 1$ if the person is male, 0 if female), and D_{2i} is the region to which the subregion belongs ($D_{2i} = 1$ if the subregion is in the South region, 0 otherwise). The estimation of the coefficient α thus gives the average consumption expenditure on the good B by a woman living in a subregion that is not located in the South. This situation is the reference situation to which the other cases will be compared.

Remark 5.5 In Chap. 2, we presented semi-log models of the type $\log Y_t = \alpha + \beta X_t + \varepsilon_t$. We have seen that the coefficient β can be interpreted as the semi-elasticity of Y_t with respect to X_t. What happens to this when we are dealing with a dummy variable and not a usual quantitative variable? Consider the following model:

$$\log Y_t = \alpha + \beta D_t + \varepsilon_t \tag{5.75}$$

where Y_t denotes the average hourly wage in euros and D_t is a dummy variable equal to 1 for women and 0 for men. For men, the model is given by $\log Y_t = \alpha + \varepsilon_t$ and for women by $\log Y_t = (\alpha + \beta) + \varepsilon_t$. Therefore, α denotes the logarithm of the average hourly wage and β is the difference in the logarithm of the average hourly wage for men and women. The anti-log of α is interpreted as the median (not average) hourly wage for men. Similarly, the anti-log of $(\alpha + \beta)$ is the median hourly wage for women.

---

[4] This is only possible if the model does not have a constant term.

Model Containing Indicator and Usual Explanatory Variables

In most cases, the regression model contains not only qualitative variables but also traditional explanatory variables, i.e., quantitative variables. Such models are called **analysis of covariance (ANCOVA)** models. They are an extension of ANOVA models in the sense that they control for the effects of quantitative variables. For this reason, the quantitative variables in ANCOVA models are called **control variables**.

If we take the previous example of the consumption expenditure on a good B, we can consider the following model:

$$Y_i = \alpha + \beta_1 D_{1i} + \beta_2 D_{2i} + \beta_3 X_i + \varepsilon_i \tag{5.76}$$

where:

- Y_i denotes the average consumption expenditure on a good B in the subregion i.
- $D_{1i} = \begin{cases} 1 \text{ if the subregion is located in the North} \\ 0 \text{ otherwise.} \end{cases}$
- $D_{2i} = \begin{cases} 1 \text{ if the subregion is located in the Southeast} \\ 0 \text{ otherwise.} \end{cases}$
- X_i designates the average wage.

The average wage is here a control variable.

Interactions

To illustrate the problem of interactions between variables, let us consider the following model:

$$Y_t = \alpha + \beta_1 D_{1t} + \beta_2 D_{2t} + \beta_3 X_t + \varepsilon_t \tag{5.77}$$

where:

- Y_t is the hourly wage in euros.
- D_{1t} is such that: $D_{1t} = \begin{cases} 1 \text{ if the person is a woman} \\ 0 \text{ otherwise.} \end{cases}$
- D_{2t} is such that: $D_{2t} = \begin{cases} 1 \text{ if the person works in the public sector} \\ 0 \text{ if the person works in the private sector.} \end{cases}$
- X_t is the level of education (in number of years).

In this model, gender (represented by D_{1t}) and employment sector (represented by D_{2t}) are qualitative variables. It is implicitly assumed that the differential effect of each of these two variables is constant, regardless of the value of the other variables. In other words, the differential effect of the variable D_{1t} is assumed to be constant in both employment sectors, and the differential effect of the variable D_{2t} is also assumed to be constant in both genders. Thus, if hourly wages are higher for men than for women, they are higher whether or not they work in the public sector. Similarly, if the hourly wage of people working in the public sector is lower than that of people working in the private sector, it is so whether they are men or women. There is therefore no interaction between the two qualitative variables. Such an assumption may seem highly restrictive, and we need to take into account the possible interactions between the variables. For example, a woman working in the public sector may earn less than a man working in the same sector. We can thus write the model (5.77):

$$Y_t = \alpha + \beta_1 D_{1t} + \beta_2 D_{2t} + \beta_3 X_t + \beta_4 (D_{1t} D_{2t}) + \varepsilon_t \tag{5.78}$$

For a woman ($D_{1t} = 1$) working in the public sector ($D_{2t} = 1$), the model is:

$$Y_t = (\alpha + \beta_1 + \beta_2 + \beta_4) + \beta_3 X_t + \varepsilon_t \tag{5.79}$$

The interpretation of the coefficients is then:

- β_1 is the differential effect of being a woman.
- β_2 is the differential effect of working in the public sector.
- β_4 is the differential effect of being a woman working in the public sector.

Let us consider a numerical example. Suppose that the estimation of the model (5.77) leads to the following results:

$$\hat{Y}_t = -0.6 - 3.4 D_{1t} - 2.7 D_{2t} + 0.7 X_t \tag{5.80}$$

This model indicates that, all other things being equal, the average hourly wage of women is €3.4 lower than that of men, and the average hourly wage of people working in the public sector is €2.7 lower than that of people working in the private sector.

Let us now assume that the estimation of the model (5.78) has led to the following results:

$$\hat{Y}_t = -0.6 - 3.4 D_{1t} - 2.7 D_{2t} + 0.7 X_t + 3.1 D_{1t} D_{2t} \tag{5.81}$$

All other things being equal, the average hourly wage of women working in the public sector is €3 lower ($-3.4 - 2.7 + 3.1 = -3$), which lies between the values -3.4 (gender difference alone) and -2.7 (employment sector difference alone).

Use of Indicator Variables for Deseasonalization

We have seen that dummy variables can be used in a variety of cases, including:

- To take account of a temporary event or exceptional phenomenon (e.g., German reunification, World War I, strikes, particular climatic phenomena, etc.)
- To take account of spatial effects (e.g., living in an urban zone or a rural area, region to which subregions belong, etc.)

– To account for characteristics (modalities or categories) of qualitative variables (e.g., gender, employment sector, political affiliation, religion, etc.)

Dummy variables can also be used to **deseasonalize** a series. To illustrate this procedure, consider a series Y_t with a quarterly frequency. To test for the existence of a seasonal effect, we run the following regression:[5]

$$Y_t = \beta_1 D_{1t} + \beta_2 D_{2t} + \beta_3 D_{3t} + \beta_4 D_{4t} + \varepsilon_t \tag{5.82}$$

where $D_{it} = 1$ for quarter i, 0 otherwise, with $i = 1, 2, 3, 4$. If there is a seasonal effect for a particular quarter, the coefficient of the dummy variable corresponding to that quarter will be significantly different from zero. As an example, if Y_t denotes a series of toy sales, it is highly likely that the coefficient β_4 assigned to the fourth quarter of the year will be significantly different from zero, reflecting the increase in toy sales at Christmas.

The deseasonalization method based on dummy variables is straightforward. It consists first in estimating the model (5.82):

$$\hat{Y}_t = \hat{\beta}_1 D_{1t} + \hat{\beta}_2 D_{2t} + \hat{\beta}_3 D_{3t} + \hat{\beta}_4 D_{4t} \tag{5.83}$$

and then calculating the series $\left(Y_t - \hat{Y}_t\right)$, which is the residual series: this series is the seasonally adjusted series.

Remark 5.6 The method described above is only valid if the series under consideration can be decomposed in an additive way, i.e., if it can be written in the form: $Y = T + C + S + \varepsilon$ where T designates the trend, C the cyclical component, S the seasonal component, and ε the residual component. This is known as an **additive decomposition scheme**. But, if the components enter multiplicatively (multiplicative decomposition scheme), i.e., $Y = T \times C \times S \times \varepsilon$, the deseasonalization method presented above is inappropriate.

Empirical Application

Consider the series of returns of the Dow Jones Industrial Average US stock index ($RDJIND$) over the period from the second quarter of 1970 to the second quarter of 2021 (source: Macrobond). We are interested in the relationship between the present value of returns and their first-lagged value. The study period includes the stock market crash of October 19, 1987, corresponding to the 71st observation. To take into account this exceptional event, let us consider the following dummy variable:

$$D_t = \begin{cases} 1 \text{ if } t = 71 \\ 0 \text{ otherwise} \end{cases} \tag{5.84}$$

[5] Note that a dummy variable is assigned to each quarter, which requires us not to introduce a constant term into the regression. We could also have written the model by introducing a constant term and only three dummy variables.

Estimating the model with the dummy variable gives:

$$\widehat{RDJIND_t} = \underset{(3.6585)}{0.0206} + \underset{(0.1028)}{0,0069} RDJIND_{t-1} - \underset{(-3.9772)}{0.3131}\, D_t \quad (5.85)$$

where the values in parentheses are the t-statistics of the estimated coefficients.

All else being equal, the stock market crash of October 1987 reduced the average value of UK index returns by -0.3131. This decrease is significant insofar as the coefficient assigned to the dummy variable is significantly different from zero.

5.3.3 Coefficient Stability Tests

It is often useful to assess the robustness of the estimated model over the entire study period, i.e., to test its **stability**. There may in fact be a **structural change** or **break** in the relationship between the dependent variable and the explanatory variables, resulting in instability of the coefficients of the model estimated over the entire period under consideration. Several causes can produce a structural change, such as the transition to the single currency, a change in exchange rate regime (from a fixed to a flexible exchange rate regime), the 1973 oil shock, the World War II, the 1987 stock market crash, the Covid-19 pandemic, etc.

There are various methods for assessing the stability of the estimated coefficients of a regression model, and we present them below.

Rolling Regressions and Recursive Residuals

General Principle

The **rolling regression** technique is very intuitive. It involves estimating the parameters of successive models in the following way. The first method consists in estimating successive models by adding one or more observations each time, starting from the beginning and going towards the end of the period. This is known as **forward regression**. The second method also involves estimating successive models by adding one or more observations each time, but starting from the end of the period and moving towards the beginning. This is known as **backward regression**.

Several graphs are then plotted to assess the stability of the various characteristics of the estimated regression, for example:

- Graph of the estimated coefficients
- Graph of the t-statistics of the estimated coefficients
- Graph of the coefficients of determination of the estimated models

It is then a matter of identifying a possible break in these graphs in order to detect a structural change. This technique is a graphical method and therefore not a statistical test in the strict sense of the term.

Recursive Residuals

Consider the usual regression model:

$$\boldsymbol{Y} = \boldsymbol{X}\boldsymbol{\beta} + \boldsymbol{\varepsilon} \tag{5.86}$$

Let us denote $\boldsymbol{x}_t$ the vector of k explanatory variables plus the constant for the *t-th* observation:

$$\boldsymbol{x}_t = (1, X_{1t}, \ldots, X_{kt})' \tag{5.87}$$

Let $\boldsymbol{X}_{t-1}$ be the matrix formed by the $(t-1)$ first rows of $\boldsymbol{X}_t$. This matrix can be used to estimate $\boldsymbol{\beta}$. Let $\hat{\boldsymbol{\beta}}_{t-1}$ be the estimator thus obtained:

$$\hat{\boldsymbol{\beta}}_{t-1} = \left(\boldsymbol{X}'_{t-1}\boldsymbol{X}_{t-1}\right)^{-1}\boldsymbol{X}'_{t-1}\boldsymbol{Y}_{t-1} \tag{5.88}$$

where $\boldsymbol{Y}_{t-1}$ is the subvector of the $(t-1)$ first elements of $\boldsymbol{Y}_t$.

It is then possible to calculate the forecast error associated with the *t-th* observation, denoted e_t:

$$e_t = Y_t - \boldsymbol{x}'_t\hat{\boldsymbol{\beta}}_{t-1} \tag{5.89}$$

The variance of this forecast error is given by (see Chap. 3):

$$V(e_t) = \sigma_\varepsilon^2\left[1 + \boldsymbol{x}'_t\left(\boldsymbol{X}'_{t-1}\boldsymbol{X}_{t-1}\right)^{-1}\boldsymbol{x}_t\right] \tag{5.90}$$

Recursive residuals can be defined as follows:

$$w_t = \frac{Y_t - \boldsymbol{x}'_t\hat{\boldsymbol{\beta}}_{t-1}}{\sqrt{1 + \boldsymbol{x}'_t\left(\boldsymbol{X}'_{t-1}\boldsymbol{X}_{t-1}\right)^{-1}\boldsymbol{x}_t}} \tag{5.91}$$

with $w_t \sim N(0, \sigma_\varepsilon^2)$. The recursive residuals are defined as the normalized forecast errors. Furthermore, the recursive residuals are a set of residuals which, if the disturbance terms are independent and of the same law, are themselves independent and of the same law. The recursive residuals thus are normally distributed since they are defined as a linear function of normal variables and the forecast given by OLS is unbiased.

To generate a sequence of recursive residuals, we proceed as follows:

- We choose a starting set of τ observations, with $\tau < T$. These may be, for example, the first τ observations of the sample (case of a forward regression). Having estimated $\hat{\boldsymbol{\beta}}_\tau$, the corresponding recursive residuals are determined:

$$w_{\tau+1} = \frac{Y_{\tau+1} - \boldsymbol{x}'_{\tau+1}\hat{\boldsymbol{\beta}}_\tau}{\sqrt{1 + \boldsymbol{x}'_{\tau+1}\left(\boldsymbol{X}'_\tau\boldsymbol{X}_\tau\right)^{-1}\boldsymbol{x}_{\tau+1}}} \tag{5.92}$$

- We increase the number of observations by one: we therefore consider the first $\tau + 1$ observations of the sample. We estimate $\hat{\boldsymbol{\beta}}_{\tau+1}$ and determine the corresponding recursive residuals:

$$w_{\tau+2} = \frac{Y_{\tau+2} - \boldsymbol{x}'_{\tau+2}\hat{\boldsymbol{\beta}}_{\tau+1}}{\sqrt{1 + \boldsymbol{x}'_{\tau+2}\left(\boldsymbol{X}'_{\tau+1}\boldsymbol{X}_{\tau+1}\right)^{-1}\boldsymbol{x}_{\tau+2}}} \tag{5.93}$$

- We repeat the previous step, each time including an additional observation.

We obtain a series of $T - \tau$ recursive residuals, defined by (5.91) for $t = \tau + 1, \ldots, T$. From these recursive residuals, Brown et al. (1975) proposed the **CUSUM** and **CUSUM of squares** tests, which allow us to test the stability of the estimated coefficients in a model. These are tests designed to test the null hypothesis of parameter stability, i.e.:

$$H_0 : \boldsymbol{\beta}_1 = \boldsymbol{\beta}_2 = \ldots = \boldsymbol{\beta}_T = \boldsymbol{\beta} \tag{5.94}$$

with:

$$\sigma^2_{\varepsilon 1} = \sigma^2_{\varepsilon 2} = \ldots = \sigma^2_{\varepsilon T} = \sigma^2_{\varepsilon} \tag{5.95}$$

where the coefficients $\boldsymbol{\beta}_t$, $t = 1, \ldots, T$, are the vectors of the regression coefficients for the period t and $\sigma^2_{\varepsilon t}$ denotes the variances of the errors for this same period.

The CUSUM Test

The first test proposed by Brown et al. (1975) is called the CUSUM test (CUmulative SUM) and is based on the cumulative sum defined by:

$$W_t = \sum_{j=\tau+1}^{t} \frac{w_j}{\hat{\sigma}_w} \tag{5.96}$$

where $t = \tau + 1, \ldots, T$ and:

$$\hat{\sigma}^2_w = \frac{1}{T - \tau} \sum_{j=\tau+1}^{T} w_j^2 \tag{5.97}$$

W_t is thus a cumulative sum that varies with t. As long as the vectors $\boldsymbol{\beta}$ are constant, the average of W_t is zero. If they vary, W_t tends to deviate from the straight line representing the null expectation. More specifically, under the null hypothesis of stability of the coefficients, W_t must lie within the interval $[-L_t, L_t]$ where:

$$L_t = \frac{a\,(2t + T - 3\tau)}{\sqrt{T - \tau}}$$

with $a = 1.143$ at the 1% significance level, $a = 0.948$ at the 5% significance level, and $a = 0.850$ at the 10% significance level.

The null hypothesis of stability is rejected if W_t cuts L_t or $-L_t$. This means that if the coefficients are not constant, there may be a disproportionate number of recursive residuals w_t of the same sign that "push" W_t out of the interval.

The CUSUM test is generally used to detect possible systematic movements in the coefficient values reflecting possible structural instability. If a break is found, the chosen specification is rejected over the whole period. On the other hand, if we wish to detect random movements (and not movements necessarily resulting from a structural modification of the coefficients), we use the CUSUM of squares test.

The CUSUM of Squares Test

The second test proposed by Brown et al. (1975) is the CUSUM of squares test based on the cumulative sums of the squares of the recursive residuals, i.e.:

$$s_t = \frac{\sum_{j=\tau+1}^{t} w_j^2}{\sum_{j=\tau+1}^{T} w_j^2}, t = \tau + 1, \ldots, T \tag{5.98}$$

The line representing the expectation of the test statistic under the null hypothesis of stability is given by:

$$E(s_t) = \frac{t - \tau}{T - \tau} \tag{5.99}$$

This expression varies from 0 (for $t = \tau$) to 1 (for $t = T$). The idea is then to study the significance of the difference between s_t and $E(s_t)$. To this end, we draw a pair of reference lines parallel to the line $E(s_t)$, one lying above it, the other below it, at a distance C. Brown et al. (1975) tabulated the values of C for various sample sizes and significance levels. The estimated coefficients are unstable if the graph of s_t intersects the previously defined reference lines. More precisely, under the null hypothesis of stability of the coefficients, s_t has a beta distribution with mean $(t - \tau)/(T - \tau)$ and is framed by the interval $\pm C + (t - \tau)/(T - \tau)$. If s_t leaves this interval at period $t = i$, this means that there is a random break that reflects the instability of the regression coefficients for this period i.

The Chow Test (1960)

The **Chow test** is very frequently used. Consider the following regression model, for $t = 1, \ldots, T$:

$$Y_t = \alpha + \beta X_t + \varepsilon_t \tag{5.100}$$

Suppose we divide the sample into two sub-samples and estimate the following models:

$$Y_t = \alpha_1 + \beta_1 X_t + \varepsilon_{1t}, \text{ for } t = 1, \ldots, \tau \tag{5.101}$$

and:

$$Y_t = \alpha_2 + \beta_2 X_t + \varepsilon_{2t}, \text{ for } t = \tau + 1, \ldots, T \tag{5.102}$$

The relationship (5.100) is based on the absence of structural change over the entire period under consideration. In other words, there is no difference between the two periods $t = 1, \ldots, \tau$ and $t = \tau + 1, \ldots, T$: the constant term and the slope coefficient remain identical. If this is indeed the case, we should have:

$$\alpha = \alpha_1 = \alpha_2 \text{ and } \beta = \beta_1 = \beta_2 \tag{5.103}$$

The Chow test consists in testing the null hypothesis:

$$H_0 : \begin{cases} \alpha_1 = \alpha_2 \\ \beta_1 = \beta_2 \end{cases} \tag{5.104}$$

against the alternative hypothesis:

$$H_1 : \begin{cases} \alpha_1 \neq \alpha_2 \\ \beta_1 \neq \beta_2 \end{cases} \tag{5.105}$$

Assuming that ε_{1t} and ε_{2t} are independent and both have normal distributions of zero mean and same variance, the Chow test is implemented as follows:

- The model (5.100) is estimated and the corresponding residual sum of squares is noted RSS_0.
- The model (5.101) is estimated and the corresponding residual sum of squares is noted RSS_1.
- The model (5.102) is estimated and the corresponding residual sum of squares is noted RSS_2.
- $RSS_a = RSS_1 + RSS_2$ is calculated.
- We calculate the test statistic:

$$F = \frac{(RSS_0 - RSS_a) / (k+1)}{RSS_a / (T - 2(k+1))} \tag{5.106}$$

where k is the number of explanatory variables (1 in our case).

Under the null hypothesis of no structural change, we have:

$$F \sim F\,(k+1, T-2(k+1)) \tag{5.107}$$

The decision rule is written:

- If $F < F\,(k+1, T-2(k+1))$, we do not reject the null hypothesis of stability of the coefficients. There is no structural change.
- If $F > F\,(k+1, T-2(k+1))$, we reject the null hypothesis of stability of the coefficients, indicating the presence of a structural change.

Remark 5.7 The Chow test can be easily generalized to the existence of more than one structural break. Thus, if we wish to test for the existence of two breaks, we will split the period into three sub-periods, the principle of the test remaining the same (the sum of the squared residuals RSS_a then being equal to the sum of the sums of the squared residuals of the three regressions corresponding to the three sub-periods).

The Chow test assumes that the date at which the structural break(s) occurs is known. Otherwise, it is possible to perform rolling regressions and to calculate the Chow test statistic for each of these regressions. The break point we are looking for then corresponds to the value for which the Chow statistic is maximum.

Empirical Application

Consider the relationship between the returns of the US Dow Jones Industrial Average index ($RDJIND$) and the Japanese Nikkei index ($RNIKKEI$). The data, taken from the Macrobond database, are quarterly over the period from the second quarter of 1978 to the second quarter of 2021 ($T = 173$). The OLS estimation of the relationship:

$$RNIKKEI_t = \alpha + \beta \times RDJIND_t + \varepsilon_t \tag{5.108}$$

over the whole period gives:

$$\widehat{RNIKKEI_t} = \underset{(-1.1688)}{-0.0079} + \underset{(9.5019)}{0.7958} RDJIND_t \tag{5.109}$$

where the figures in parentheses correspond to the t-statistics of the estimated coefficients. In addition, we have the following statistics: $R_0^2 = 0.3455$ and $RSS_0 = 1.2654$.

As our study period includes the stock market crash of October 19, 1987, it is pertinent to question the stability of the estimated relationship.

Rolling Regressions

To get a rough idea of the stability of the estimated coefficients, we perform rolling regressions by adding an observation each time. We then graphically represent the

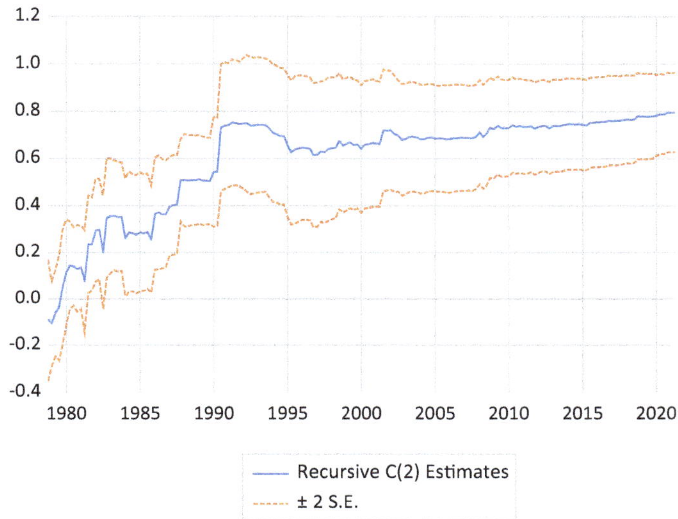

Fig. 5.1 Rolling regressions. Change in the slope coefficient

estimated coefficients corresponding to each of the regressions estimated. Figure 5.1 shows the change in the slope coefficient. The dotted curves correspond to plus or minus twice the standard deviation of the estimated coefficient. The graph shows some instability of the slope coefficient, which is more marked in the first part of the sample.

The same type of analysis can be carried out for the constant term. Figure 5.2 also shows some instability of the coefficient, with, in particular, a change in sign over the study period.

CUSUM and CUSUM of Squares Tests

To assess the stability of the estimated relationship, let us calculate the recursive residuals using the method previously presented. Figure 5.3 plots the series of the recursive residuals, along with two curves representing more or less twice the standard deviation of recursive residuals at each date. If some residuals lie outside the band formed by these two curves, this is an indication of instability. We observe that such a phenomenon appears on several occasions, particularly in the first part of the sample.

Figure 5.4 corresponds to the application of the CUSUM test for the 5% significance level. It can be seen that the series of the cumulative sum of recursive residuals remains within the interval formed by the two lines, suggesting there is no structural instability in the relationship over the period under consideration.

Figure 5.5 corresponds to the application of the CUSUM of squares for the 5% significance level. This graph highlights that the cumulative sum of squares falls

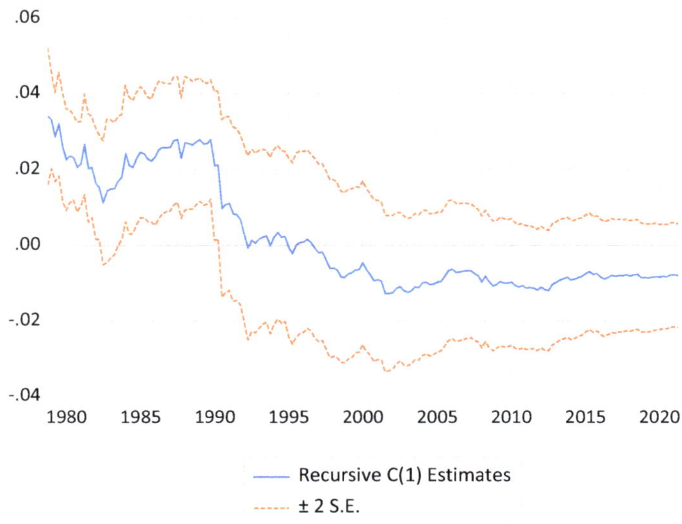

Fig. 5.2 Rolling regressions. Change in the constant term

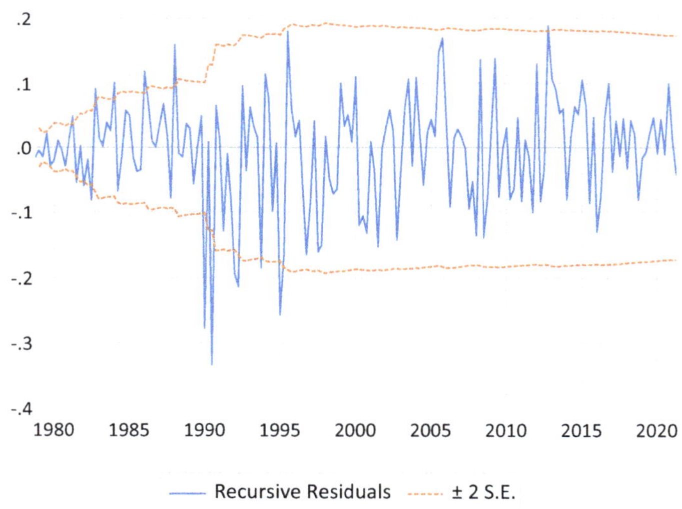

Fig. 5.3 Recursive residuals

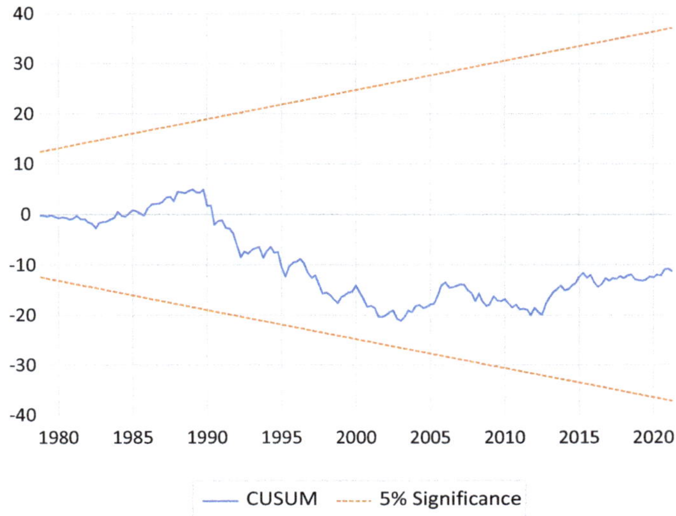

Fig. 5.4 CUSUM test

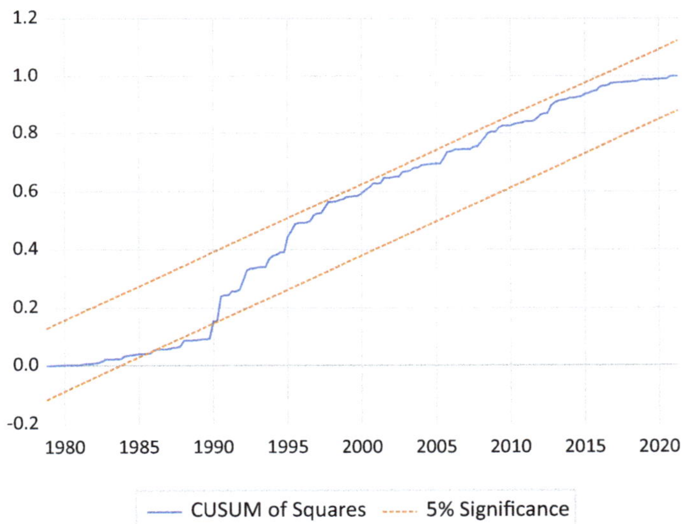

Fig. 5.5 CUSUM of squares test

outside the interval delimited by the two lines around the 1987 stock market crash, indicating some instability (random break) in the parameters or variance.

Chow Test

To investigate whether the stock market crash of October 1987 caused a structural break in the relationship between the returns of the two indices under consideration, let us apply the Chow test. To this end, we estimate two regressions: a regression over the period 1978.2–1987.3 (before the crash) and a regression over the period 1987.4–2021.2 (after the crash). The results are given below.

Over the period 1978.2–1987.3, i.e., $t = 1, \ldots, 70$:

$$\widehat{RNIKKEI_t} = \underset{(3.3720)}{0.0279} + \underset{(3.8941)}{0.4072} RDJIND_t \tag{5.110}$$

with $R_1^2 = 0.2964$ and $RSS_1 = 0.0782$.

Over the period 1987.4–2021.2, i.e., $t = 71, \ldots, 173$:

$$\widehat{RNIKKEI_t} = \underset{(-1.9608)}{-\,0.0159} + \underset{(8.7515)}{0.8728} RDJIND_t \tag{5.111}$$

with $R_2^2 = 0.3654$ and $RSS_2 = 1.1258$.

So we have: $RSS_a = 0.0782 + 1.1258 = 1.2040$.

It is possible to calculate the Chow test statistic:

$$F = \frac{(1.2654 - 1.2040)\,/\,(1+1)}{1.2040/(173 - 2\,(1+1))} = 4.3128 \tag{5.112}$$

The Fisher table gives us, at the 5% significance level: $F\,(2, 169) = 2.997$. The calculated value of the test statistic being higher than the critical value, the null hypothesis of stability of the estimated coefficients is rejected at the 5% significance level. There is indeed a break in the fourth quarter of 1987. This result was expected in view of the differences obtained in the estimates over the two sub-periods: the constant term is positive in the first sub-period and negative in the second, and the slope coefficient is more than twice as high in the second sub-period as in the first.

It is possible to recover the results of the Chow test by introducing a dummy variable and running a single regression. Consider the following model:

$$RNIKKEI_t = \alpha + \beta \times RDJIND_t + \gamma D_t + \delta\,(D_t \times RDJIND_t) + \varepsilon_t \tag{5.113}$$

with $D_t = \begin{cases} 0 \text{ over the period 1978.2–1987.3} \\ 1 \text{ over the period 1987.4–2021.2} \end{cases}$.

Thus, over the period 1978.2–1987.3, the model is written:

$$RNIKKEI_t = \alpha + \beta \times RDJIND_t + \varepsilon_t \tag{5.114}$$

and over the period 1987.4–2021.2:

$$RNIKKEI_t = (\alpha + \gamma) + (\beta + \delta) \times RDJIND_t + \varepsilon_t \tag{5.115}$$

In Eq. (5.113), the coefficient δ indicates how much the slope coefficient of the second period differs from that of the first period. Estimating this relationship yields:

$$\begin{aligned}\widehat{RNIKKEI}_t &= \underset{(1.8619)}{0.0279} + \underset{(2.1502)}{0.4072} RDJIND_t - \underset{(-2.6195)}{0.0439}\, D_t \\ &\quad + \underset{(2.2140)}{0.4656}\,(D_t \times RDJIND_t)\end{aligned} \tag{5.116}$$

All coefficients are significantly different from zero (at the 10% significance level for the constant term), suggesting that the relationship between the two series of returns is different over the two sub-periods. From this estimation, we deduce the relationship over the 1978.2–1987.3 period:

$$\widehat{RNIKKEI}_t = 0.0279 + 0.4072 \times RDJIND_t \tag{5.117}$$

and the relationship over the period 1987.4–2021.2:

$$\begin{aligned}\widehat{RNIKKEI}_t &= (0.0279 - 0.0439) + (0.4072 + 0.4656)\, RDJIND_t \\ &= -0.0160 + 0.8728 RDJIND_t\end{aligned} \tag{5.118}$$

We naturally find the results obtained when implementing the Chow test. We see that the coefficients γ and δ are significantly different from zero. We deduce that the regressions over the two sub-periods differ not only in the constant term but also in the slope coefficient. The findings therefore confirm the results of the Chow test.

Conclusion

In this chapter, we have considered that two of the assumptions of the regression model concerning the explanatory variables are violated: the assumption of independence between the explanatory variables and the error term, on the one hand, and the assumption of independence between the explanatory variables, on the other. We have also studied third problem relating to the explanatory variables, namely, the question of the instability of the estimated model. So far, we have considered models in which the dependent variable is a function of one or more explanatory variables at the same date, i.e., at the same moment in time. Frequently, however, the explanatory variables include lagged variables or the lagged endogenous variable. These are referred to as dynamic models, as opposed to static models. These models are the subject of the next two chapters.

The Gist of the Chapter

Random explanatory variables	Estimation method: instrumental variables (IV)
	IV estimator $\hat{\boldsymbol{\beta}}_{IV} = \left(\boldsymbol{Z}'\boldsymbol{X}\right)^{-1}\boldsymbol{Z}'\boldsymbol{Y}$
	$\boldsymbol{Z}$: matrix of instrumental variables
Multicollinearity	Explanatory variables not independent of each other
Detection	Calculation of correlation coefficients
	Klein (1962) test
	Farrar and Glauber (1967) test
	Eigenvalue method
	Calculation of variance inflation factors (VIF)
Solution	Ridge regression
Structural changes	
Constrained estimation	Constrained least squares
Consideration	Indicator variables (dummy)
Tests	CUSUM and CUSUM of squares (recursive residuals)
	Chow (1960)

Further Reading

In addition to the references cited in this chapter concerning, in particular, collinearity detection techniques or tests to detect breaks, the readers can extend their knowledge through the selected reading below.

For developments relating to multicollinearity and model selection, see the chapter by Leamer (1983) in the book edited by Griliches and Intriligator (1983). Readers may also refer to Belsley et al. (1980) and to various econometric textbooks such as Judge et al. (1988).

Concerning the constrained least squares method, interested readers may refer to the following textbooks: Judge et al. (1985, 1988), Gouriéroux and Monfort (2008), and Greene (2020). For more information on the use of dummy variables, see Fox (1997) and Kennedy (2008). Regarding seasonal adjustment methods, an interesting reference is provided by Diebold (2012). Readers can also consult Johnston and Dinardo (1996) for further discussion on recursive residuals. We have not dealt with models with random coefficients in this book; interested readers may consult Swamy (1971). Similarly, a key reference on regime-switching models is Goldfeld and Quandt (1972).

Appendix: Demonstration of the Formula for Constrained Least Squares Estimators

In order to determine the constrained least squares estimator, we need to solve a minimization program of sum of squared residuals:

$$Min\left(\boldsymbol{Y}-\boldsymbol{X}\hat{\boldsymbol{\beta}}_0\right)'\left(\boldsymbol{Y}-\boldsymbol{X}\hat{\boldsymbol{\beta}}_0\right)=Min\left(\boldsymbol{e}'\boldsymbol{e}\right) \tag{5.119}$$

under the constraint: $\boldsymbol{R}\hat{\boldsymbol{\beta}}_0=\boldsymbol{r}$.

We define the Lagrange function:

$$\mathcal{L}=\left(\boldsymbol{Y}-\boldsymbol{X}\hat{\boldsymbol{\beta}}_0\right)'\left(\boldsymbol{Y}-\boldsymbol{X}\hat{\boldsymbol{\beta}}_0\right)-2\boldsymbol{\lambda}'\left(\boldsymbol{R}\hat{\boldsymbol{\beta}}_0-\boldsymbol{r}\right) \tag{5.120}$$

where $\boldsymbol{\lambda}$ is a column vector formed by the q Lagrange multipliers. We calculate the partial derivatives:

$$\frac{\partial\mathcal{L}}{\partial\hat{\boldsymbol{\beta}}_0}=-2\boldsymbol{X}'\boldsymbol{Y}+2\boldsymbol{X}'\boldsymbol{X}\hat{\boldsymbol{\beta}}_0-2\boldsymbol{R}'\boldsymbol{\lambda} \tag{5.121}$$

and:

$$\frac{\partial\mathcal{L}}{\partial\boldsymbol{\lambda}}=-2\left(\boldsymbol{R}\hat{\boldsymbol{\beta}}_0-\boldsymbol{r}\right) \tag{5.122}$$

Canceling these partial derivatives, we have:

$$\boldsymbol{X}'\boldsymbol{X}\hat{\boldsymbol{\beta}}_0-\boldsymbol{X}'\boldsymbol{Y}-\boldsymbol{R}'\boldsymbol{\lambda}=0 \tag{5.123}$$

and:

$$\boldsymbol{R}\hat{\boldsymbol{\beta}}_0-\boldsymbol{r}=0 \tag{5.124}$$

Let us multiply each member of (5.123) by $\boldsymbol{R}\left(\boldsymbol{X}'\boldsymbol{X}\right)^{-1}$:

$$\boldsymbol{R}\hat{\boldsymbol{\beta}}_0-\boldsymbol{R}\left(\boldsymbol{X}'\boldsymbol{X}\right)^{-1}\boldsymbol{X}'\boldsymbol{Y}-\boldsymbol{R}\left(\boldsymbol{X}'\boldsymbol{X}\right)^{-1}\boldsymbol{R}'\boldsymbol{\lambda}=0 \tag{5.125}$$

Hence:

$$\boldsymbol{\lambda}=\left(\boldsymbol{R}\left(\boldsymbol{X}'\boldsymbol{X}\right)^{-1}\boldsymbol{R}'\right)^{-1}\left(\boldsymbol{r}-\boldsymbol{R}\hat{\boldsymbol{\beta}}\right) \tag{5.126}$$

with $\hat{\boldsymbol{\beta}} = \left(\boldsymbol{X}'\boldsymbol{X}\right)^{-1} \boldsymbol{X}'\boldsymbol{Y}$ denoting the OLS estimator of the unconstrained model. It is then sufficient to replace $\boldsymbol{\lambda}$ by its value in (5.123):

$$\hat{\boldsymbol{\beta}}_0 = \left(\boldsymbol{X}'\boldsymbol{X}\right)^{-1} \boldsymbol{X}'\boldsymbol{Y} + \left(\boldsymbol{X}'\boldsymbol{X}\right)^{-1} \boldsymbol{R}' \left(\boldsymbol{R}\left(\boldsymbol{X}'\boldsymbol{X}\right)^{-1} \boldsymbol{R}'\right)^{-1} \left(\boldsymbol{r} - \boldsymbol{R}\hat{\boldsymbol{\beta}}\right) \tag{5.127}$$

Hence:

$$\hat{\boldsymbol{\beta}}_0 = \hat{\boldsymbol{\beta}} + \left(\boldsymbol{X}'\boldsymbol{X}\right)^{-1} \boldsymbol{R}' \left(\boldsymbol{R}\left(\boldsymbol{X}'\boldsymbol{X}\right)^{-1} \boldsymbol{R}'\right)^{-1} \left(\boldsymbol{r} - \boldsymbol{R}\hat{\boldsymbol{\beta}}\right) \tag{5.128}$$

which defines the constrained least squares estimator.

6 Distributed Lag Models

In the previous chapters, we essentially considered models in which the variables were all expressed at the same instant of time. However, it is common for models to include lagged variables, i.e., variables that are not all expressed at the same period. These are known as **dynamic models**. There are two main categories:

- Models including present and lagged values of explanatory variables; these are **distributed lag models**.
- Models in which the lagged values of the dependent variable intervene among the explanatory variables; in this case, we speak of **autoregressive models**.[1]

This chapter proposes a study of the first category of models. Autoregressive models will be treated in depth in the following chapter dealing with time series models. We have thus chosen to divide the presentation of dynamic models into two chapters, the distinction residing in whether or not the lagged dependent variable is among the explanatory variables.

6.1 Why Introduce Lags? Some Examples

In economics, the present value of the dependent variable often depends on the past values of the explanatory variables. In other words, the influence of the explanatory variables is only exerted after a certain **lag**. Let us take a few examples to illustrate this.

[1] It is possible to introduce a nuance in the terminology. We generally speak of autoregressive models when only the lagged values of the dependent variable are present as explanatory variables. We speak of **autoregressive distributed lag (ARDL) models** when the lagged values of the dependent variable are among the explanatory variables in addition to the lagged values of the usual explanatory variables.

V. Mignon, *Principles of Econometrics*, Classroom Companion: Economics,
https://doi.org/10.1007/978-3-031-52535-3_6

Consider, as a first example, the **consumption function**. A simple way of illustrating the consideration of lags is to refer to Duesenberry's (1949) ratchet effect. According to this approach, consumption depends on income of the same period, but also on the highest income achieved in the past. This introduces an irreversibility of consumption decisions over time, in the sense that the attainment of a higher income permanently modifies consumption habits. Another way of taking into account the influence of the past is to explain consumption not only by income in the same period, but also by lagged consumption. Such a formulation, widely used in econometric studies on consumption, makes it possible to model consumption habits. We can write, noting income R and consumption C:[2]

$$C_t = \alpha + \beta_1 R_t + \phi_1 C_{t-1} \tag{6.1}$$

By replacing C_{t-1} with:

$$C_{t-1} = \alpha + \beta_1 R_{t-1} + \phi_1 C_{t-2} \tag{6.2}$$

we can write:

$$C_t = \alpha + \beta_1 R_t + \phi_1 \left(\alpha + \beta_1 R_{t-1} + \phi_1 C_{t-2}\right) \tag{6.3}$$

that is:

$$C_t = \alpha \left(1 + \phi_1\right) + \beta_1 \left(R_t + \phi_1 R_{t-1}\right) + \phi_1^2 C_{t-2} \tag{6.4}$$

Replacing C_{t-2} with $\alpha + \beta_1 R_{t-2} + \phi_1 C_{t-3}$ and so on, we get:

$$C_t = \beta_1 \sum_{i=0}^{\infty} \phi_1^i R_{t-i} + \frac{\alpha}{1 - \phi_1} \tag{6.5}$$

Thus, in addition to income of the current period, all past income has an influence on present consumption. The reaction of consumption to a change in income is therefore spread, i.e., staggered, over time. The slower the reaction, the closer the coefficient ϕ_1 is to 1. This coefficient represents the **degree of inertia** of consumption. The model (6.5) is a distributed lag model in the sense that the explanatory variable (income) has a distributed impact over time on the dependent variable (consumption).

Another possible illustration, again in the field of consumption, is provided by Friedman's (1957) permanent income model. According to this theory, consumption in a given period depends not just on income of the same period, but on all income anticipated in future periods. Since future incomes are unknown, they need to be

---

[2] We ignore the error term here to simplify the notations and calculations to follow.

approximated in order to estimate such a model. Friedman proposes to approximate permanent income by current income and all past income, with observed income assigned a decreasing weight over time. Under these conditions, an increase in an individual's permanent income affects consumption over time. In other words, an increase in income is not immediately reflected in consumption. A model describing such a situation can, for example, be written as:

$$C_t = \mu + \delta_0 R_t + \delta_1 R_{t-1} + \delta_2 R_{t-2} + \varepsilon_t \tag{6.6}$$

where R denotes income and C consumption. In this model, the present and lagged values of one and two periods of income are involved in explaining present consumption, meaning that an increase in income is spread, or distributed, over three periods. The model (6.6) is called a distributed lag model because the explanatory variable exerts a time-distributed influence on the dependent variable.

A second example illustrating the spread over time of the influence of explanatory variables is given by the **investment function**. In line with the accelerator model, investment reacts immediately to changes in demand, i.e.:

$$I_t = \nu \Delta Y_t \tag{6.7}$$

where I_t denotes investment at date t and $\Delta Y_t = Y_t - Y_{t-1}$ represents the change in output perceived as the variation in demand, ν being the acceleration coefficient. In line with this formulation, a change in demand generates an immediate increase in investment: there is no lag between the change in demand and the reaction of investment. Such a formulation is too restrictive in the sense that it leads to too abrupt variations in investment, and that there are lags in the adjustment of investment to changes in demand. These limitations led to the flexible accelerator model in which the capital stock K is linked to a weighted average of current and past output, with the weight assigned to past output decreasing over time:

$$K_t = \phi (1 - \lambda) \left(Y_t + \lambda Y_{t-1} + \lambda^2 Y_{t-2} + \ldots + \lambda^h Y_{t-h} + \ldots \right) \tag{6.8}$$

where the weight λ is between 0 and 1. After a few simple calculations[3] and remembering that investment is equal to the change in the capital stock ($I_t = K_t - K_{t-1}$), the accelerator model can be written:

$$I_t = \lambda \nu \sum_{i=0}^{\infty} (1 - \lambda)^i \, \Delta Y_{t-i} \tag{6.9}$$

[3] See classic textbooks on macroeconomics or economic dynamics, for example, Blanchard and Fischer (1989) and Dowrick et al. (2008).

This shows that investment reacts in a distributed way to changes in demand, not adjusting immediately as was the case in the simple accelerator model. It is therefore a distributed lag model.

These examples illustrate that a variety of factors can justify the existence of lags and the use of distributed lag models. Lags can have a number of causes, including but not limited to:

- The existence of memory or **inertia** phenomena. To take the example of consumption, agents do not immediately modify their consumption following an increase in income. There is inertia due, for example, to consumption habits.
- Technological or technical reasons. An increase in capital expenditure may have staggered effects on investment, due in particular to the existence of production delays. Similarly, the reaction of a variable to an economic policy is often spread over several periods or only appears after a certain time lag.
- Institutional or political reasons. As an example, certain contractual obligations may contribute to the occurrence of lags.
- One of the main reasons lies in expectations. Variables are often a function of agents' expectations, which are themselves frequently based on the past.

6.2 General Formulation and Definitions of Distributed Lag Models

Noting h the number of lags, a **distributed lag model** is written generally as follows:

$$Y_t = \mu + \delta_0 X_t + \delta_1 X_{t-1} + \ldots + \delta_h X_{t-h} + \varepsilon_t \tag{6.10}$$

The number of lags h can be finite or infinite. An infinite lag model is used when the lagged effects of the explanatory variables are likely to be very long-lasting. Finite lag models are preferred when the effect of a change in X no longer has an influence on Y after a relatively small number of periods.

To simplify the notations, let us introduce the **lag operator** L such that:

$$LX_t = X_{t-1} \tag{6.11}$$

The lag operator thus transforms a variable into its past value. More generally, we have:

$$L^i X_t = X_{t-i} \tag{6.12}$$

Let us define the **lag polynomial** $D(L)$ of degree h such that:

$$D(L) = \delta_0 + \delta_1 L + \ldots + \delta_h L^h \tag{6.13}$$

The distributed lag model (6.10) is then written as:

$$Y_t = \mu + D(L)X_t + \varepsilon_t \tag{6.14}$$

The coefficient δ_0 measures the variation of Y_t following the variation of X_t:

$$\delta_0 = \frac{\Delta Y_t}{\Delta X_t} \tag{6.15}$$

δ_0 is called the **short-term multiplier** or impact multiplier of X. The partial sums of the coefficients $\delta_i, i = 1, \ldots, h$, define the **cumulative multipliers**. Thus, the cumulative effect τ periods after a shock occurring at period t is given by $\sum_{i=0}^{\tau} \delta_i$.

The polynomial:

$$D(1) = \delta_0 + \delta_1 + \ldots + \delta_h \tag{6.16}$$

equal to the sum of all coefficients $\delta_i, i = 1, \ldots, h$, measures the effect, in the long term, of a variation in X on the value of Y. $D(1)$ is called the **long-term multiplier** or equilibrium multiplier.

It is possible to normalize the coefficients $\delta_i, i = 1, \ldots, h$, by dividing them by their sum $D(1)$. The partial sums of these normalized δ_i coefficients measure the proportion of the total effect of a change in X reached after a certain period.

Let us consider a numerical example to illustrate this. Consider the model (6.6), by giving values to the coefficients:

$$C_t = \mu + 0.4R_t + 0.2R_{t-1} + 0.1R_{t-2} \tag{6.17}$$

The short-term multiplier is 0.4: following a one-unit increase in income, individuals increase their consumption in the same period by 0.4 units. The long-term multiplier is $0.4+0.2+0.1 = 0.7$: following a one-unit increase in income, the individual increases consumption by 0.4 units in the same period, by 0.2 units in the following period, and by 0.1 units in the period after that. In the long term, the total effect of a one-unit increase in income is an increase in consumption of 0.7 units. Let us now calculate the standardized coefficients, dividing each coefficient by 0.7. We obtain 0.57, 0.29, and 0.14, respectively. This means that 57% of the total effect of a change in income is felt in the same period, 86% after one period, and 100% after two periods.

Another useful concept is that of **median lag**: this is the number of periods required for 50% of the total effect to be reached. The notion of **mean lag**, on the other hand, allows us to grasp the time period corresponding to the mean value of the coefficients[4] $\delta_i, i = 1, \ldots, h$. It is defined by the weighted average of the

[4] The concepts of median and mean lags only really make sense if the coefficients are of the same sign.

coefficients, i.e.:

$$\bar{D} = \frac{\sum_{i=0}^{h} i\delta_i}{\sum_{i=0}^{h} \delta_i} = \frac{\delta_1 + 2\delta_2 + \ldots + h\delta_h}{\delta_0 + \delta_1 + \delta_2 + \ldots + \delta_h} = \frac{D'(1)}{D(1)} \tag{6.18}$$

where D' denotes the derivative of D.

6.3 Determination of the Number of Lags and Estimation

6.3.1 Determination of the Number of Lags

There are several procedures for determining the number of lags h in a distributed lag model:

$$Y_t = \mu + \delta_0 X_t + \delta_1 X_{t-1} + \ldots + \delta_h X_{t-h} + \varepsilon_t \tag{6.19}$$

- A first technique is to perform significance tests on the coefficients. For example, we can perform a Fisher test, testing the nullity of coefficients associated with lags of order greater than h.
- A second technique relies on the use of various criteria: the adjusted coefficient of determination, the Akaike information criterion (AIC), Schwarz information criterion (SIC), Hannan-Quinn information criterion (HQ), etc. We select the value of h that maximizes the adjusted coefficient of determination or the one that minimizes the AIC, SIC, and HQ criteria:

$$AIC(h) = \log \frac{RSS_h}{T} + \frac{2h}{T} \tag{6.20}$$

$$SIC(h) = \log \frac{RSS_h}{T} + \frac{h \log T}{T} \tag{6.21}$$

$$HQ(h) = \log \frac{RSS_h}{T} + 2\frac{h \log(\log T)}{T} \tag{6.22}$$

where RSS_h denotes the sum of squared residuals of the model with h lags and T is the number of observations.[5]

[5] It has been assumed here that the constant c is equal to 1 in the expression of the HQ criterion.

Of course, each technique has its advantages and disadvantages. In particular, we know that the AIC criterion tends to overestimate the value of h, while the SIC criterion is more parsimonious.

6.3.2 The Question of Estimating Distributed Lag Models

In addition to determining the number of lags, estimating a distributed lag model poses a second problem. It is theoretically possible to estimate such a model by OLS if the explanatory variable is assumed to be non-random. However, the greater the number of lags, the higher the risk of multicollinearity between the lagged explanatory variables. Under these conditions, it is known that the estimation of the coefficients is imprecise, as coefficient standard deviations tend to be too high.

To overcome this limitation, assumptions are made about the structure of the lags in order to reduce the number of parameters to be estimated. A distinction is made between models with a finite number of distributed lags and models with an infinite number of distributed lags.

6.4 Finite Distributed Lag Models: Almon Lag Models

Finite distributed lag models are **polynomial distributed lag (PDL) models**, also known as **Almon lag models** (see Almon, 1962).

Almon's technique avoids directly estimating the coefficients δ_i, since it consists in assuming that the true lag distribution can be approximated by a polynomial of order q:

$$\delta_i = \alpha_0 + \alpha_1 i + \alpha_2 i^2 + \ldots + \alpha_q i^q = \sum_{j=0}^{q} \alpha_j i^j \tag{6.23}$$

with $h > q$.

Consider, as an example, that the polynomial is of second order ($q = 2$). Then we have:

- $\delta_0 = \alpha_0$
- $\delta_1 = \alpha_0 + \alpha_1 + \alpha_2$
- $\delta_2 = \alpha_0 + 2\alpha_1 + 4\alpha_2$
- $\delta_3 = \alpha_0 + 3\alpha_1 + 9\alpha_2$
- $\ldots$
- $\delta_h = \alpha_0 + h\alpha_1 + h^2\alpha_2$

Let us plug these values into (6.19):

$$Y_t = \mu + \alpha_0 X_t + (\alpha_0 + \alpha_1 + \alpha_2) X_{t-1} + \ldots + \left(\alpha_0 + h\alpha_1 + h^2\alpha_2\right) X_{t-h} + \varepsilon_t \tag{6.24}$$

that is:

$$\begin{aligned} Y_t = \mu &+ \alpha_0 (X_t + X_{t-1} + \ldots + X_{t-h}) \\ &+ \alpha_1 (X_{t-1} + 2X_{t-2} + \ldots + hX_{t-h}) \\ &+ \alpha_2 \left(X_{t-1} + 4X_{t-2} + \ldots + h^2 X_{t-h}\right) + \varepsilon_t \end{aligned} \tag{6.25}$$

The "new" explanatory variables are linear combinations of the lagged explanatory variables. Thus, a regression of Y on these "new" explanatory variables yields estimates of the coefficients α, which, in turn, allows us to determine the coefficients δ.

More generally, in matrix form, we can write for h lags and a polynomial of degree q:

$$\begin{pmatrix} \delta_0 \\ \delta_1 \\ \delta_2 \\ \vdots \\ \delta_h \end{pmatrix} = \begin{pmatrix} 1 & 0 & 0 & \cdots\cdots & 0 \\ 1 & 1 & 1 & \cdots\cdots & 1 \\ 1 & 2 & 2^2 & \cdots\cdots & 2^q \\ \vdots & & & & \\ 1 & h & h^2 & \cdots\cdots & h^q \end{pmatrix} \begin{pmatrix} \alpha_0 \\ \alpha_1 \\ \alpha_2 \\ \vdots \\ \alpha_h \end{pmatrix} \tag{6.26}$$

Let us note $\boldsymbol{W}$ the matrix:

$$\boldsymbol{W} = \begin{pmatrix} 1 & 0 & 0 & \cdots\cdots & 0 \\ 1 & 1 & 1 & \cdots\cdots & 1 \\ 1 & 2 & 2^2 & \cdots\cdots & 2^q \\ \vdots & & & & \\ 1 & h & h^2 & \cdots\cdots & h^q \end{pmatrix} \tag{6.27}$$

The matrix form of (6.19) being given by:

$$\boldsymbol{Y} = \boldsymbol{I}\mu + \boldsymbol{X}\boldsymbol{\delta} + \boldsymbol{\varepsilon} \tag{6.28}$$

we can write:

$$\boldsymbol{Y} = \boldsymbol{I}\mu + \boldsymbol{X}\boldsymbol{W}\boldsymbol{\alpha} + \boldsymbol{\varepsilon} \tag{6.29}$$

where $\boldsymbol{\delta} = \begin{pmatrix} \delta_0 \\ \delta_1 \\ \delta_2 \\ \vdots \\ \delta_h \end{pmatrix}$ and $\boldsymbol{\alpha} = \begin{pmatrix} \alpha_0 \\ \alpha_1 \\ \alpha_2 \\ \vdots \\ \alpha_h \end{pmatrix}$.

It is then possible to estimate the regression (6.29) by OLS to obtain the estimator $\hat{\boldsymbol{\alpha}}$ of $\boldsymbol{\alpha}$ and to deduce the estimator $\hat{\boldsymbol{\delta}}$ of $\boldsymbol{\delta}$ from (6.26).

The method just described assumes that the degree q of the polynomial used for the approximation is known. In practice, this is not the case and q needs to be determined. One possible technique is to start with a high value, $q = h - 1$, and test the significance of the associated coefficient (α_{h-1}) by means of a t-test. The degree of the polynomial is then progressively reduced until a significant coefficient appears.

6.5 Infinite Distributed Lag Models

In infinite distributed lag models, the effect of the explanatory variable is unlimited in time. It is assumed, however, that the recent past has more influence than the distant past, and that the weight of past observations tends to decrease steadily over time.

Generally speaking, an **infinite distributed lag model** is written as:

$$Y_t = \mu + \sum_{i=0}^{\infty} \delta_i X_{t-i} + \varepsilon_t \tag{6.30}$$

or:

$$Y_t = \mu + \sum_{i=0}^{\infty} \delta_i L^i X_t + \varepsilon_t \tag{6.31}$$

In order to estimate the model, it is necessary to reduce it to a model with a finite number of parameters to be estimated. To this end, a particular form is imposed on the structure of the coefficients δ_i. The two most commonly used forms are based on the **Koyck approach** and the **Pascal approach**.

6.5.1 The Koyck Approach

The Koyck Transformation

Under the assumption that the coefficients δ_i are of the same sign, Koyck (1954) assumes that the lags decrease geometrically:

$$\delta_i = \lambda^i \delta_0 \tag{6.32}$$

where $i = 0, 1, 2, \ldots$ and $0 < \lambda < 1$.

Since $\lambda < 1$, the relationship (6.32) expresses the fact that the coefficients δ_i decrease as we move further into the past: more recent observations are assigned higher weights than past observations. The closer λ is to 1, the slower the rate of decrease of the coefficients, and the closer λ is to 0, the faster that rate.

Substituting (6.32) into (6.30), we have:

$$Y_t = \mu + \delta_0 X_t + \lambda\delta_0 X_{t-1} + \lambda^2\delta_0 X_{t-2} + \ldots + \lambda^i\delta_0 X_{t-i} + \ldots + \varepsilon_t \tag{6.33}$$

or:

$$Y_t = \mu + \delta_0 \left(X_t + \lambda X_{t-1} + \lambda^2 X_{t-2} + \ldots + \lambda^i X_{t-i} + \ldots \right) + \varepsilon_t \tag{6.34}$$

The associated polynomial $D(L)$ is written as:

$$D(L) = \delta_0 + \lambda\delta_0 L + \lambda^2\delta_0 L^2 + \ldots + \lambda^i\delta_0 L^i + \ldots \tag{6.35}$$

We can rewrite (6.34) as follows:

$$Y_t = \mu + D(L) X_t + \varepsilon_t \tag{6.36}$$

or:

$$D(L)^{-1} Y_t = D(L)^{-1}\mu + D(L)^{-1} D(L) X_t + D(L)^{-1}\varepsilon_t \tag{6.37}$$

Knowing that:

$$D(L) = \delta_0 \left(1 + \lambda L + \lambda^2 L^2 + \ldots + \lambda^i L^i + \ldots \right) \tag{6.38}$$

represents the sum of the terms of a geometric sequence, we have:

$$D(L) = \frac{\delta_0}{(1 - \lambda L)} \tag{6.39}$$

and therefore:

$$D(L)^{-1} = \frac{(1 - \lambda L)}{\delta_0} \tag{6.40}$$

Substituting into (6.37), we get:

$$(1 - \lambda L)\, Y_t = (1 - \lambda L)\, \mu + \delta_0 X_t + (1 - \lambda L)\, \varepsilon_t \tag{6.41}$$

That is:

$$Y_t - \lambda Y_{t-1} = (1 - \lambda)\, \mu + \delta_0 X_t + \varepsilon_t - \lambda\varepsilon_{t-1} \tag{6.42}$$

Hence:

$$Y_t = \lambda Y_{t-1} + (1 - \lambda)\,\mu + \delta_0 X_t + \varepsilon_t - \lambda\varepsilon_{t-1} \tag{6.43}$$

This gives an autoregressive model with autocorrelated errors of order 1. This transformation, from a distributed lag model (Eq. (6.30)) to an autoregressive model (Eq. (6.43)), is called the **Koyck transformation**. It significantly reduces the number of parameters to be estimated. Indeed, if we compare Eqs. (6.30) and (6.43), it appears that, instead of estimating the constant term μ and an infinite number of parameters δ_i, we now only need to estimate three parameters: the constant term μ, δ_0, and λ. The risk of multicollinearity is consequently greatly reduced, if not eliminated.

A few remarks are in order. Firstly, the Koyck transformation shows that we can move from a distributed lag model to an autoregressive model. The endogenous lagged variable, Y_{t-1}, now appears as an explanatory variable of Y_t, which has important implications in terms of estimation. We know that one of the basic assumptions of the OLS method is that the matrix of explanatory variables is non-random. Such an assumption is violated here since Y_{t-1}, like Y_t, is a random variable. However, this assumption can be reformulated by writing that the matrix of explanatory variables can contain random variables, provided that they are not correlated with the error term (see Chap. 3). It will therefore be necessary to check this characteristic during the estimation phase; we will return to this point when discussing estimation methods (see below).

Secondly, the error term of the model (6.43) is $\varepsilon_t - \lambda\varepsilon_{t-1}$, and no longer only ε_t as was the case in the original model (6.30). Let us posit $\eta_t = \varepsilon_t - \lambda\varepsilon_{t-1}$. It appears that while the ε_t are indeed non-autocorrelated, this is not the case for the η_t, a characteristic which must be taken into account during the estimation phase (see below).

Thirdly, it is possible to define median and mean lags in the Koyck approach, which makes it possible to quantify the speed with which the dependent variable Y_t responds to a unit variation in the explanatory variable X_t. The median lag corresponds to the number of periods required for 50% of the total effect of a unit change in the explanatory variable X_t on Y_t to be reached. It can be shown that, in the Koyck model, the median lag is given by $\log 2/\log\lambda$. Thus, the higher the value of λ, the greater the median lag and the lower the speed of adjustment. On the other hand, the mean lag is defined by:

$$\bar{D} = \frac{\sum\limits_{i=0}^{h} i\delta_i}{\sum\limits_{i=0}^{h} \delta_i} \tag{6.44}$$

or, in the case of the Koyck model:

$$\bar{D} = \frac{\lambda}{1-\lambda} \tag{6.45}$$

The median and mean lags can thus be used to assess the speed with which Y_t adjusts following a unit variation in X_t.

Estimation: The Instrumental Variables Method

As previously mentioned, it is not possible to apply the OLS method directly to estimate the Koyck model. This stems from two reasons:

- The lagged endogenous variable is among the explanatory variables, so the matrix of explanatory variables is not non-random.
- The error term $\eta_t = \varepsilon_t - \lambda\varepsilon_{t-1}$ of the Koyck model exhibits autocorrelation.

If we wish to apply OLS to the Koyck model, we need to ensure that the lagged endogenous variable Y_{t-1} is independent of the error term η_t. However, such an assumption does not hold. Indeed, in accordance with (6.43), ε_{t-1} has an impact on ε_t. Similarly, if we write Eq. (6.43) in $t-1$, it is clear that ε_{t-1} has an impact on Y_{t-1}. There is thus a link between ε_t and Y_{t-1}.

The consequence of this dependence between the lagged endogenous variable and the error term ε_t is that the OLS estimators are no longer consistent. In other words, even if the sample size grows indefinitely, the OLS estimators do not approach their true population values. We know that, in such a case, it is possible to use the **instrumental variables method** whose estimator is given by (see Chap. 5):

$$\hat{\boldsymbol{\beta}}_{IV} = \left(\boldsymbol{Z}'\boldsymbol{X}\right)^{-1}\boldsymbol{Z}'\boldsymbol{Y} \tag{6.46}$$

where $\boldsymbol{Z}$ is the instrument matrix.

In the case of the Koyck model (Eq. (6.43)):

$$Y_t = \lambda Y_{t-1} + (1-\lambda)\mu + \delta_0 X_t + \varepsilon_t - \lambda\varepsilon_{t-1} \tag{6.47}$$

only one instrument needs to be found, since only the variable Y_{t-1} needs to be instrumented (the variable X_t is indeed independent of the error term, by assumption). We frequently use X_{t-1} as the instrument of Y_{t-1}. We then have the following matrix $\boldsymbol{Z}$:

$$\boldsymbol{Z} = \begin{pmatrix} 1 & X_1 & X_0 \\ 1 & X_2 & X_1 \\ \vdots & \vdots & \vdots \\ 1 & X_T & X_{T-1} \end{pmatrix} \tag{6.48}$$

and the estimator of the instrumental variables is written:[6]

$$\hat{\boldsymbol{\beta}}_{IV} = \begin{pmatrix} T & \sum X_t & \sum Y_{t-1} \\ \sum X_t & \sum X_t^2 & \sum X_t Y_{t-1} \\ \sum X_{t-1} & \sum X_t X_{t-1} & \sum X_{t-1} Y_{t-1} \end{pmatrix}^{-1} \begin{pmatrix} \sum Y_t \\ \sum X_t Y_t \\ \sum X_{t-1} Y_t \end{pmatrix} \tag{6.49}$$

Remark 6.1 It is not always easy to find the "right" instrumental variables. In these circumstances, the instrumental variables method may be of limited practical interest, and it is preferable to resort to the maximum likelihood method. In the case of the Koyck model, the essential role of the method of instrumental variables is to obtain a consistent estimator of $\boldsymbol{\beta}$ to serve as the initial value of an iterative procedure, such as the maximum likelihood method.

Remark 6.2 (The Sargan Test) Sargan (1964) developed a **test of instrument validity**. The test can be described sequentially as follows:

- Split the variables appearing in the regression model into two groups: the group of variables independent of the error term (noted $X_1, X_2, \ldots, X_{k1}$) and the group of variables that are not independent of the error term (noted $W_1, W_2, \ldots, W_{k2}$).
- Note $Z_1, Z_2, \ldots, Z_{k3}$ the instruments chosen for the variables W, with $k3 \geq k2$.
- Estimate the parameters of the model by the instrumental variables method, i.e., $\hat{\boldsymbol{\beta}}_{IV} = \left(\boldsymbol{Z}'\boldsymbol{X}\right)^{-1} \boldsymbol{Z}'\boldsymbol{Y}$, and deduce the estimated series of residuals e_t.
- Regress the residuals e_t on a constant, the variables X and the variables Z. Determine the coefficient of determination R^2 of the estimated regression.
- Calculate the Sargan test statistic:

$$S = (T - k - 1)R^2 \tag{6.50}$$

  where T is the number of observations and k is the number of variables in the original regression model.
- Under the null hypothesis of validity of all instruments, the statistic S follows a Chi-squared distribution with r degrees of freedom, where $r = k3 - k2$. If the calculated value of the statistic S is less than the theoretical Chi-squared value, the null hypothesis of instrument validity is not rejected. If the calculated value of the statistic S is greater than the theoretical Chi-squared value, the null hypothesis is rejected, meaning that at least one instrument is not valid in the sense that it is not independent of the error term. In the latter case, the estimators of the instrumental variables are not valid.

---

[6] The sums run from 1 to T.

The Partial Adjustment Model

The **partial adjustment model** is an example of an application of the Koyck model (see in particular Nerlove, 1958). This model includes lagged endogenous variables among the explanatory variables. The underlying idea is that, due to the presence of rigidities or various constraints, the dependent variable cannot reach the desired value in a single period. In other words, the adjustment to the desired value takes some time. Generally speaking, the partial adjustment model is written:

$$Y_t^* = \alpha + \beta X_t + \varepsilon_t \tag{6.51}$$

where Y_t^* denotes the desired level of the dependent variable Y_t and X_t is an explanatory variable. As the variable Y_t^* is unobservable, we express it as a function of Y_t by using a partial adjustment mechanism of the type:

$$Y_t - Y_{t-1} = \lambda \left(Y_t^* - Y_{t-1}\right) \tag{6.52}$$

where $0 \leq \lambda \leq 1$ is called the **adjustment coefficient**. The variation $(Y_t - Y_{t-1})$ corresponds to the observed variation, $\left(Y_t^* - Y_{t-1}\right)$ being the desired variation. Substituting (6.51) into (6.52) gives:

$$Y_t - Y_{t-1} = \lambda \left(\alpha + \beta X_t + \varepsilon_t - Y_{t-1}\right) \tag{6.53}$$

that is:

$$Y_t = (1 - \lambda) Y_{t-1} + \lambda\alpha + \lambda\beta X_t + \lambda\varepsilon_t \tag{6.54}$$

This partial adjustment model has a similar structure to the Koyck model, the error term being simpler since it is only multiplied by the constant λ.

The Adaptive Expectations Model

The **adaptive expectations model** is another example of an application of the Koyck model. In this type of model, the values of the explained variable are a function, not of the observed values of the explanatory variables, but of the anticipated or expected values. Generally speaking, we can write an adaptive expectations model as follows:

$$Y_t = \alpha + \beta X_t^* + \varepsilon_t \tag{6.55}$$

where X_t^* denotes the expected value of the explanatory variable X_t. As the variable X_t^* is generally not directly observable, we assume an adaptive training process for expectations of the type:

$$X_t^* - X_{t-1}^* = \lambda \left(X_t - X_{t-1}^*\right) \tag{6.56}$$

with $0 \leq \lambda \leq 1$, λ is called the **expectation coefficient**. If $\lambda = 0$, then $X_t^* = X_{t-1}^*$, which means that expectations remain identical from period to period (static expectations). If $\lambda = 1$, then $X_t^* = X_t$, which implies that the anticipated value is equal to the observed value (naive expectations).

In line with the adaptive expectations hypothesis, expectations are revised each period according to the information provided by the last value actually taken by the variable. Low values of λ indicate large adjustments in expectations, while high values imply slow changes.

We can rewrite (6.56) as follows:

$$X_t^* = \lambda X_t + (1 - \lambda) X_{t-1}^* \tag{6.57}$$

Substituting (6.57) into (6.55), we obtain:

$$Y_t = \alpha + \beta \left[\lambda X_t + (1 - \lambda) X_{t-1}^*\right] + \varepsilon_t \tag{6.58}$$

This model can be reduced to a Koyck model. Let us write the model (6.55) in $(t - 1)$ and multiply each member by $(1 - \lambda)$. This gives us:

$$(1 - \lambda) Y_{t-1} = (1 - \lambda) \alpha + (1 - \lambda) \beta X_{t-1}^* + (1 - \lambda) \varepsilon_{t-1} \tag{6.59}$$

By subtracting Eqs. (6.58) and (6.59), we get:

$$Y_t - (1 - \lambda) Y_{t-1} = \alpha - (1 - \lambda) \alpha + \lambda\beta X_t + \varepsilon_t - (1 - \lambda) \varepsilon_{t-1} \tag{6.60}$$

that is:

$$Y_t = \lambda\alpha + \lambda\beta X_t + (1 - \lambda) Y_{t-1} + \varepsilon_t - (1 - \lambda) \varepsilon_{t-1} \tag{6.61}$$

This gives us a structure similar to that of the Koyck model.

Remark 6.3 It is possible to combine partial adjustment and adaptive expectations models. The dependent variable is then the desired level of the variable Y_t, the explanatory variable being the expected value of the variable X_t. The result is a model in which the endogenous variable lagged by one period, but also by two periods, is included among the explanatory variables. An economic illustration of such a model is provided by Friedman's permanent income model.

6.5.2 The Pascal Approach

The Pascal approach is another technique aimed at imposing a particular form on the structure of the coefficients δ_i in order to obtain a model with a finite number of parameters to be estimated. Such an approach was adopted by Solow (1960) and makes it possible to account for a distribution such that the coefficients are initially low, increase until they reach a maximum, and then decrease (a kind of bell curve). With this approach, the coefficients δ_i are distributed as follows:

$$\delta_i = (1 - \lambda)^{r+1} C_{r+i}^{i} \lambda^i \tag{6.62}$$

where C_{r+i}^{i} is the coefficient of Newton's binomial, $0 \leq \lambda \leq 1$ and $r \in N$.

The Pascal approach is a generalization of the Koyck approach. If we posit $r = 0$, we find the geometric distribution of Koyck.

Using Eq. (6.30), the distributed lag model is expressed as follows:

$$Y_t = \mu + \sum_{i=0}^{\infty} (1 - \lambda)^{r+1} C_{r+i}^{i} \lambda^i X_{t-i} + \varepsilon_t \tag{6.63}$$

The associated $D(L)$ polynomial is written as:

$$D(L) = (1 - \lambda)^{r+1} \sum_{i=0}^{\infty} C_{r+i}^{i} \lambda^i L^i \tag{6.64}$$

which can also be expressed as:

$$D(L) = \frac{\delta_0}{(1 - \lambda L)^{r+1}} \tag{6.65}$$

The model (6.63) becomes:

$$Y_t = \mu + D(L) X_t + \varepsilon_t \tag{6.66}$$

or:

$$D(L)^{-1} Y_t = D(L)^{-1} \mu + D(L)^{-1} D(L) X_t + D(L)^{-1} \varepsilon_t \tag{6.67}$$

- For $r = 0$, then we have:

$$D(L) = \frac{\delta_0}{(1 - \lambda L)} \tag{6.68}$$

  and we find the Koyck model.
- For $r = 1$, we have:

$$D(L) = \frac{\delta_0}{(1 - \lambda L)^2} \tag{6.69}$$

  or:

$$D(L)^{-1} = \frac{(1 - \lambda L)^2}{\delta_0} \tag{6.70}$$

Substituting in (6.67), we get:

$$(1-\lambda L)^2 Y_t = (1-\lambda L)^2 \mu + \delta_0 X_t + (1-\lambda L)^2 \varepsilon_t \tag{6.71}$$

Noting that $(1-\lambda L)^2 = \left(1 - 2\lambda L + \lambda^2 L^2\right)$, we get:

$$Y_t = 2\lambda Y_{t-1} - \lambda^2 Y_{t-2} + \left(1 - 2\lambda + \lambda^2\right)\mu + \delta_0 X_t + \varepsilon_t - 2\lambda\varepsilon_{t-1} + \lambda^2\varepsilon_{t-2} \tag{6.72}$$

which corresponds to a second-order autoregressive model.

- For $r = 2$, we have:

$$D(L)^{-1} = \frac{(1-\lambda L)^3}{\delta_0} \tag{6.73}$$

Substituting in (6.67), we get:

$$\begin{aligned} Y_t = {} & 3\lambda Y_{t-1} - 3\lambda^2 Y_{t-2} + \lambda^3 Y_{t-3} + \left(1 - 3\lambda + 3\lambda^2 - \lambda^3\right)\mu \\ & + \delta_0 X_t + \varepsilon_t - 3\lambda\varepsilon_{t-1} + 3\lambda^2\varepsilon_{t-2} - \lambda^3\varepsilon_{t-3} \end{aligned} \tag{6.74}$$

which corresponds to an autoregressive model of order 3.

Generally speaking, the autoregressive form associated with the distributed lag model in which the coefficients are distributed according to (6.62) has $(r+1)$ lagged endogenous variables whose associated coefficients are a function of λ.

Remark 6.4 In order to determine the value of r, Maddala and Rao (1971) suggest adopting a sweeping approach: we give ourselves a set of possible values for r and select the value that maximizes the adjusted coefficient of determination.

6.6 Autoregressive Distributed Lag Models

6.6.1 Writing the ARDL Model

In **autoregressive distributed lag (ARDL) models**, the lagged values of the dependent variable are added to the present and past values of the "usual" explanatory variables in the set of explanatory variables.[7]

Generally speaking, an autoregressive distributed lag model is written:

$$Y_t = \mu + \phi_1 Y_{t-1} + \ldots + \phi_p Y_{t-p} + \delta_0 X_t + \delta_1 X_{t-1} + \ldots + \delta_h X_{t-h} + \varepsilon_t \tag{6.75}$$

---

[7] We will not deal in detail with ARDL models in this book. For a more exhaustive presentation, readers can refer to Greene (2020).

that is:

$$Y_t = \mu + \sum_{i=1}^{p} \phi_i Y_{t-i} + \sum_{j=0}^{h} \delta_j X_{t-j} + \varepsilon_t \tag{6.76}$$

where ε_t is a non-autocorrelated homoskedastic process. By introducing the lag operator L, we can write:

$$\Phi(L) Y_t = \mu + D(L) X_t + \varepsilon_t \tag{6.77}$$

with $\Phi(L) = 1 - \phi_1 L - \phi_2 L^2 - \ldots - \phi_p L^p$ and $D(L) = \delta_0 + \delta_1 L + \ldots + \delta_h L^h$.

Such an autoregressive distributed lag model is denoted $ARDL(p, h)$. We observe that the Koyck model is a special case of the $ARDL(p, h)$ model in which $p = 1$ and $h = 0$.

$ARDL(p, h)$ models can be estimated by the OLS method as long as the error term ε_t is assumed to have the "good" statistical properties. Because of this characteristic, the OLS estimator is an efficient estimator.

6.6.2 Calculation of ARDL Model Weights

Let us write the distributed lag form of the ARDL model (6.77). To do this, divide each term of (6.77) by the autoregressive lag polynomial $\Phi(L)$:

$$Y_t = \frac{\mu}{\Phi(L)} + \frac{D(L)}{\Phi(L)} X_t + \frac{\varepsilon_t}{\Phi(L)} \tag{6.78}$$

which can also be written:

$$Y_t = \frac{\mu}{1 - \phi_1 - \ldots - \phi_p} + \sum_{j=0}^{\infty} \alpha_j X_{t-j} + \sum_{l=0}^{\infty} \theta_l \varepsilon_{t-l} \tag{6.79}$$

where the coefficients α_j, $j = 0, 1, \ldots, \infty$, are the terms associated with the ratio of the polynomials $D(L)$ and $\Phi(L)$. Thus, α_0 is the coefficient of 1 in $\frac{D(L)}{\Phi(L)}$, α_1 is the coefficient of L in $\frac{D(L)}{\Phi(L)}$, α_2 is the coefficient of L^2 in $\frac{D(L)}{\Phi(L)}$, and so on. Similarly, the coefficients θ_l, $l = 0, 1, \ldots, \infty$, are the terms associated with the ratio $\frac{1}{\Phi(L)}$.

The model (6.79) has a very general lag structure and is referred to as a **rational lag model** by Jorgenson (1966). The long-term multiplier associated with such a model is given by:

$$\sum_{j=0}^{\infty} \alpha_j = \frac{D(1)}{\Phi(1)} \tag{6.80}$$

6.7 Empirical Application

Consider the following two series:

- The returns of the Hang Seng Index of the Hong Kong Stock Exchange: RHK
- The returns of the Japanese index $NIKKEI$ 225: $RNIKKEI$

The data are weekly and cover the period from the week of December 1, 1969, to that of July 5, 2021, i.e., a number of observations $T = 2\,693$ (data source: Macrobond). Suppose we wish to explain the returns of the Hang Seng Index by the present and lagged returns of the Japanese index. The dependent variable is therefore RHK and the explanatory variables are the present and lagged values of $RNIKKEI$. We seek to estimate the following distributed lag model:

$$RHK_t = \mu + \delta_0 RNIKKEI_t + \delta_1 RNIKKEI_{t-1} + \ldots + \delta_h RNIKKEI_{t-h} + \varepsilon_t \tag{6.81}$$

Let us start by determining the number of lags to take into account. To do this, we estimate the model (6.81) for various values of h and select the one that minimizes the information criteria. Table 6.1 shows the values taken by the three criteria AIC, SIC, and Hannan-Quinn (HQ) for values of h ranging from 1 to 6. These results lead us to select a number of lags h equal to 1 according to the SIC and HQ criteria and 2 for the AIC criterion. For reasons of parsimony, and given that two out of three criteria favor a number of lags equal to 1, we choose $h = 1$.[8]

Let us assume a geometric distribution for the lags (Koyck model). We thus seek to estimate the following model:

$$RHK_t = \lambda RHK_{t-1} + (1 - \lambda)\,\mu + \delta_0 RNIKKEI_t + \varepsilon_t - \lambda\varepsilon_{t-1} \tag{6.82}$$

Since errors are autocorrelated in this model, we estimate it by applying the Newey-West correction (see Chap. 4). The results obtained are shown in Table 6.2.

Table 6.1 Determining the number of lags

h	AIC	SIC	HQ
1	−3.7672	**−3.7607**	**−3.7649**
2	**−3.7673**	−3.7585	−3.7641
3	−3.7666	−3.7556	−3.7626
4	−3.7655	−3.7524	−3.7608
5	−3.7649	−3.7495	−3.7593
6	−3.7656	−3.7480	−3.7593

Values in bold correspond to values minimizing information criteria

[8] Note further that the values taken by the AIC criterion for $h = 1$ and $h = 2$ are almost identical.

Table 6.2 OLS model estimation with the Newey-West correction

Dependent variable: RHK				
Variable	Coefficient	Std. error	t-Statistic	Prob.
C	0.001262	0.000742	1.700514	0.0891
RHK(-1)	0.113356	0.024923	4.548237	0.0000
RNIKKEI	0.484588	0.034404	14.08505	0.0000
R-squared	0.126952	Mean dependent var	0.001940	
Adjusted R-squared	0.126303	S.D. dependent var	0.039058	
S.E. of regression	0.036508	Akaike info criterion	−3.781467	
Sum squared resid	3.583954	Schwarz criterion	−3.774894	
Log likelihood	5092.855	Hannan-Quinn criterion	−3.779090	
F-statistic	195.5068	Durbin-Watson stat	1.998281	
Prob(F-statistic)	0.000000	Wald F-statistic	117.1121	
Prob(Wald F-statistic)	0.000000			

We have $\hat{\lambda} = 0.1134$. This value is small, which means that the decay rate of the coefficients of the distributed lag model is rapid. In other words, the influence of past values of $RNIKKEI$ on RHK decreases rapidly. The model can also be written as:

$$RHK_t = \mu + \delta_0 RNIKKEI_t + \lambda\delta_0 RNIKKEI_{t-1} + \lambda^2\delta_0 RNIKKEI_{t-2} + \ldots + \lambda^i\delta_0 RNIKKEI_{t-i} + \ldots + \varepsilon_t \tag{6.83}$$

Knowing that: $\left(1 - \hat{\lambda}\right)\hat{\mu} = 0.0013$, we deduce:

$$\hat{\mu} = \frac{0.0013}{1 - 0.1134} = 0.0014 \tag{6.84}$$

The estimation of the model (6.83) is therefore given by:

$$\widehat{RHK}_t = 0.0014 + 0.4846 RNIKKEI_t + 0.1134 \times 0.4846 RNIKKEI_{t-1} + 0.1134^2 \times 0.4846 RNIKKEI_{t-2} + \ldots \tag{6.85}$$

that is:

$$\widehat{RHK}_t = 0.0014 + 0.4846 RNIKKEI_t + 0.0549 RNIKKEI_{t-1} + 0.0062 RNIKKEI_{t-2} + \ldots \tag{6.86}$$

We can see that the value of the coefficients associated with the variable $RNIKKEI$ decreases rapidly as the number of lags increases. We can calculate the median lag, given by $\log 2/\log\hat{\lambda}$, i.e., 0.3184: following a unit variation of

$RNIKKEI$, 50% of the total variation of RHK is achieved in just over a day and a half. As the value of $\hat{\lambda}$ is small, so is the median lag, highlighting a rapid adjustment. It is also possible to calculate the mean lag:

$$\bar{D} = \frac{\hat{\lambda}}{1-\hat{\lambda}} = 0.1278 \tag{6.87}$$

The mean lag is around 0.13: it takes around half a day for the effect of a variation in $RNIKKEI$ to be reflected in RHK, which is rapid.

Conclusion

This chapter has introduced a first category of dynamic models: distributed lag models. There is a second category of dynamic models, generally referred to as time series models, in which the lagged endogenous variable is one of the explanatory variables. These are the subject of the next chapter, which presents the basics of time series econometrics.

The Gist of the Chapter

Distributed lag model	
Definition	$Y_t = \mu + \delta_0 X_t + \delta_1 X_{t-1} + \ldots + \delta_h X_{t-h} + \varepsilon_t$
Short-term multiplier	$\delta_0 = \frac{\Delta Y_t}{\Delta X_t}$
Long-term multiplier	$D(1) = \delta_0 + \delta_1 + \ldots + \delta_h$
Lag form	
Almon	$\delta_i = \alpha_0 + \alpha_1 i + \alpha_2 i^2 + \ldots + \alpha_q i^q = \sum_{j=0}^{q} \alpha_j i^j$, $h > q$
Koyck	$\delta_i = \lambda^i \delta_0$, $i = 0, 1, 2, \ldots$ and $0 < \lambda < 1$
Pascal	$\delta_i = (1-\lambda)^{r+1} C_{r+i}^{i} \lambda^i$, where $0 \leq \lambda \leq 1$ and $r \in N$
ARDL model	$Y_t = \mu + \sum_{i=1}^{p} \phi_i Y_{t-i} + \sum_{j=0}^{h} \delta_j X_{t-j} + \varepsilon_t$
Lag operator	$L^i X_t = X_{t-i}$

Further Reading

In addition to the references cited in the chapter, readers interested in distributed lag models can consult Nerlove (1958) and Griliches (1967). A detailed presentation can also be found in Davidson and MacKinnon (1993) and Gujarati et al. (2017).

An Introduction to Time Series Models 7

Time series econometrics is a branch of econometrics that has undergone many developments over the last 40 years.[1] We offer here an introduction to time series models. After laying down a number of definitions, we focus on the essential concept of stationarity. We present the Dickey-Fuller unit root test for testing the non-stationary nature of a time series. We then expose the basic models of time series – the autoregressive moving-average models (ARMA models) – and the related Box and Jenkins methodology. A multivariate extension is proposed through the presentation of VAR (vector autoregressive) models. Finally, we present the concepts of non-stationary time series econometrics by studying the notions of cointegration and error-correction models.

7.1 Some Definitions

7.1.1 Time Series

A **time series** is a sequence of real numbers, indexed by relative integers such as time. For each instant of time, the value of the quantity under study Y_t is called a **random variable**.[2] The set of values Y_t when t varies is called a **random process** (or **stochastic process**): $\{Y_t, t \in Z\}$. A time series is thus the realization of a random process.

An example of a time series is provided by Fig. 7.1 which represents Standard and Poor's 500 stock index (denoted SP) at monthly frequency over the period

[1] This chapter takes up a number of developments appearing in the work by Lardic and Mignon (2002), which interested readers may refer to for further details.

[2] There are continuous and discrete random variables. We are only interested in discrete variables here.

V. Mignon, *Principles of Econometrics*, Classroom Companion: Economics,
https://doi.org/10.1007/978-3-031-52535-3_7

Fig. 7.1 Standard and Poor's 500 stock index series, 1980.01–2021.06

Table 7.1 Standard and Poor's 500 stock index series

	SP
1980.01	386.0129
1980.02	395.7329
1980.03	353.9680
1980.04	344.3516
...	...
2021.03	3997.9635
2021.04	4199.2766
2021.05	4192.7108
2021.06	4246.8837

Data source: Robert Shiller's website (www.econ.yale.edu/~shiller)

from January 1980 to June 2021. The first and last values of this series are given in Table 7.1: for each month, we have a value of the stock market index.

As the class of random processes is very large, time series analysis initially focused on a particular class of processes: **stationary random processes**. These processes are characterized by the fact that their statistical properties do not change over time.

7.1.2 Second-Order Stationarity

The notion of stationarity of a time series was briefly discussed in the first chapter. We have seen that, when working with time series, it is necessary to study their characteristics in terms of stationarity before analyzing and attempting to model them. Here we present only the concept of **second-order stationarity** or **weak stationarity**, which is the notion of stationarity usually retained in time series econometrics.[3]

Definition 7.1 A process Y_t is second-order stationary if:

- (1) $E\left(Y_t^2\right) < \infty\ \forall t \in Z$.
- (2) $E\left(Y_t\right) = m\ \forall t \in Z$.
- (3) $Cov\left(Y_t, Y_{t+h}\right) = \gamma_h,\ \forall t, h \in Z$, where γ is the autocovariance function of the process.

Condition (1) means that the process is of second order: second-order moments, such as variance, are finite and independent of time. Condition (2) means that the expectation of the process is constant over time (mean stationarity). Condition (3) reflects the fact that the covariance between two periods t and $t + h$ is solely a function of the time difference, h. Note that the variance $\sigma_Y^2 = Cov\left(Y_t, Y_t\right) = \gamma_0$ is also independent of time. The fact that the variance is constant over time reflects the property of homoskedasticity.

In the remainder of the chapter, the term stationary will refer to the concept of second-order stationarity.

7.1.3 Autocovariance Function, Autocorrelation Function, and Partial Autocorrelation Function

We have already mentioned the notions of autocovariance function and autocorrelation function in Chap. 4. Let us recall here the definitions of these central concepts for the study of time series.

Definition 7.2 Let Y_t be a random process with finite variance. The **autovariance function** γ_h of Y_t is defined as:

$$\gamma_h = Cov\left(Y_t, Y_{t+h}\right) = E\left[\left[Y_t - E\left(Y_t\right)\right]\left[Y_{t+h} - E\left(Y_{t+h}\right)\right]\right] \tag{7.1}$$

The autocovariance function measures the covariance between two values of the same series Y_t separated by a certain time h.

[3] For a more detailed study of stationarity and a definition of the various concepts, see in particular Lardic and Mignon (2002).

Theorem 7.1 *The autocovariance function of a stationary process Y_t has the following properties:*

- $\gamma_0 = Cov\left(Y_t, Y_t\right) = E\left[\left[Y_t - E\left(Y_t\right)\right]^2\right] = V\left(Y_t\right) = \sigma_Y^2 \geq 0$
- $|\gamma_h| \leq \gamma_0$
- $\gamma_h = \gamma_{-h}$: *the autocovariance function is an even function,* $Cov\left(Y_t, Y_{t+h}\right) = Cov\left(Y_t, Y_{t-h}\right)$.

Remark 7.1 We restrict ourselves here to the analysis of series in the time domain. However, it is possible to study a series in the spectral or frequency domain. The analog of the autocovariance function in the spectral domain is called the **spectral density**. This book does not deal with spectral analysis. Interested readers should refer to Hamilton (1994) or Greene (2020).

Definition 7.3 Let Y_t be a stationary process. The **autocorrelation function** ρ_h is defined as:

$$\rho_h = \frac{\gamma_h}{\gamma_0}, h \in Z \tag{7.2}$$

The autocorrelation function measures the temporal links between the various components of the series Y_t. Specifically:

$$\rho_h = \frac{Cov\left(Y_t, Y_{t+h}\right)}{\sigma_{Y_t}\sigma_{Y_{t+h}}} = \frac{\gamma_h}{\sqrt{\gamma_0}\sqrt{\gamma_0}} = \frac{\gamma_h}{\gamma_0} \tag{7.3}$$

By virtue of the definitions of covariance and standard deviations, we can write:

$$\rho_h = \frac{\sum_{t=1}^{T-h}(Y_t - \bar{Y})(Y_{t+h} - \bar{Y})}{\sqrt{\sum_{t=1}^{T-h}(Y_t - \bar{Y})^2}\sqrt{\sum_{t=1}^{T-h}(Y_{t-h} - \bar{Y})^2}} \tag{7.4}$$

where $\bar{Y}$ is the mean of the series Y_t calculated on $(T - h)$ observations:

$$\bar{Y} = \frac{1}{T - h}\sum_{t=1}^{T-h} Y_t \tag{7.5}$$

To simplify the calculations, and since in practice only a sample is available, we can define the **sampling autocorrelation function** (or estimated autocorrelation function) as follows:

$$\hat{\rho}_h = \frac{\sum_{t=1}^{T-h} (Y_t - \bar{Y})(Y_{t+h} - \bar{Y})}{\sum_{t=1}^{T} (Y_t - \bar{Y})^2} \tag{7.6}$$

where $\bar{Y}$ represents the mean of the series Y_t calculated over T observations:

$$\bar{Y} = \frac{1}{T} \sum_{t=1}^{T} Y_t \tag{7.7}$$

For a sufficiently large number of observations T, expressions (7.4) and (7.6) give very similar results.

Remark 7.2 The graph of the sampling autocorrelation function is called a **correlogram**. An example is shown in Fig. 7.2, with the number of lags on the x-axis and the value of the autocorrelation function on the y-axis.

Theorem 7.2 *The autocorrelation function of a stationary process* Y_t *has the following properties:*

- $\rho_0 = 1$
- $|\rho_h| \leq \rho_0$
- $\rho_h = \rho_{-h}$: *even function.*

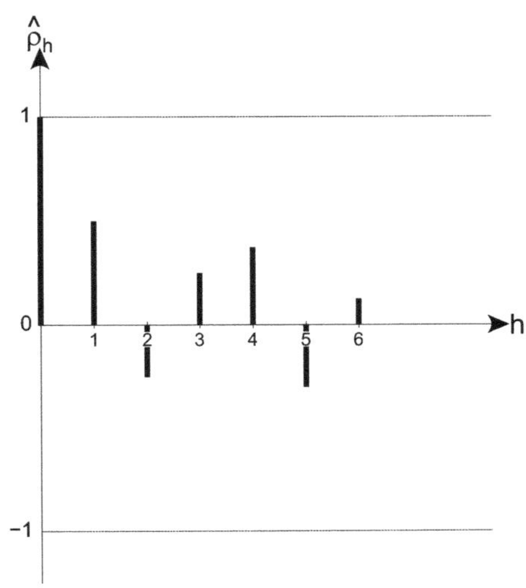

Fig. 7.2 Example of a correlogram

The practical interest of the autocorrelation function can be found in particular in the study of ARMA processes (see below). Another fundamental function in the study of time series is the partial autocorrelation function. We have already mentioned the notion of partial correlation coefficient in Chap. 3.

The **partial autocorrelation function** measures the correlation between Y_t and Y_{t-h}, the influence of the variables Y_{t-h+i} (for $i < h$) having been removed. Let ρ_h and ϕ_{hh} be the autocorrelation and partial autocorrelation functions of Y_t, respectively. Let P_h be the symmetric matrix formed by the $(h-1)$ first autocorrelations of Y_t:

$$P_h = \begin{bmatrix} 1 & \rho_1 & \dots & \rho_{h-1} \\ . & 1 & & \\ . & & . & \\ . & & . & \\ . & & & . \\ \rho_{h-1} & & & 1 \end{bmatrix} \tag{7.8}$$

The partial autocorrelation function is given by:

$$\phi_{hh} = \frac{|P_h^*|}{|P_h|} \tag{7.9}$$

where $|P_h|$ is the determinant of the matrix P_h. The matrix P_h^* is given by:

$$P_h^* = \begin{bmatrix} 1 & \rho_1 & . & . & \rho_{h-2} & \rho_1 \\ . & 1 & & & & . \\ . & & . & & & . \\ . & & & . & & . \\ . & & & & 1 & . \\ \rho_{h-1} & & & & & \rho_h \end{bmatrix} \tag{7.10}$$

P_h^* is thus the matrix P_h in which the last column has been replaced by the vector $[\rho_1 .... \rho_h]'$.

The partial autocorrelation function is written as:

$$\phi_{ii} = \begin{cases} \rho_1 & \text{if } i = 1 \\ \dfrac{\rho_i - \sum\limits_{j=1}^{i-1} \phi_{i-1,j}\rho_{i-j}}{1 - \sum\limits_{j=1}^{i-1} \phi_{i-1,j}\rho_j} & \text{for } i = 2, \dots, h \end{cases} \tag{7.11}$$

and

$$\phi_{ij} = \phi_{i-1,j} - \phi_{ii}\phi_{i-1,i-j} \text{ for } i = 2, \dots, h \text{ and } j = 1, \dots, i-1. \tag{7.12}$$

This algorithm is known as the **Durbin algorithm** (Durbin, 1960). It is based on the **Yule-Walker equations** (see below). The partial autocorrelation coefficients are given by the autocorrelation coefficients and by a set of recursive equations.

7.2 Stationarity: Study of the Autocorrelation Function and Unit Root Test

7.2.1 Study of the Autocorrelation Function

In addition to the graphical representation of the series itself, a first idea concerning the stationarity or not of a series can be provided by the autocorrelation function. We know that the autocorrelation function of a stationary time series decreases very rapidly. If no autocorrelation coefficient is significantly different from zero, we say that the process has **no memory**. It is therefore stationary, as in the case of white noise. If, for example, only the first-order autocorrelation is significant, the process is said to have a **short memory**. Conversely, the autocorrelation function of a non-stationary time series decreases very slowly, indicating a strong dependence between observations.

Figures 7.3 and 7.4 represent the correlogram of a stationary series. It can be seen that the autocorrelation function decreases very rapidly (here it is cancelled out from the fourth lag). Similarly, Fig. 7.5 relates to a stationary series: the autocorrelation function decreases sinusoidally, but the decay of the envelope curve is exponential, testifying to a very rapid decrease in the autocorrelation function. Conversely, the correlograms in Figs. 7.6 and 7.7 relate to a non-stationary series insofar as it appears that the autocorrelation function decreases very slowly.

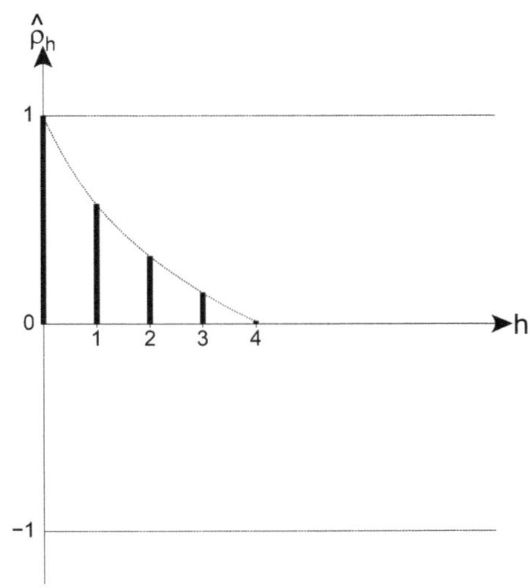

Fig. 7.3 Correlogram of a stationary series

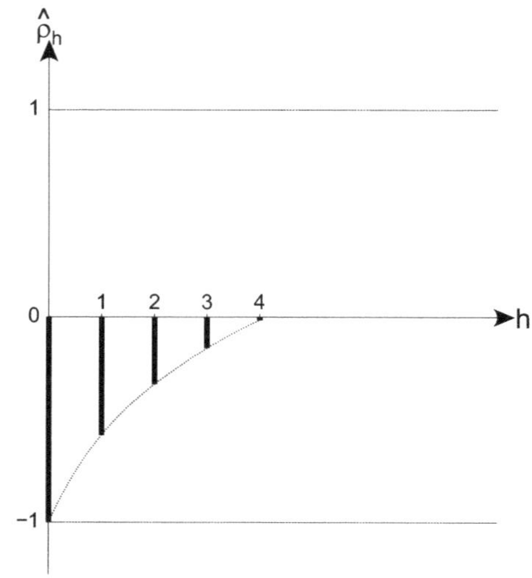

Fig. 7.4 Correlogram of a stationary series

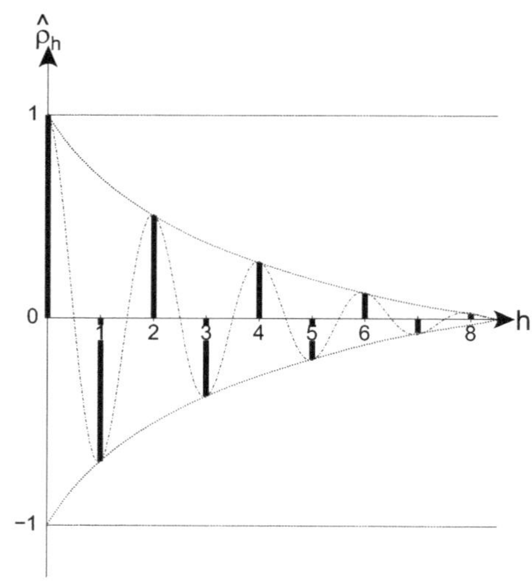

Fig. 7.5 Correlogram of a stationary series

As an illustration, consider Standard and Poor's 500 stock index (denoted SP) monthly over the period from January 1980 to June 2021 (Fig. 7.1). Figure 7.8 reproduces the dynamics of this same series in logarithms, noted LSP. These graphs highlight the existence of an overall upward trend illustrating that the mean of the series varies over time: the US stock market index series appears to be non-stationary in the mean.

Let us differentiate the series LSP by applying the first-difference operator:

$$\Delta LSP_t = LSP_t - LSP_{t-1} = RSP_t \tag{7.13}$$

Fig. 7.6 Correlogram of a non-stationary series

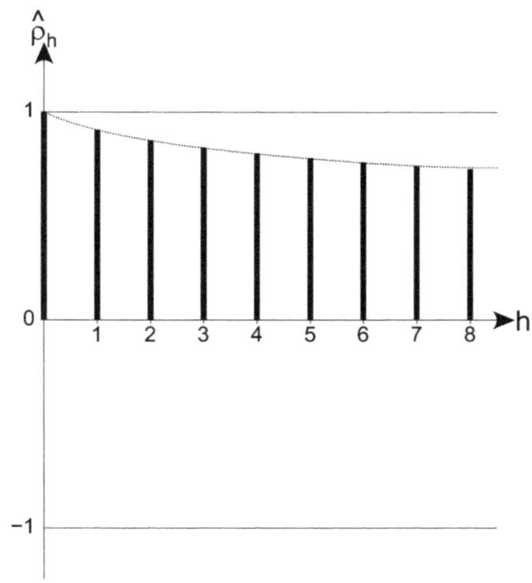

Fig. 7.7 Correlogram of a non-stationary series

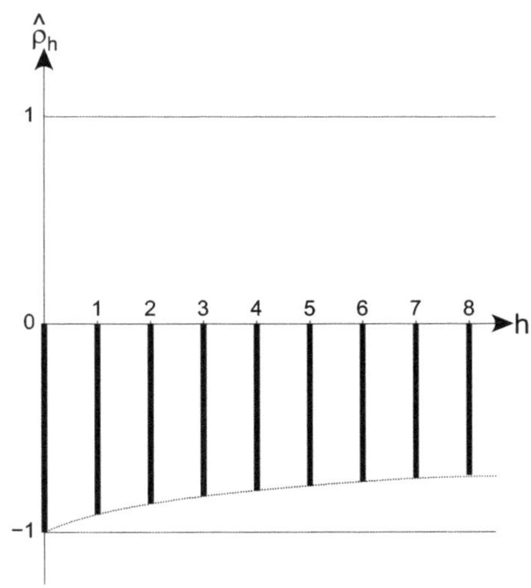

RSP_t represents the series of returns of the US stock index over the period from February 1980 to June 2021 (one observation, corresponding to January 1980, is lost at the beginning of the period due to the differentiation operation). This series is shown in Fig. 7.9: the upward trend in the mean has been suppressed by the differentiation operation, indicating that the series of returns is a priori stationary in the mean.

Let us confirm these intuitions by examining the correlograms of the LSP and RSP series. The correlogram of LSP is plotted in Fig. 7.10 and that of RSP in Fig. 7.11. The vertical dotted lines on the graphs of the autocorrelation and partial autocorrelation functions define the bounds of the confidence interval. Each

Fig. 7.8 Logarithm of Standard and Poor's 500 stock index, 1980.01–2021.06

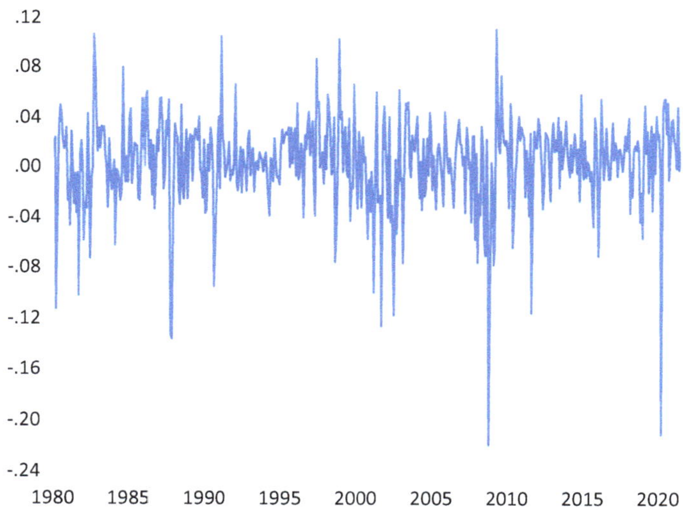

Fig. 7.9 Standard and Poor's 500 returns, 1980.02–2021.06

value (autocorrelation or partial autocorrelation) that falls outside this confidence interval is significantly different from zero. We can see from Fig. 7.10 that the autocorrelation function of the series LSP decreases very slowly (the values taken by the autocorrelation function are given in column AC for lags ranging from 1 to 20). All the values of the autocorrelation function are also outside the confidence interval; they are significantly different from zero. The column $Q - Stat$

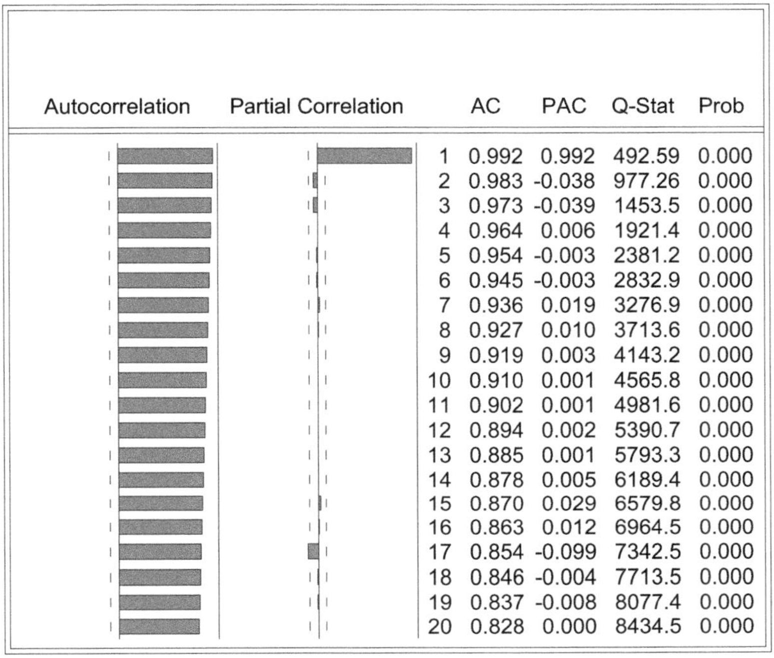

	AC	PAC	Q-Stat	Prob
1	0.992	0.992	492.59	0.000
2	0.983	-0.038	977.26	0.000
3	0.973	-0.039	1453.5	0.000
4	0.964	0.006	1921.4	0.000
5	0.954	-0.003	2381.2	0.000
6	0.945	-0.003	2832.9	0.000
7	0.936	0.019	3276.9	0.000
8	0.927	0.010	3713.6	0.000
9	0.919	0.003	4143.2	0.000
10	0.910	0.001	4565.8	0.000
11	0.902	0.001	4981.6	0.000
12	0.894	0.002	5390.7	0.000
13	0.885	0.001	5793.3	0.000
14	0.878	0.005	6189.4	0.000
15	0.870	0.029	6579.8	0.000
16	0.863	0.012	6964.5	0.000
17	0.854	-0.099	7342.5	0.000
18	0.846	-0.004	7713.5	0.000
19	0.837	-0.008	8077.4	0.000
20	0.828	0.000	8434.5	0.000

Fig. 7.10 Correlogram of the series LSP

gives the values of the Ljung-Box statistic used to test the null hypothesis of no autocorrelation (see Chap. 4) for a number of lags ranging from 1 to 20. We see that the value of this statistic for 20 lags is 8 434,5, which is higher than the critical value of the Chi-squared distribution with 20 degrees of freedom (31.41 at the 5% significance level): the null hypothesis of no autocorrelation is consequently rejected. These elements confirm the intuition about the non-stationary nature of the series LSP. On the other hand, we notice that the autocorrelation function of RSP no longer shows any particular structure, which pleads in favor of the stationarity of the series. Of course, this intuition must be confirmed by the application of **unit root tests** (see below). However, the Ljung-Box statistic for 20 lags is 36.589, which is slightly higher than the critical value (31.41 at the 5% significance level), leading to the rejection of the null hypothesis of no autocorrelation.

7.2.2 TS and DS Processes

Economic and financial series are very often non-stationary series. We are interested here in non-stationarity in the mean. We have seen that non-stationarity can be identified graphically through the graph of the series and the correlogram. Since

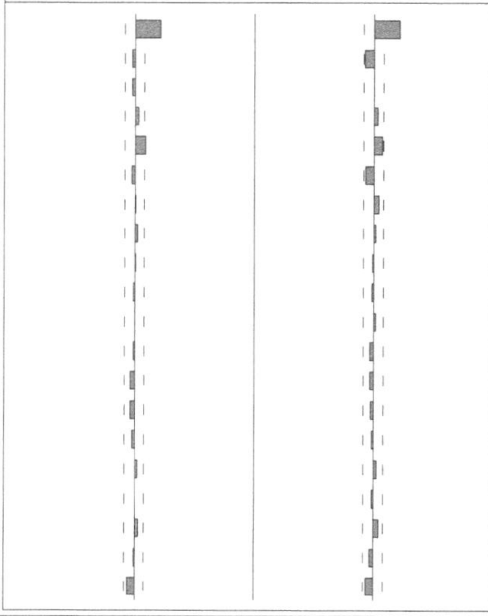

	AC	PAC	Q-Stat	Prob
1	0.225	0.225	25.466	0.000
2	-0.022	-0.077	25.712	0.000
3	-0.020	0.003	25.918	0.000
4	0.036	0.041	26.564	0.000
5	0.090	0.075	30.672	0.000
6	-0.027	-0.067	31.046	0.000
7	0.010	0.043	31.092	0.000
8	0.030	0.017	31.544	0.000
9	0.010	-0.007	31.590	0.000
10	-0.010	-0.013	31.640	0.000
11	0.005	0.020	31.650	0.001
12	-0.014	-0.031	31.755	0.002
13	-0.038	-0.032	32.491	0.002
14	-0.034	-0.017	33.092	0.003
15	-0.017	-0.008	33.245	0.004
16	0.030	0.031	33.723	0.006
17	0.001	-0.010	33.723	0.009
18	0.034	0.048	34.339	0.011
19	-0.007	-0.026	34.361	0.017
20	-0.065	-0.058	36.589	0.013

Fig. 7.11 Correlogram of the series RSP

Nelson and Plosser (1982), cases of non-stationarity in the mean have been analyzed using two types of processes:

- **TS (trend stationary) processes** which are characterized by non-stationarity of a deterministic nature
- **DS (difference stationary) processes** whose non-stationarity is stochastic (random) in nature

Non-stationarity has fundamental consequences for econometrics. If it is stochastic in nature, the usual asymptotic properties of estimators are no longer valid, and it is necessary to develop a particular asymptotic theory. Moreover, in a multivariate framework, applying the usual econometric methods to non-stationary series can lead to the estimation of regressions that seem statistically correct, but which in reality make no sense at all. In other words, the links highlighted between the variables appearing in these regressions are spurious; this is the classic problem of **spurious regressions** (see below).

Characteristics of TS Processes

Generally speaking, a **TS process** Y_t can be written:

$$Y_t = f_t + \varepsilon_t \tag{7.14}$$

where f_t is a deterministic function of time and ε_t is a stationary process. In the simple case where f_t is a polynomial function of order 1, we have:

$$Y_t = \gamma + t\,\beta + \varepsilon_t \tag{7.15}$$

where t denotes time.

For simplicity, further assume that $\varepsilon_t \sim WN\left(0, \sigma_\varepsilon^2\right)$. Let us determine the expectation, variance, and autocovariance function of this process in order to identify its characteristics. Calculating the expectation yields:

$$E[Y_t] = E\left[\gamma + t\,\beta + \varepsilon_t\right] \tag{7.16}$$

Hence, since $E[\varepsilon_t] = 0$:

$$E[Y_t] = \gamma + t\,\beta \tag{7.17}$$

Now let us calculate the variance:

$$V[Y_t] = E\left[Y_t - E\left[Y_t\right]\right]^2 = E\left[\varepsilon_t\right]^2 = V[\varepsilon_t] \tag{7.18}$$

Hence:

$$V[Y_t] = \sigma_\varepsilon^2 \tag{7.19}$$

Finally, let us determine the autocovariance function of the process Y_t:

$$Cov[Y_t, Y_s] = E[(Y_t - E[Y_t])(Y_s - E[Y_s])] = E[\varepsilon_t \varepsilon_s] \tag{7.20}$$

Hence:

$$Cov[Y_t, Y_s] = 0 \quad \forall\, t \neq s \tag{7.21}$$

Thus, the expectation of a TS process exhibits a deterministic trend: the process is non stationary in the mean, the non-stationarity being of a deterministic type. On the other hand, its variance is constant over time, showing that a TS process is stationary in variance. Finally, its autocovariance function is independent of time. Thus, by virtue of (7.19), the long-term forecast error has a finite variance σ_ε^2. In other words, the long-term behavior of Y_t is deterministic, which is the main characteristic of TS processes. In this type of modeling, the effects of a shock on Y_t are transitory (ε_t being assumed stationary and invertible): following a shock, the series returns to its long-term level represented here by the trend.

Remark 7.3 A TS process is a process that can be made stationary (i.e., detrended) by a regression on a deterministic trend.

Characteristics of DS Processes

A **DS process** is a non-stationary process that can be stationarized by applying a difference filter $\Delta = (1 - L)^d$ where L is the lag operator and d is a positive integer called the **differentiation** or **integration parameter**:

$$(1 - L)^d \, Y_t = \beta + \varepsilon_t \tag{7.22}$$

where ε_t is a stationary process. Often $d = 1$ and the DS process is written as:

$$Y_t - Y_{t-1} = \beta + \varepsilon_t \tag{7.23}$$

$Y_t - Y_{t-1} = \Delta Y_t$ is stationary: in a DS process, the difference of the series is stationary.

Remark 7.4 If ε_t is white noise, the process:

$$Y_t = Y_{t-1} + \beta + \varepsilon_t \tag{7.24}$$

is known as a **random walk with drift** if $\beta \neq 0$. If $\beta = 0$, it is referred to as a **random walk without drift**. A random walk is thus characterized by the presence of a **unit root** (the coefficient assigned to Y_{t-1} is equal to 1)[4] and by the fact that ε_t is white noise.

In order to highlight the main characteristics of a DS process, let us reason by recurrence:

$$Y_1 = Y_0 + \beta + \varepsilon_1 \tag{7.25}$$

$$Y_2 = Y_1 + \beta + \varepsilon_2 = Y_0 + 2\beta + \varepsilon_1 + \varepsilon_2 \tag{7.26}$$

Proceeding in this way, we have:

$$Y_t = Y_0 + t\,\beta + \sum_{j=1}^{t} \varepsilon_j \tag{7.27}$$

where Y_0 denotes the first term of the series Y_t.

Unlike the error term in Eq. (7.15) of a TS process, the error term in the DS process (Eq. (7.27)) corresponds to an accumulation of random shocks $\left(\sum_{j=1}^{t} \varepsilon_j\right)$. This remark is fundamental as it means that a shock at a given date has permanent consequences.

---

[4] By introducing the lag operator L, we can write $(1 - L)\, Y_t = \beta + \varepsilon_t$. If we posit $1 - L = 0$, we deduce $L = 1$, hence the name of unit root.

Let us examine the statistical characteristics of DS processes, assuming that ε_t is a white noise process. Consider the calculation of the expectation:

$$E[Y_t] = E\left[Y_0 + t\,\beta + \sum_{j=1}^{t} \varepsilon_j\right] \tag{7.28}$$

Hence:

$$E[Y_t] = Y_0 + t\,\beta \tag{7.29}$$

Now let us determine the variance of the process:

$$V[Y_t] = E\,[Y_t - E\,[Y_t]]^2 = E\left[Y_0 + t\,\beta + \sum_{j=1}^{t} \varepsilon_j - Y_0 - t\,\beta\right]^2 \tag{7.30}$$

that is:

$$V[Y_t] = V\left[\sum_{j=1}^{t} \varepsilon_j\right] \tag{7.31}$$

So we have:

$$V[Y_t] = t\,\sigma_\varepsilon^2 \tag{7.32}$$

Finally, let us calculate the autocovariance function:

$$Cov[Y_t, Y_s] = E[(Y_t - E[Y_t])(Y_s - E[Y_s])] = E\left[\left(\sum_{j=1}^{t} \varepsilon_j\right)\left(\sum_{j=1}^{s} \varepsilon_j\right)\right] \tag{7.33}$$

Hence:

$$Cov[Y_t, Y_s] = \min(t, s)\,\sigma_\varepsilon^2 \qquad s \neq t \tag{7.34}$$

The expectation and variance of a DS process are time-dependent. The DS process is thus characterized by non-stationarity of a deterministic nature via the expectation but also by a non-stationarity of a stochastic nature through the disturbances whose variance follows a linear trend. For a DS process, the variance of the forecast error is not constant, but increases with the horizon. Thus, each random shock has a lasting effect on the behavior of the series.

Because of their very different characteristics, it is crucial to be able to distinguish between the two types of processes, TS and DS. This distinction can be made by means of unit root tests, such as the Dickey-Fuller test, which we present below.

7.2.3 The Dickey-Fuller Test

To determine whether a series is stationary or not, unit root tests are applied. There are numerous unit root tests (see in particular Lardic and Mignon, 2002). We present here only the **test of Dickey and Fuller** (1979, 1981) aimed at testing the null hypothesis of non-stationarity against the alternative hypothesis of stationarity. We thus test:

- H_0: the series is non-stationary, i.e., it has at least one unit root.
- H_1: the series is stationary, i.e., it has no unit root.

Simple Dickey-Fuller (DF) Test

Dickey and Fuller (1979) consider three basic models for the series Y_t, $t = 1, \ldots, T$:

- Model [1]: model without constant or deterministic trend:

$$(1 - \rho L) Y_t = \varepsilon_t \tag{7.35}$$

  that is:

$$Y_t = \rho Y_{t-1} + \varepsilon_t \tag{7.36}$$

- Model [2]: model with constant without deterministic trend:

$$(1 - \rho L)(Y_t - \mu) = \varepsilon_t \tag{7.37}$$

  that is:

$$Y_t = \rho Y_{t-1} + \mu (1 - \rho) \varepsilon_t \tag{7.38}$$

- Model [3]: model with constant and deterministic trend:

$$(1 - \rho L)(Y_t - \alpha - \beta t) = \varepsilon_t \tag{7.39}$$

  that is:

$$Y_t - \alpha - \beta t - \rho Y_{t-1} + \alpha\rho + \beta (t - 1) = \varepsilon_t \tag{7.40}$$

hence:

$$Y_t = \rho Y_{t-1} + \alpha(1-\rho) + \beta\rho + \beta(1-\rho)t + \varepsilon_t \tag{7.41}$$

In each of the three models, it is assumed that $\varepsilon_t \sim WN(0, \sigma_\varepsilon^2)$.

If $\rho = 1$, this means that one of the roots of the lag polynomial is equal to 1. In this case, there is a **unit root** and Y_t is a non-stationary process.

We test the null hypothesis of non-stationarity, i.e., the presence of a unit root ($\rho = 1$), against the alternative hypothesis of no unit root ($|\rho| < 1$).

Let us write more precisely the null and alternative hypotheses for each of the three models considered:

- Model [1]:

$$\begin{cases} H_0 : \rho = 1 \Leftrightarrow Y_t = Y_{t-1} + \varepsilon_t \\ H_1 : |\rho| < 1 \Leftrightarrow Y_t = \rho Y_{t-1} + \varepsilon_t \end{cases} \tag{7.42}$$

  Under the null hypothesis, Y_t follows a random walk process without drift. Under the alternative hypothesis, Y_t follows an autoregressive process of order 1 ($AR(1)$).
- Model [2]:

$$\begin{cases} H_0 : \rho = 1 \Leftrightarrow Y_t = Y_{t-1} + \varepsilon_t \\ H_1 : |\rho| < 1 \Leftrightarrow Y_t = \rho Y_{t-1} + \gamma + \varepsilon_t \text{ with } \gamma = \mu(1-\rho) \end{cases} \tag{7.43}$$

  The null hypothesis corresponds to a random walk process without drift. Under the alternative hypothesis, Y_t follows an $AR(1)$ process with drift.
- Model [3]:

$$\begin{cases} H_0 : \rho = 1 \Leftrightarrow Y_t = Y_{t-1} + \beta + \varepsilon_t \\ H_1 : |\rho| < 1 \Leftrightarrow Y_t = \rho Y_{t-1} + \lambda + \delta t + \varepsilon_t \\ \text{with } \lambda = \alpha(1-\rho) + \rho\beta \text{ and } \delta = \beta(1-\rho) \end{cases} \tag{7.44}$$

  Under the null hypothesis, Y_t follows a random walk with drift. Under the alternative hypothesis, Y_t is a TS process. It can be made stationary by calculating the deviations from the trend estimated by OLS.

To facilitate the application of the test, models [1], [2], and [3] are in practice estimated in the following form:[5]

[5] The first-difference models allow us to reduce to usual tests of significance of the coefficients, the critical values being tabulated by Dickey and Fuller (see below).

– Model [1]:

$$\Delta Y_t = \phi Y_{t-1} + \varepsilon_t \tag{7.45}$$

– Model [2]:

$$\Delta Y_t = \gamma + \phi Y_{t-1} + \varepsilon_t \tag{7.46}$$

– Model [3]:

$$\Delta Y_t = \lambda + \delta t + \phi Y_{t-1} + \varepsilon_t \tag{7.47}$$

with $\phi = \rho - 1$ and ε_t is white noise. We test the null hypothesis $\phi = 0$ (non-stationarity) against the alternative hypothesis $\phi < 0$ (stationarity). To this end, the t-statistic for the coefficient ϕ is calculated. This statistic is compared with the values tabulated by Dickey and Fuller (see Table 7.2). As the critical values are negative, the decision rule is reversed:

– If the calculated value of the t-statistic associated with ϕ is lower than the critical value, the null hypothesis is rejected, the series is stationary.

Table 7.2 Critical values of the Dickey-Fuller test for $\rho = 1$

T	*1%*	*5%*	*10%*
	Model [1]		
100	−2.60	−1.95	−1.61
250	−2.58	−1.95	−1.62
500	−2.58	−1.95	−1.62
∞	−2.58	−1.95	−1.62
	Model [2]		
100	−3.51	−2.89	−2.58
250	−3.46	−2.88	−2.57
500	−3.44	−2.87	−2.57
∞	−3.43	−2.86	−2.57
	Model [3]		
100	−4.04	−3.45	−3.15
250	−3.99	−3.43	−3.13
500	−3.98	−3.42	−3.13
∞	−3.96	−3.41	−3.12

Model [1]: model without constant or deterministic trend. Model [2]: model with constant, without trend. Model [3]: model with constant and trend

- If the calculated value of the t-statistic associated with ϕ is higher than the critical value, the null hypothesis is not rejected, the series is non-stationary.

The models used in the DF test are restrictive in that ε_t is assumed to be white noise. However, this assumption is very often questioned due to autocorrelation and/or heteroskedasticity. To solve this problem, Dickey and Fuller proposed a parametric correction leading to the **augmented Dickey-Fuller test**.

Augmented Dickey-Fuller (ADF) Test

To account for possible autocorrelation of errors, lags are introduced on the endogenous variable.[6] As before, three models are distinguished:

- Model [1]:

$$\Delta Y_t = \phi Y_{t-1} + \sum_{j=1}^{p} \phi_j \Delta Y_{t-j} + \varepsilon_t \tag{7.48}$$

- Model [2]:

$$\Delta Y_t = \gamma + \phi Y_{t-1} + \sum_{j=1}^{p} \phi_j \Delta Y_{t-j} + \varepsilon_t \tag{7.49}$$

- Model [3]:

$$\Delta Y_t = \lambda + \delta t + \phi Y_{t-1} + \sum_{j=1}^{p} \phi_j \Delta Y_{t-j} + \varepsilon_t \tag{7.50}$$

Again, we test the null hypothesis $\phi = 0$ against the alternative hypothesis $\phi < 0$. The t-statistic of the coefficient ϕ is compared to the critical values tabulated by Dickey and Fuller (see Table 7.2). The null hypothesis of unit root is rejected if the calculated value is less than the critical value.

It should be noted that the application of the ADF test requires us to choose the number of lags p – called the **truncation parameter** of the ADF test – to

[6] One of the causes of error autocorrelation lies in the omission of explanatory variables. The correction provided by Dickey and Fuller thus consists in adding explanatory variables represented by the lagged values of the endogenous variable.

be introduced so that the residuals are indeed white noise. Several methods are available for making this choice, including:

- The study of partial autocorrelations of the series ΔY_t. We select for p the lag corresponding to the last partial autocorrelation significantly different from zero.
- The estimation of several processes for different values of p. We retain the model that minimizes the information criteria of Akaike, Schwarz, or Hannan-Quinn.
- The use of the procedure suggested by Campbell and Perron (1991) consisting in setting a maximum value for p, noted p_{max}. We then estimate the regression model of the ADF test and test the significance of the coefficient associated with the term $\Delta Y_{t-p_{\max}}$. If this coefficient is significant, we select this value $p_{\max}$ for p. If the coefficient associated with $\Delta Y_{t-p_{\max}}$ is not significant, we re-estimate the ADF regression model for a value of p equal to $p_{max} - 1$ and test the significance of the coefficient relating to the term $\Delta Y_{t-p_{\max}-1}$ and so on.

Sequential Testing Strategy

It is fundamental to note that the unit root test should not be performed on all three models. Instead, the Dickey-Fuller test should be applied to just one of the three models. In practice, we adopt a three-step sequential strategy.

- *Step 1*. We estimate the general model with constant and trend:

$$\Delta Y_t = \alpha + \beta t + \phi Y_{t-1} + \sum_{j=1}^{p} \phi_j \Delta Y_{t-j} + \varepsilon_t \tag{7.51}$$

  We start by testing the significance of the trend by referring to the Dickey-Fuller tables (see Table 7.3). Two cases may arise:
  - If the trend is not significant, we go on to Step 2.
  - If the trend is significant, we keep the model and test the null hypothesis of unit root by comparing the t-statistic of ϕ with the values tabulated by Dickey and Fuller (see Table 7.2). We then have two possibilities:

Table 7.3 Critical values of constant and trend, Dickey-Fuller tests

	Model [2]			*Model [3]*					
	Constant			*Constant*			*Trend*		
T	*1%*	*5%*	*10%*	*1%*	*5%*	*10%*	*1%*	*5%*	*10%*
100	3.22	2.54	2.17	3.78	3.11	2.73	3.53	2.79	2.38
250	3.19	2.53	2.16	3.74	3.09	2.73	3.49	2.79	2.38
500	3.18	2.52	2.16	3.72	3.08	2.72	3.48	2.78	2.38
∞	3.18	2.52	2.16	3.71	3.08	2.72	3.46	2.78	2.38

Model [2]: model with constant, without deterministic trend. Model [3]: model with constant and trend

  - If we do not reject the null hypothesis, Y_t is non-stationary. In this case, it must be differentiated and the test procedure must be repeated on the series in first difference.
  - If the null hypothesis is rejected, Y_t is stationary. In this case, the test procedure stops and we can work directly on the series Y_t.
- *Step 2*. This step should only be applied if the trend in the previous model is not significant. We estimate model [2]:

$$\Delta Y_t = \alpha + \phi\, Y_{t-1} + \sum_{j=1}^{p} \phi_j \Delta Y_{t-j} + \varepsilon_t \tag{7.52}$$

  and begin by testing the significance of the constant by referring to the Dickey-Fuller tables (see Table 7.3):
  - If the constant is not significant, we go to Step 3.
  - If the constant is significant, we test the null hypothesis of unit root by comparing the t-statistic of ϕ with the values tabulated by Dickey and Fuller (see Table 7.2). We then have two possibilities:
    - If we do not reject the null hypothesis, Y_t is non-stationary. In this case, it must be differentiated and the test procedure must be repeated on the series in first difference.
    - If the null hypothesis is rejected, Y_t is stationary. In this case, the test procedure stops and we can work directly on the series Y_t.
- *Step 3*. This step should only be applied if the constant in the previous model is not significant. We estimate model [1]:

$$\Delta Y_t = \phi\, Y_{t-1} + \sum_{j=1}^{p} \phi_j \Delta Y_{t-j} + \varepsilon_t \tag{7.53}$$

  and test the null hypothesis of unit root using Dickey-Fuller critical values (see Table 7.2):
  - If the null hypothesis is not rejected, Y_t is non-stationary. In this case, it must be differentiated and the test procedure must be repeated on the series in first difference.
  - If the null hypothesis is rejected, Y_t is stationary. In this case, the test procedure stops and we can work directly on the series Y_t.

Remark 7.5 If, after applying this procedure, we find that Y_t is non-stationary, this means that the series contains at least one unit root. In this case, we should repeat the Dickey-Fuller tests on the series in first difference. If ΔY_t is found to be non-stationary, the procedure should be applied again on the series in second difference and so on.

Remark 7.6 A non-stationary series is also called an **integrated series**. For example, if Y_t is non-stationary and ΔY_t is stationary, then Y_t is integrated of order

1: it must be differentiated once to make it stationary. ΔY_t is integrated of order 0: there is no need to differentiate it to make it stationary. An integrated series of order 0 is thus a stationary series.

Definition 7.4 A series Y_t is integrated of order d, which we note $Y_t \sim I(d)$, if it is necessary to differentiate it d times to make it stationary. In other words, $Y_t \sim I(d)$ if and only if $(1-L)^d Y_t \sim I(0)$. d is called the **integration** parameter.

Empirical Application

Consider the series SP of Standard and Poor's 500 stock index over the period from January 1980 to June 2021. The logarithmic series is denoted LSP, with RSP standing for the series of returns. Our aim is to apply the Dickey-Fuller test strategy.

Let us first study the stationarity of the series LSP. We test the null hypothesis of non-stationarity of the series LSP (presence of unit root) against the alternative hypothesis of stationarity (absence of unit root). To this end, we begin by estimating the model with constant and trend:

$$\Delta LSP_t = RSP_t = \lambda + \delta t + \phi\, LSP_{t-1} + \sum_{j=1}^{p} \phi_j RSP_{t-j} + \varepsilon_t \tag{7.54}$$

Estimating this model involves determining the value of the truncation parameter p. As previously mentioned, this choice can be guided by the graph of the partial autocorrelation function of the series RSP (Fig. 7.11). As shown, only the first partial autocorrelation lies outside the confidence interval. In other words, only the first partial autocorrelation is significantly different from zero, which leads us to take a value of p equal to 1. Another technique involves estimating the model (7.54) for different values of p and selecting the value that minimizes the information criteria. Table 7.4 shows the values taken by the AIC, SIC, and HQ information criteria for values of p ranging from 1 to 12. Minimizing the SIC and HQ criteria leads us to choose $p = 1$, while the AIC criterion tends to select $p = 2$. For reasons of parsimony, and insofar as two out of three criteria favor a value of p equal to 1, we choose $p = 1$.[7]

As a result, we estimate the following model:

$$RSP_t = \lambda + \delta t + \phi\, LSP_{t-1} + \phi_1 RSP_{t-1} + \varepsilon_t \tag{7.55}$$

The results are set out in Table 7.5. We start by testing the significance of the trend (noted $@TREND$("1980M01")) by referring to the Dickey-Fuller tables. The critical value of the trend in a model with constant and trend for 500 observations being 2.78 (see Table 7.3), we have $1.9612 < 2.78$: we do not reject the null hypothesis of non-significance of the trend. We then proceed to the next step, which

[7] For robustness, we also conducted the analysis with two lags. The results are identical to those presented here.

Table 7.4 Choosing the truncation parameter p

p	AIC	SIC	HQ
1	−3.814483	**−3.780663**	**−3.80121**
2	**−3.815117**	−3.772842	−3.798526
3	−3.811125	−3.760395	−3.791216
4	−3.809856	−3.750671	−3.786628
8	−3.808701	−3.715695	−3.772199
12	−3.79428	−3.667454	−3.744505

Values in bold correspond to values minimizing information criteria

Table 7.5 ADF test on LSP. Model with constant and trend

Null hypothesis: LSP has a unit root				
Exogenous: constant, linear trend				
Lag length: 1 (automatic – based on SIC, maxlag = 17)				
			t-Statistic	Prob.*
Augmented Dickey-Fuller test statistic			−2.037919	0.5786
Test critical values	1%level		−3.976591	
	5% level		−3.418870	
	10% level		−3.131976	
*MacKinnon (1996) one-sided p-values				
Augmented Dickey-Fuller test equation				
Dependent variable: D(LSP)				
Method: least squares				
Sample: 1980M01 2021M06				
Included observations: 498				
Variable	Coefficient	Std. error	t-Statistic	Prob.
LSP(−1)	−0.013371	0.006561	−2.037919	0.0421
D(LSP(−1))	0.232493	0.043764	5.312378	0.0000
C	0.084074	0.039810	2.111899	0.0352
@TREND("1980M01")	5.82E-05	2.97E-05	1.961221	0.0504
R-squared	0.058922	Mean dependent var	0.004844	
Adjusted R-squared	0.053207	S.D. dependent var	0.036778	
S.E. of regression	0.035787	Akaike info criterion	−3.814483	
Sum squared resid	0.632659	Schwarz criterion	−3.780663	
Log likelihood	953.8063	Hannan-Quinn criterion	−3.801210	
F-statistic	10.31000	Durbin-Watson stat	1.968341	
Prob(F-statistic)	0.000001			

consists in estimating the model with constant, without trend:

$$RSP_t = \lambda + \phi\, LSP_{t-1} + \phi_1 RSP_{t-1} + \varepsilon_t \tag{7.56}$$

The results are given in Table 7.6. We test the significance of the constant. The critical value, at the 5% significance level, of the constant in a model with constant

Table 7.6 ADF test on LSP. Model with constant, without trend

Null hypothesis: LSP has a unit root				
Exogenous: constant				
Lag length: 1 (automatic – based on SIC, maxlag = 17)				
		t-Statistic	Prob.*	
Augmented Dickey-Fuller test statistic			−0.584842	0.8709
Test critical values:	1%level		−3.443254	
	5% level		−2.867124	
	10% level		−2.569806	
*MacKinnon (1996) one-sided p-values				
Augmented Dickey-Fuller test equation				
Dependent variable: D(LSP)				
Method: least squares				
Sample: 1980M01 2021M06				
Included observations: 498				
Variable	Coefficient	Std. error	t-Statistic	Prob.
LSP(−1)	−0.001445	0.002472	−0.584842	0.5589
D(LSP(−1))	0.226533	0.043784	5.173865	0.0000
C	0.013992	0.017597	0.795109	0.4269
R-squared	0.051595	Mean dependent var	0.004844	
Adjusted R-squared	0.047763	S.D. dependent var	0.036778	
S.E. of regression	0.035889	Akaike info criterion	−3.810743	
Sum squared resid	0.637585	Schwarz criterion	−3.785378	
Log likelihood	951.8751	Hannan-Quinn criterion	−3.800788	
F-statistic	13.46438	Durbin-Watson stat	1.965871	
Prob(F-statistic)	0.000002			

without trend is 2.52 (see Table 7.3). Since $0.7951 < 2.52$, we do not reject the null hypothesis that the constant is insignificant. Finally, we estimate the model without constant or trend:

$$RSP_t = \phi\, LSP_{t-1} + \phi_1 RSP_{t-1} + \varepsilon_t \tag{7.57}$$

The results in Table 7.7 allow us to proceed with the unit root test, i.e., the test of the null hypothesis $\phi = 0$ against the alternative hypothesis $\phi < 0$. The calculated value of the ADF statistic is 2.2448 and the critical value is -1.95 at the 5% significance level (Table 7.2). Since $2.2448 > -1.95$, we do not reject the null hypothesis of non-stationarity of the series LSP. We deduce that LSP is non-stationary and characterized by the presence of at least one unit root.

To determine the order of integration of LSP, we differentiate it:

$$\Delta LSP_t = LSP_t - LSP_{t-1} = RSP_t \tag{7.58}$$

Table 7.7 ADF test on LSP. Model without constant or trend

Null hypothesis: LSP has a unit root				
Exogenous: none				
Lag length: 1 (automatic – based on SIC, maxlag = 17)				
			t-Statistic	Prob.*
Augmented Dickey-Fuller test statistic			2.244829	0.9944
Test critical values:	1%level		−2.569614	
	5% level		−1.941460	
	10% level		−1.616272	
*MacKinnon (1996) one-sided p-values				
Augmented Dickey-Fuller test equation				
Dependent variable: D(LSP)				
Method: least squares				
Sample: 1980M01 2021M06				
Included observations: 498				
Variable	Coefficient	Std. error	t-Statistic	Prob.
LSP(−1)	0.000511	0.000228	2.244829	0.0252
D(LSP(−1))	0.225714	0.043756	5.158492	0.0000
R-squared	0.050383	Mean dependent var	0.004844	
Adjusted R-squared	0.048469	S.D. dependent var	0.036778	
S.E. of regression	0.035876	Akaike info criterion	−3.813483	
Sum squared resid	0.638399	Schwarz criterion	−3.796573	
Log likelihood	951.5572	Hannan-Quinn criterion	−3.806846	
Durbin-Watson stat	1.965761			

and we perform the ADF test on the series RSP. The null hypothesis that RSP is non-stationary is tested against the alternative hypothesis of stationarity. We adopt the same sequential strategy as before, first estimating the model with constant and trend:

$$\Delta RSP_t = \lambda + \delta t + \phi\, RSP_{t-1} + \sum_{j=1}^{p} \phi_j \Delta RSP_{t-j} + \varepsilon_t \tag{7.59}$$

The endogenous variable is the series of changes in returns, in other words, the second difference of the LSP series. In order to determine the truncation parameter p, we have estimated this model for various values of p and selected the one that minimizes the information criteria. The application of this methodology leads us to choose a number of lags p equal to 0, which corresponds to the case of a simple Dickey-Fuller test. Consequently, we estimate the following model:

$$\Delta RSP_t = \lambda + \delta t + \phi\, RSP_{t-1} + \varepsilon_t \tag{7.60}$$

Table 7.8 ADF test on RSP. Model without constant or trend

Null hypothesis: RSP has a unit root				
Exogenous: none				
Lag length: 0 (automatic – based on SIC, maxlag = 17)				
		t-Statistic	Prob.*	
Augmented Dickey-Fuller test statistic			−17.48229	0.0000
Test critical values:	1%level		−2.569614	
	5% level		−1.941460	
	10% level		−1.616272	
*MacKinnon (1996) one-sided p-values				
Augmented Dickey-Fuller test equation				
Dependent variable: D(RSP)				
Method: least squares				
Sample: 1980M01 2021M06				
Included observations: 498				
Variable	Coefficient	Std. error	t-Statistic	Prob.
D(LSP(−1))	−0.761108	0.043536	−17.48229	0.0000
R-squared	0.380786	Mean dependent var	−3.10E-05	
Adjusted R-squared	0.380786	S.D. dependent var	0.045776	
S.E. of regression	0.036022	Akaike info criterion	−3.807390	
Sum squared resid	0.644885	Schwarz criterion	−3.798935	
Log likelihood	949.0402	Hannan-Quinn criterion	−3.804072	
Durbin-Watson stat	1.969349			

and start by testing the significance of the trend. The results (not reported here) give us a calculated t-statistic associated with the trend equal to 0.1925. As this value is lower than the critical value of 2.78, we do not reject the null hypothesis that the trend is not significant. We therefore estimate the model with constant, without trend. The results lead to a t-statistic associated with the constant equal to 2.3093, below the critical value of 2.52. We finally estimate the model with no constant or trend, the results of which are shown in Table 7.8.

The calculated value of the ADF statistic being equal to -17.4823 and the critical value at the 5% significance level being -1.95, we have: $-17.4823 < -1.95$. We therefore reject the null hypothesis of non-stationarity of the series RSP. We deduce that RSP is stationary, i.e., integrated of order 0. It follows that the series LSP is integrated of order 1, since it has to be differentiated once to make it stationary.

7.3 ARMA Processes

ARMA (autoregressive moving-average) processes were introduced by Box and Jenkins (1970). Such processes are sometimes referred to as a-theoretical in that their purpose is to model a time series in terms of its past values and the present and past values of the error term (noise). In other words, they do not refer to any underlying economic theory. We begin by presenting the definition of ARMA processes before describing the four-step methodology of Box and Jenkins.

7.3.1 Definitions

Autoregressive Processes

Definitions

Definition 7.5 An **autoregressive process of order** p, denoted $AR(p)$, is a stationary process Y_t verifying a relation of the type:

$$Y_t - \phi_1 Y_{t-1} - \quad \cdots \quad - \phi_p Y_{t-p} = \varepsilon_t \tag{7.61}$$

where $\phi_i \quad (i = 1, \ldots, p)$ are real numbers and $\varepsilon_t \sim WN(0, \sigma_\varepsilon^2)$.

By introducing the lag operator L, the relation (7.61) can also be written as:

$$(1 - \phi_1 L - \quad \cdots \quad - \phi_p L^p)\, Y_t = \varepsilon_t \tag{7.62}$$

or:

$$\Phi(L)\, Y_t = \varepsilon_t \tag{7.63}$$

with: $\Phi(L) = 1 - \phi_1 L - \quad \cdots \quad - \phi_p L^p$.

Remark 7.7 In time series models, the error term ε_t is often called **innovation**. This name derives from the fact that it is the only new information involved in the process at date t.

Autocorrelations and Yule-Walker Equations

The autocorrelations of a process $AR(p)$ can be calculated by multiplying each member of Eq. (7.61) by Y_{t-h} $(h > 0)$. Then taking the expectation of the variables and dividing by γ_0, we obtain the following relationship:

$$\frac{1}{\gamma_0}\left[E[Y_t Y_{t-h}] - \phi_1 E[Y_{t-1} Y_{t-h}] - \quad \cdots \quad - \phi_p E[Y_{t-p} Y_{t-h}]\right] = \frac{1}{\gamma_0} E[\varepsilon_t Y_{t-h}] \tag{7.64}$$

Since ε_t is white noise, we have: $E[\varepsilon_t Y_{t-h}] = 0$. We deduce that:

$$\frac{1}{\gamma_0}\left[\gamma_h - \phi_1 \gamma_{h-1} - \ldots - \phi_p \gamma_{h-p}\right] = 0 \tag{7.65}$$

Hence, noting $\rho_h = \frac{\gamma_h}{\gamma_0}$ the autocorrelation function:

$$\rho_h - \phi_1 \rho_{h-1} - \ldots - \phi_p \rho_{h-p} = 0 \tag{7.66}$$

The autocorrelation function of a process $AR(p)$ is finally given by:

$$\rho_h = \sum_{i=1}^{p} \phi_i \rho_{h-i} \qquad \forall\, h > 0 \tag{7.67}$$

The autocorrelations of a process $AR(p)$ are thus described by a linear recurrence equation of order p. Writing this relation for different values of h ($h = 1, 2, \ldots, p$), we obtain the **Yule-Walker equations**:

$$\begin{pmatrix} \rho_1 \\ \rho_2 \\ \vdots \\ \rho_p \end{pmatrix} = \begin{pmatrix} 1 & \rho_1 \; \rho_2 \; \ldots & \rho_{p-1} \\ \rho_1 & 1 & \vdots \\ \vdots & & \rho_1 \\ \rho_{p-1} & & 1 \end{pmatrix} \begin{pmatrix} \phi_1 \\ \phi_2 \\ \vdots \\ \phi_p \end{pmatrix} \tag{7.68}$$

These equations allow us to obtain the autocorrelation coefficients as a function of the autoregressive coefficients and vice versa.

Partial Autocorrelations

It is possible to calculate the partial autocorrelations of the AR process from the Yule-Walker equations and the autocorrelations. For this, we use the **algorithm of Durbin** (1960):

$$\begin{cases} \phi_{11} = \rho_1 & \text{algorithm initialization} \\ \phi_{hh} = \dfrac{\rho_h - \sum_{j=1}^{h-1} \phi_{h-1,j}\,\rho_{h-j}}{1 - \sum_{j=1}^{h-1} \phi_{h-1,j}\,\rho_j} & \text{for } h = 2, 3, \ldots \\ \phi_{hj} = \phi_{h-1,j} - \phi_{hh}\phi_{h-1,h-j} & \text{for } h = 2, 3 \ldots \text{ and } j = 1, \ldots, h-1 \end{cases} \tag{7.69}$$

Property 7.1 *For a process $AR(p)$, $\phi_{hh} = 0 \;\; \forall\, h > p$. In other words, for a process $AR(p)$, the partial autocorrelations cancel out from rank $p + 1$.*

This property is fundamental in that it allows us to identify the order p of AR processes (see below).

Moving-Average Processes

Definitions

Definition 7.6 A **moving-average process of order** q, denoted $MA(q)$, is a stationary process Y_t verifying a relationship of the type:

$$Y_t = \varepsilon_t - \theta_1 \varepsilon_{t-1} - \cdots - \theta_q \varepsilon_{t-q} \tag{7.70}$$

where the θ_i $(i = 1, \ldots, q)$ are real numbers and $\varepsilon_t \sim WN(0, \sigma_\varepsilon^2)$.

By introducing the lag operator L, the relation (7.70) can be written:

$$Y_t = (1 - \theta_1 L - \cdots - \theta_q L^q)\, \varepsilon_t \tag{7.71}$$

or:

$$Y_t = \Theta(L)\varepsilon_t \tag{7.72}$$

with $\Theta(L) = 1 - \theta_1 L - \cdots \theta_q L^q$.

Autocovariances and Autocorrelations

The autocovariance function of a process $MA(q)$ is given by:

$$\begin{aligned} \gamma_h &= E[Y_t Y_{t-h}] \\ &= E\left[(\varepsilon_t - \theta_1 \varepsilon_{t-1} - \cdots - \theta_q \varepsilon_{t-q})(\varepsilon_{t-h} - \cdots - \theta_q \varepsilon_{t-h-q})\right] \end{aligned} \tag{7.73}$$

Some simple calculations lead to the following expression:

$$\gamma_h = \begin{cases} (-\theta_h + \theta_1\theta_{h+1} + \cdots + \theta_{q-h}\theta_q)\sigma_\varepsilon^2 & \text{if } h = 1, \ldots, q \\ 0 & \text{if } h > q \end{cases} \tag{7.74}$$

If $h = 0$, we obtain the variance of the process:

$$\gamma_0 = \sigma_Y^2 = (1 + \theta_1^2 + \cdots + \theta_q^2)\sigma_\varepsilon^2 \tag{7.75}$$

We deduce the autocorrelation function $\rho_h = \frac{\gamma_h}{\gamma_0}$:

$$\rho_h = \begin{cases} \frac{-\theta_h + \theta_1\theta_{h+1} + \cdots + \theta_{q-h}\theta_q}{1 + \theta_1^2 + \cdots + \theta_q^2} & \text{if } \quad 1 \le h \le q \\ 0 & \text{if } \quad h > q \end{cases} \tag{7.76}$$

Property 7.2 *For a process $MA(q)$, $\rho_h = 0$ for $h > q$. In other words, the autocorrelations cancel from rank $q + 1$, when the true data generating process is a $MA(q)$.*

This fundamental property allows us to identify the order q of MA processes.

Partial Autocorrelations

In order to calculate the partial autocorrelations of an MA process, we use the Durbin algorithm. However, the partial autocorrelation function of a process $MA(q)$ has no particular property and its expression is relatively complicated.

Autoregressive Moving-Average Processes: ARMA(p,q)

These processes are a natural extension of AR and MA processes. They are mixed processes – in the sense that they simultaneously incorporate AR and MA components – which allows for a more parsimonious description of the data.

Definitions

Definition 7.7 A stationary process Y_t follows an $ARMA(p, q)$ process if:

$$Y_t - \phi_1 Y_{t-1} - \quad \cdots \quad - \phi_p Y_{t-p} = \varepsilon_t - \theta_1 \varepsilon_{t-1} \quad \cdots \quad \theta_q \varepsilon_{t-q} \tag{7.77}$$

where the coefficients ϕ_i $(i = 1, \ldots, p)$ and θ_j $(j = 1, \ldots, q)$ are real numbers and $\varepsilon_t \sim WN(0, \sigma_\varepsilon^2)$.

By introducing the lag operator L, the relation (7.77) is written as:

$$\Phi(L)\, Y_t = \Theta(L)\, \varepsilon_t \tag{7.78}$$

with $\Phi(L) = 1 - \phi_1 L - \quad \cdots \quad - \phi_p L^p$ and $\Theta(L) = 1 - \theta_1 L - \cdots \theta_q L^q$.

Autocorrelations

To calculate the autocorrelations of an ARMA process, we proceed as in the case of AR processes. We obtain the following expression:

$$\rho_h = \sum_{i=1}^{p} \phi_i \, \rho_{h-i} \qquad \forall\, h > q \tag{7.79}$$

The autocorrelation function of ARMA processes satisfies the same difference equation as that of AR processes.

Partial Autocorrelations

The partial autocorrelation function of ARMA processes has no simple expression. It depends on the order of each part (p and q) and the value of the parameters. It is most frequently characterized either by a decreasing exponential form or by a damped oscillatory form.

7.3.2 The Box and Jenkins Methodology

In order to determine the appropriate ARMA process for modeling the time series under consideration, Box and Jenkins suggested a four-step methodology: identification, estimation, validation, and forecasting. Let us briefly review these different steps.

Step 1: Identification of ARMA Processes

The purpose of this first step is to find the values of the parameters p and q of the ARMA processes. To this end, we rely on the study of the autocorrelation and partial autocorrelation functions.

Autocorrelation Function

We start by calculating the autocorrelation coefficients from the expression (7.6):

$$\hat{\rho}_h = \frac{\sum_{t=1}^{T-h} (Y_t - \bar{Y})(Y_{t+h} - \bar{Y})}{\sum_{t=1}^{T} (Y_t - \bar{Y})^2} \tag{7.80}$$

for various values of h: $h = 1, 2, \ldots, H$. Box and Jenkins suggest retaining a maximum number of lags $H = \frac{T}{4}$ where T is the number of observations in the series. After evaluating the function $\hat{\rho}_h$, we test the statistical significance of each autocorrelation coefficient using Bartlett's result that $\hat{\rho}_h$ follows a normal distribution. Thus, to test the null hypothesis that the autocorrelations are not significantly different from zero, i.e., $\rho_h = 0$, we calculate the value of the t-statistic[8] $t_{\hat{\rho}_h} = \frac{\hat{\rho}_h}{\hat{\sigma}(\hat{\rho}_h)}$ which we compare with the critical value read from the Student's t distribution table. The decision rule is:

- If $|t_{\hat{\rho}_h}| < t(T - l)$, we do not reject the null hypothesis: ρ_h is not significant.
- If $|t_{\hat{\rho}_h}| \geq t(T - l)$, we reject the null hypothesis: ρ_h is significantly different from zero,

where $t(T - l)$ is the value of the Student's t distribution with $(T - l)$ degrees of freedom, l being the number of estimated parameters.

This test enables us to identify the order q of the MA processes, since we know that the autocorrelations of a process $MA(q)$ cancel out from rank $q + 1$.

[8] Bartlett showed that the standard deviation is given by $\hat{\sigma}(\hat{\rho}_h) = \left[\frac{1}{T}\left(1 + 2\sum_{i=1}^{h-1} \hat{\rho}_i^2\right)\right]^{1/2}$.

Example 7.1 Suppose that the application of the t-test on autocorrelations yields $\rho_1 \neq 0$ and $\rho_2 = \ldots = \rho_H = 0$. The process identified is then an $MA(1)$ since the autocorrelations cancel out from rank $q + 1$, with $q = 1$.

Partial Autocorrelation Function

It is also possible to construct a test of the null hypothesis that the partial autocorrelations are not significantly different from zero, i.e., $\phi_{hh} = 0$. For large samples, the partial autocorrelations follow a normal distribution with mean zero and variance $1/T$. To test the null hypothesis of nullity of the partial autocorrelations, we calculate the test statistic: $t_{\hat{\phi}_{hh}} = \frac{\widehat{\phi_{hh}}}{\sqrt{1/T}}$. The value obtained is compared to the critical value read from the Student's t distribution table. The decision rule is:

- If $|t_{\hat{\phi}_{hh}}| < t(T - l)$, we do not reject the null hypothesis: ϕ_{hh} is not significantly different from zero.
- If $|t_{\hat{\phi}_{hh}}| \geq t(T - l)$, we reject the null hypothesis: ϕ_{hh} is significantly different from zero.,

where $t(T - l)$ is the value of the Student's t distribution with $(T - l)$ degrees of freedom, l being the number of estimated parameters.

This test enables us to identify the order p of the AR processes, since we know that the partial autocorrelations of a process $AR(p)$ cancel out from rank $p + 1$.

Example 7.2 Suppose that the application of the t-test on partial autocorrelations yields $\phi_{11} \neq 0$ and $\phi_{22} = \ldots = \phi_{HH} = 0$. The process identified is then an $AR(1)$ since the partial autocorrelations cancel out from rank $p + 1$, with $p = 1$.

At the end of this identification stage, one or more models have been selected. It is now necessary to estimate each selected model, which is the object of the second step of the Box and Jenkins procedure.

Step 2: Estimation of ARMA Processes

After identifying the values p and q of one or more ARMA processes, the next step is to estimate the coefficients associated with the autoregressive and moving-average terms. In some cases, notably for $AR(p)$ processes with no autocorrelation of the errors, it is possible to apply the OLS method. More generally, we use the maximum likelihood method or nonlinear least squares. We will not describe these estimation techniques here and refer readers to Gouriéroux and Monfort (2008) or Greene (2020).

Step 3: Validation of ARMA Processes

At the beginning of this step, we have several ARMA processes whose parameters have been estimated. We now need to validate these models in order to distinguish between them. To do this, we apply tests on the coefficients and on the residuals:

- With regard to the coefficients, these are the usual significance tests (t-tests). As these tests are identical to those presented in the previous chapters, we will not repeat them here. Let us simply note that if some of the estimated coefficients are not significant, the estimation must be repeated by deleting the variable(s) associated with the non-significant coefficients.
- With regard to the residuals, the aim is to test whether they have the "good" statistical properties. In particular, we need to test whether the residuals are homoskedastic and not autocorrelated.

If several models are validated, the validation step should continue with a comparison between these models.

Tests on Residuals

The purpose of these tests is to verify that the residuals $e_t = \frac{\widehat{\Phi}(L)}{\widehat{\Theta}(L)} Y_t$ do follow a white noise process. To this end, we apply tests of absence of autocorrelation and tests of homoskedasticity. These various tests have already been presented in detail in Chap. 4 and remain valid in the context of ARMA processes. Thus, in order to test the null hypothesis of no autocorrelation, the Breusch-Godfrey, Box-Pierce, or Ljung-Box tests can be applied. Similarly, to test the null hypothesis of homoskedasticity, the tests of Goldfeld and Quandt, Glejser, Breusch-Pagan, White, or the ARCH test can be implemented.

The tests most commonly used in time series econometrics are the Box-Pierce or Ljung-Box tests with regard to absence of autocorrelation, and the ARCH test with regard to homoskedasticity. It is worth clarifying the number of degrees of freedom associated with the Box-Pierce and Ljung-Box tests. Under the null hypothesis of no autocorrelation, these two statistics have a Chi-squared distribution with $(H - p - q)$ degrees of freedom, where H is the maximum number of lags considered for calculating autocorrelations, p is the order of the autoregressive part, and q is the order of the moving-average part.

Once the various tests have been applied, several models can be validated. It remains for us to compare them in an attempt to select the most "adequate" model. To this end, various model selection criteria can be used.

Model Selection Criteria

There are several types of criteria that can be used to compare validated models:

- *Standard criteria*: they are based on the calculation of the forecast error that we seek to minimize. In this context, the most frequently used criteria are:
  - The **mean absolute error**:

$$MAE = \frac{1}{T} \sum_t |e_t| \tag{7.81}$$

– The **root mean squared error**:

$$RMSE = \sqrt{\frac{1}{T}\sum_t e_t^2} \tag{7.82}$$

– The **mean absolute percent error**:

$$MAPE = 100\frac{1}{T}\sum_t \left|\frac{e_t}{Y_t}\right| \tag{7.83}$$

where T is the number of observations in the series Y_t studied and e_t are the residuals.

The lower the value taken by these criteria, the closer the estimated model is to the observations.

– *Information criteria*: we have already presented them in Chap. 3. The most widely used criteria are those of Akaike, Schwarz, and, to a lesser extent, Hannan-Quinn:

– The **Akaike information criterion** (1969):[9]

$$AIC = \log\widehat{\sigma}_\varepsilon^2 + \frac{2(p+q)}{T} \tag{7.84}$$

– The **Schwarz information criterion** (1978):

$$SIC = \log\widehat{\sigma}_\varepsilon^2 + (p+q)\frac{\log T}{T} \tag{7.85}$$

– The **Hannan-Quinn information criterion** (1979):[10]

$$HQ = \log\widehat{\sigma}_\varepsilon^2 + 2(p+q)\frac{\log(\log T)}{T} \tag{7.86}$$

We seek to minimize these various criteria. Their application allows us to select a model among the various validated ARMA processes.

Step 4: Prediction of ARMA Processes

The final step in the Box and Jenkins methodology is the prediction step. Consider a process $ARMA(p,q)$:

$$\Phi(L)\,Y_t = \Theta(L)\,\varepsilon_t \tag{7.87}$$

[9] See also Akaike (1969, 1974).

[10] It is assumed here that the constant c in the expression of the HQ criterion is equal to 1.

and let $\widehat{Y}_{t+h}$ denote the forecast made at t for the date $t + h$, with h denoting the forecast horizon. By definition, we have the following expression:

$$\widehat{Y}_{t+h} = E[Y_{t+h}|I_t] \tag{7.88}$$

where I_t is the set of information available at date t, i.e., $I_t = (Y_1, Y_2, \ldots, Y_t, \varepsilon_1, \varepsilon_2, \ldots, \varepsilon_t)$. The expectation here is taken in the sense of conditional expectation: it represents the best forecast of the series Y conditionally on the set of available information. In the linear case, it is a regression function.

Let us take the example of a process $ARMA(1, 1)$

$$Y_t = \phi_1 Y_{t-1} + \varepsilon_t - \theta_1 \varepsilon_{t-1} \tag{7.89}$$

with $|\phi_1| < 1$ and $|\theta_1| < 1$. Let us calculate the forecasts for various horizons.

- $Y_{t+1} = \phi_1 Y_t + \varepsilon_{t+1} - \theta_1 \varepsilon_t$
  $\widehat{Y}_{t+1} = E[Y_{t+1}|I_t] = \phi_1 Y_t - \theta_1 \varepsilon_t$
- $Y_{t+2} = \phi_1 Y_{t+1} + \varepsilon_{t+2} - \theta_1 \varepsilon_{t+1}$
  $\widehat{Y}_{t+2} = E[Y_{t+2}|I_t] = \phi_1 \widehat{Y}_{t+1}$

We deduce the following relationship giving the series of recursive forecasts:

$$\widehat{Y}_{t+h} = \phi_1 \widehat{Y}_{t+h-1} \quad \forall\, h > 1 \tag{7.90}$$

We can calculate the **forecast error**, $e_{t+h} = Y_{t+h} - \widehat{Y}_{t+h}$, and construct a **prediction interval**:

$$\widehat{Y}_{t+h} \pm u \times \sigma_{e_{t+h}} \tag{7.91}$$

assuming that the residuals follow a Gaussian white noise process, with u being the value of the standard normal distribution at the selected significance level (at the 5% level, $u = 1.96$). It is then possible to impart a certain degree of confidence to the forecast if the value of the dependent variable, for the horizon considered, lies within the prediction interval.

7.3.3 Empirical Application

Consider again the series RSP of the returns of Standard and Poor's stock index at monthly frequency over the period from February 1980 to June 2021. As we have previously shown, this series is stationary and can, therefore, be modeled by an ARMA-type process. To this end, let us take up the four steps of the Box and Jenkins methodology.

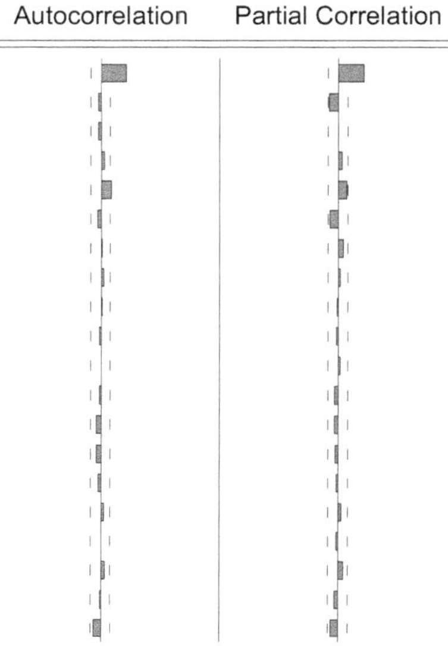

	AC	PAC	Q-Stat	Prob
1	0.225	0.225	25.466	0.000
2	-0.022	-0.077	25.712	0.000
3	-0.020	0.003	25.918	0.000
4	0.036	0.041	26.564	0.000
5	0.090	0.075	30.672	0.000
6	-0.027	-0.067	31.046	0.000
7	0.010	0.043	31.092	0.000
8	0.030	0.017	31.544	0.000
9	0.010	-0.007	31.590	0.000
10	-0.010	-0.013	31.640	0.000
11	0.005	0.020	31.650	0.001
12	-0.014	-0.031	31.755	0.002
13	-0.038	-0.032	32.491	0.002
14	-0.034	-0.017	33.092	0.003
15	-0.017	-0.008	33.245	0.004
16	0.030	0.031	33.723	0.006
17	0.001	-0.010	33.723	0.009
18	0.034	0.048	34.339	0.011
19	-0.007	-0.026	34.361	0.017
20	-0.065	-0.058	36.589	0.013

Fig. 7.12 Correlogram of the series RSP

Step 1: Identification

In order to identify the orders p and q, let us consider the graph of autocorrelations and partial autocorrelations of the series RSP. Examining Fig. 7.12 shows that:

- The first autocorrelation falls outside the confidence interval, being significantly different from zero. From order 2 onwards, the autocorrelations cancel out. We deduce $q = 1$.
- The first partial autocorrelation lies outside the confidence interval, and is significantly different from zero. From order 2 onwards, the partial autocorrelations cancel out. We deduce $p = 1$.

At the end of this step, we identify three processes: $AR(1)$, $MA(1)$, and $ARMA(1, 1)$. We can now estimate each of these models.

Step 2: Estimation

We estimate the three processes identified: $AR(1)$ (Table 7.9), $MA(1)$ (Table 7.10), and $ARMA(1, 1)$ (Table 7.11).

Table 7.9 Estimation of the process $AR(1)$

Dependent variable: RSP				
Sample: 1980M01 2021M06				
Included observations: 498				
Variable	Coefficient	Std. error	t-Statistic	Prob.
C	0.004854	0.002402	2.020679	0.0439
AR(1)	0.225054	0.036249	6.208603	0.0000
R-squared	0.050841	Mean dependent var		0.004844
Adjusted R-squared	0.047006	S.D. dependent var		0.036778
S.E. of regression	0.035904	Akaike info criterion		−3.809845
Sum squared resid	0.638091	Schwarz criterion		−3.784480
Log likelihood	951.6514	Hannan-Quinn criterion		−3.799890
F-statistic	13.25728	Durbin-Watson stat		1.964289
Prob(F-statistic)	0.000002			

Table 7.10 Estimation of the process $MA(1)$

Dependent variable: RSP				
Sample: 1980M01 2021M06				
Included observations: 498				
Variable	Coefficient	Std. error	t-Statistic	Prob.
C	0.004849	0.002277	2.129807	0.0337
MA(1)	0.245678	0.035607	6.899709	0.0000
R-squared	0.055787	Mean dependent var		0.004844
Adjusted R-squared	0.051972	S.D. dependent var		0.036778
S.E. of regression	0.035810	Akaike info criterion		−3.815048
Sum squared resid	0.634766	Schwarz criterion		−3.789683
Log likelihood	952.9471	Hannan-Quinn criterion		−3.805094
F-statistic	14.62309	Durbin-Watson stat		2.004813
Prob(F-statistic)	0.000001			

Step 3: Validation

Tests of Significance of Coefficients

Let us first proceed to the significance of the coefficients in each of the three estimated models:

- $AR(1)$ process: the first-order autoregressive coefficient is significantly different from zero as its t-statistic 6.2086 is higher than the critical value 1.96 at the 5% significance level. The $AR(1)$ model is therefore a candidate for the modeling of RSP.
- $MA(1)$ process: the first-order moving-average coefficient is significantly different from zero as its t-statistic 6.8997 is higher than the critical value 1.96 at the 5% significance level. The $MA(1)$ model is therefore a candidate for modeling RSP.

Table 7.11 Estimation of the process $ARMA(1)$

Dependent variable: RSP				
Sample: 1980M01 2021M06				
Included observations: 498				
Variable	Coefficient	Std. error	t-Statistic	Prob.
C	0.004848	0.002286	2.120727	0.0344
AR(1)	−0.049048	0.180326	−0.271997	0.7857
MA(1)	0.291703	0.175861	1.658713	0.0978
R-squared	0.055927	Mean dependent var		0.004844
Adjusted R-squared	0.050194	S.D. dependent var		0.036778
S.E. of regression	0.035844	Akaike info criterion		−3.811180
Sum squared resid	0.634672	Schwarz criterion		−3.777360
Log likelihood	952.9839	Hannan-Quinn criterion		−3.797907
F-statistic	9.754938	Durbin-Watson stat		1.999172
Prob(F-statistic)	0.000003			

- $ARMA(1, 1)$ process: the t-statistic associated with the autoregressive and moving-average coefficients being less than 1.96 in absolute value, none of the coefficients is significantly different from zero. We can therefore reject the $ARMA(1, 1)$ model.

At the end of this first phase of the validation stage, two processes are candidates for modeling the series RSP: the $AR(1)$ and the $MA(1)$ processes.

Tests on Residuals

We now apply the tests to the residuals of the $AR(1)$ and $MA(1)$ models. We start with the Ljung-Box test of absence of autocorrelation. The results are shown in Figs. 7.13 and 7.14. These figures first show that the autocorrelations of the residuals lie within the confidence interval for each of the two models, suggesting the absence of autocorrelation. Let us calculate the Ljung-Box statistic for a maximum number of lags H of 20:

- For the residuals of the $AR(1)$ model, we have $LB(20) = 14.333$. Under the null hypothesis of no autocorrelation, this statistic follows a Chi-squared distribution with $(H - p - q) = (20 - 1 - 0) = 19$ degrees of freedom. At the 5% significance level, the critical value of the Chi-squared distribution with 19 degrees of freedom is 30.144. Since $14.333 < 30.144$, we do not reject the null hypothesis of no autocorrelation of residuals. The model $AR(1)$ therefore remains a candidate.
- For the residuals of the $MA(1)$ model, we have $LB(20) = 11.388$. Under the null hypothesis of no autocorrelation, this statistic has a Chi-squared distribution with $(H - p - q) = (20-0-1) = 19$ degrees of freedom, the corresponding critical value being 30.144 at the 5% significance level. We find that $11.388 < 30.144$,

Q-statistic probabilities adjusted for 1 ARMA term

Autocorrelation	Partial Correlation		AC	PAC	Q-Stat	Prob
		1	0.018	0.018	0.1574	
		2	-0.073	-0.073	2.8443	0.092
		3	-0.026	-0.023	3.1718	0.205
		4	0.023	0.019	3.4385	0.329
		5	0.098	0.094	8.2691	0.082
		6	-0.054	-0.055	9.7296	0.083
		7	0.010	0.027	9.7806	0.134
		8	0.029	0.024	10.195	0.178
		9	0.006	0.001	10.213	0.250
		10	-0.014	-0.018	10.317	0.325
		11	0.011	0.024	10.377	0.408
		12	-0.008	-0.019	10.410	0.494
		13	-0.030	-0.032	10.879	0.539
		14	-0.025	-0.022	11.192	0.595
		15	-0.018	-0.020	11.359	0.658
		16	0.037	0.028	12.063	0.674
		17	-0.014	-0.014	12.169	0.732
		18	0.040	0.051	12.980	0.738
		19	0.000	-0.001	12.980	0.793
		20	-0.051	-0.045	14.333	0.764

Fig. 7.13 Correlogram of the residuals of the AR(1) model

implying that the null hypothesis of no autocorrelation of residuals is not rejected. The $MA(1)$ model therefore remains a candidate.

Let us now apply the ARCH test to check that the residuals of both models are indeed homoskedastic. This test involves regressing the squared residuals on a constant and their ℓ past values:

$$e_t^2 = a_0 + \sum_{i=1}^{\ell} a_i e_{t-i}^2 \tag{7.92}$$

and testing the null hypothesis of homoskedasticity:

$$H_0 : a_1 = a_2 = \ldots = a_\ell = 0 \tag{7.93}$$

against the alternative hypothesis of conditional heteroskedasticity, which states that at least one of the coefficients a_i, $i = 1, \ldots, \ell$, is significantly different from zero. Under the null hypothesis of homoskedasticity, the test statistic TR^2, where T is the number of observations and R^2 is the coefficient of determination associated with the regression (7.92), follows a Chi-squared distribution with ℓ degrees of freedom.

Q-statistic probabilities adjusted for 1 ARMA term

Autocorrelation	Partial Correlation		AC	PAC	Q-Stat	Prob
		1	-0.003	-0.003	0.0034	
		2	-0.017	-0.017	0.1453	0.703
		3	-0.020	-0.020	0.3502	0.839
		4	0.018	0.017	0.5058	0.918
		5	0.098	0.098	5.4056	0.248
		6	-0.054	-0.053	6.8662	0.231
		7	0.016	0.020	7.0039	0.320
		8	0.024	0.027	7.3081	0.398
		9	0.007	0.002	7.3338	0.501
		10	-0.014	-0.021	7.4324	0.592
		11	0.010	0.022	7.4854	0.679
		12	-0.010	-0.018	7.5323	0.754
		13	-0.030	-0.034	8.0006	0.785
		14	-0.022	-0.020	8.2520	0.827
		15	-0.021	-0.020	8.4747	0.863
		16	0.038	0.031	9.2276	0.865
		17	-0.017	-0.013	9.3782	0.897
		18	0.040	0.047	10.189	0.895
		19	-0.005	-0.002	10.200	0.925
		20	-0.048	-0.046	11.388	0.910

Fig. 7.14 Correlogram of the residuals of the MA(1) model

Table 7.12 ARCH test results

	$AR(1)$	$MA(1)$
a_0	0.0012 (6.9184)	0.0012 (6.9287)
a_1	0.0891 (1.9891)	0.0941 (2.1032)
TR^2	3.9413	4.4021

Values in parentheses are t-statistics of the estimated coefficients

We have estimated the relationship (7.92) on the squares of the residuals of each of the two models considered, using a number of lags $\ell = 1$. The results are shown in Table 7.12. For both models, the critical value to which the TR^2 test statistic must be compared is that of the Chi-squared distribution with 1 degree of freedom, i.e., 3.841 at the 5% significance level. It can be seen that, for both models, the calculated value of the test statistic is higher than the critical value. The null hypothesis of homoskedasticity is consequently rejected at the 5% significance level.

In summary, the residuals of the $AR(1)$ and $MA(1)$ models are not autocorrelated, but (slightly) heteroskedastic. Both models therefore pass the validation stage from the point of view of the absence of autocorrelation, but not from the point of view of the homoskedasticity property. This result is not surprising insofar as the study concerns financial series that are known to exhibit heteroskedasticity due to their time-varying volatility.

Table 7.13 Model comparison criteria

	$AR(1)$	$MA(1)$
$RMSE$	0.035794	**0.035702**
MAE	0.025478	**0.025390**
$MAPE$	**357.1195**	397.9688
AIC	−3.809845	**−3.815048**
SIC	−3.784480	**−3.789683**
HQ	−3.799890	**−3.805094**

Values in bold correspond to values minimizing model selection criteria

Model Selection Criteria

To conclude the validation step, let us compare the two models using model selection criteria. Table 7.13 summarizes the values obtained. These results show that the $AR(1)$ model minimizes only the MAPE criterion, the other five criteria being minimized by the $MA(1)$ model. If one of the two models has to be chosen, the $MA(1)$ model should be preferred for the returns of the US stock market index. It is then possible to forecast returns based on the $MA(1)$ process. Strictly speaking, neither model should be retained insofar as they do not pass the validation step due to heteroskedasticity in the errors.

7.4 Extension to the Multivariate Case: VAR Processes

The **$VAR(p)$ (vector autoregressive) processes** are a generalization of the autoregressive processes to the multivariate case. They were introduced by Sims (1980) as an alternative to structural macroeconometric models, i.e., **simultaneous equations models** (see Chap. 8). According to Sims (1980), these macroeconometric models can be criticized on several points, including the existence of a priori restrictions on parameters that are too strong in relation to what the theory predicts, the simultaneity of the relationships, the assumed exogeneity of certain variables, and poor predictive quality. VAR models have been developed in response to these various criticisms. Their essential characteristic is that they no longer distinguish between endogenous and exogenous variables, in the sense that all the variables in the model have the same status.

VAR models have been the subject of many developments, and we will only present their general structure here (for more details, readers can refer to Hamilton, 1994; Lardic and Mignon, 2002; or Greene, 2020).

7.4.1 Writing the Model

Introductory Example

Consider two stationary variables Y_{1t} and Y_{2t}. Each variable is a function of its own past values, but also of the past and present values of the other variables. Suppose we have $p = 4$. The $VAR(4)$ model describing these two variables is written as:

$$\begin{cases} Y_{1t} = a_1 + \sum_{i=1}^{4} b_{1i} Y_{1t-i} + \sum_{j=1}^{4} c_{1j} Y_{2t-j} - d_1 Y_{2t} + \varepsilon_{1t} \\ Y_{2t} = a_2 + \sum_{i=1}^{4} b_{2i} Y_{1t-i} + \sum_{j=1}^{4} c_{2j} Y_{2t-j} - d_2 Y_{1t} + \varepsilon_{2t} \end{cases} \tag{7.94}$$

where ε_{1t} and ε_{2t} are two uncorrelated white noise processes.

This model involves estimating 20 coefficients. The number of parameters to be estimated grows rapidly with the number of lags, as pN^2, where p is the number of lags and N the number of variables in the model.

In matrix form, the $VAR(4)$ process is written:

$$\boldsymbol{B}\boldsymbol{Y}_t = \boldsymbol{\Phi}_0 + \sum_{i=1}^{4} \boldsymbol{\Phi}_i \boldsymbol{Y}_{t-i} + \boldsymbol{\varepsilon}_t \tag{7.95}$$

with:

$$\boldsymbol{B} = \begin{pmatrix} 1 & d_1 \\ d_2 & 1 \end{pmatrix} \quad \boldsymbol{\Phi}_0 = \begin{pmatrix} a_1 \\ a_2 \end{pmatrix} \quad \boldsymbol{Y}_t = \begin{pmatrix} Y_{1t} \\ Y_{2t} \end{pmatrix}$$
$$\boldsymbol{\Phi}_i = \begin{pmatrix} b_{1i} & c_{1i} \\ b_{2i} & c_{2i} \end{pmatrix} \quad \boldsymbol{\varepsilon}_t = \begin{pmatrix} \varepsilon_{1t} \\ \varepsilon_{2t} \end{pmatrix} \tag{7.96}$$

Then we simply multiply each term of (7.95) by $\boldsymbol{B}^{-1}$, assuming $\boldsymbol{B}$ invertible, to obtain the usual form of the VAR model.

General Formulation

We generalize the previous example to the case where $\boldsymbol{Y}_t$ contains N variables and for any order of lags p. A $VAR(p)$ process with N variables is written in matrix form:

$$\boldsymbol{Y}_t = \boldsymbol{\Phi}_0 + \boldsymbol{\Phi}_1 \boldsymbol{Y}_{t-1} + \cdots + \boldsymbol{\Phi}_p \boldsymbol{Y}_{t-p} + \boldsymbol{\varepsilon}_t \tag{7.97}$$

$$\boldsymbol{Y}_t = \begin{pmatrix} Y_{1t} \\ \vdots \\ Y_{Nt} \end{pmatrix} \quad \boldsymbol{\varepsilon}_t = \begin{pmatrix} \varepsilon_{1t} \\ \vdots \\ \varepsilon_{Nt} \end{pmatrix} \quad \boldsymbol{\Phi}_0 = \begin{pmatrix} a_1^0 \\ \vdots \\ a_N^0 \end{pmatrix}$$
$$\boldsymbol{\Phi}_p = \begin{pmatrix} a_{1p}^1 & a_{1p}^2 & \cdots & a_{1p}^N \\ \vdots & \vdots & & \vdots \\ a_{Np}^1 & a_{Np}^2 & \cdots & a_{Np}^N \end{pmatrix} \tag{7.98}$$

where $\boldsymbol{\varepsilon}_t$ is white noise with variance-covariance matrix $\boldsymbol{\Sigma}_\varepsilon$.

We can also write:

$$(\boldsymbol{I} - \boldsymbol{\Phi}_1 L - \boldsymbol{\Phi}_2 L^2 - \cdots - \boldsymbol{\Phi}_p L^p)\boldsymbol{Y}_t = \boldsymbol{\Phi}_0 + \boldsymbol{\varepsilon}_t \tag{7.99}$$

or:

$$\boldsymbol{\Phi}(L)\boldsymbol{Y}_t = \boldsymbol{\Phi}_0 + \boldsymbol{\varepsilon}_t \tag{7.100}$$

with $\boldsymbol{\Phi}(L) = \boldsymbol{I} - \sum_{i=1}^{p} \boldsymbol{\Phi}_i L^i$.

More formally, the following definition is used.

Definition 7.8 $\boldsymbol{Y}_t$ follows a $\boldsymbol{VAR(p)}$ **process** if and only if there exist white noise $\boldsymbol{\varepsilon}_t$ ($\boldsymbol{\varepsilon}_t \sim WN(0, \boldsymbol{\Sigma}_\varepsilon)$), $\boldsymbol{\Phi}_0 \in R^N$, and p matrices $\boldsymbol{\Phi}_1, \ldots, \boldsymbol{\Phi}_p$ such that:

$$\boldsymbol{Y}_t - \sum_{i=1}^{p} \boldsymbol{\Phi}_i \boldsymbol{Y}_{t-i} = \boldsymbol{\Phi}_0 + \boldsymbol{\varepsilon}_t \tag{7.101}$$

or:

$$\boldsymbol{\Phi}(L)\boldsymbol{Y}_t = \boldsymbol{\Phi}_0 + \boldsymbol{\varepsilon}_t \tag{7.102}$$

where $\boldsymbol{\Phi}_0$ is the identity matrix ($\boldsymbol{I}$) and:

$$\boldsymbol{\Phi}(L) = \boldsymbol{I} - \sum_{i=1}^{p} \boldsymbol{\Phi}_i L^i \tag{7.103}$$

7.4.2 Estimation of the Parameters of a $VAR(p)$ Process and Validation

The parameters of the VAR process can only be estimated on stationary time series.[11] Two estimation techniques are possible: estimation of each equation of the VAR model by OLS or estimation by the maximum likelihood technique. The estimation of a VAR model involves choosing the number of lags p. To determine this value, the information criteria can be used. The procedure consists in estimating a number of VAR models for an order p ranging from 0 to h where h is the maximum

[11] Strictly speaking, it is possible to estimate VAR processes in which non-stationary variables are involved using OLS. In this case, the estimators are super-consistent, but they are no longer asymptotically normal, which poses a problem for statistical inference since the usual tests can no longer be implemented.

lag. We select the lag p that minimizes the information criteria AIC, SIC, and HQ[12] defined as follows:

$$AIC = \log \det \hat{\boldsymbol{\Sigma}}_{\varepsilon} + \frac{2N^2 p}{T} \tag{7.104}$$

$$SIC = \log \det \hat{\boldsymbol{\Sigma}}_{\varepsilon} + N^2 p \frac{\log T}{T} \tag{7.105}$$

$$HQ = \log \det \hat{\boldsymbol{\Sigma}}_{\varepsilon} + 2N^2 p \frac{\log(\log T)}{T} \tag{7.106}$$

where N is the number of variables in the system, T is the number of observations, and $\hat{\boldsymbol{\Sigma}}_{\varepsilon}$ is an estimator of the variance-covariance matrix of the residuals, *det* denoting its determinant.

Remark 7.8 It is also possible to perform maximum likelihood ratio tests to validate the number of lags p selected. Generally speaking, we test:

H_0 : $\boldsymbol{\Phi}_{p+1} = 0$: process $VAR(p)$

H_1 : $\boldsymbol{\Phi}_{p+1} \neq 0$: process $VAR(p+1)$

The technique involves estimating a constrained model ($VAR(p)$) and an unconstrained model ($VAR(p+1)$) and performing the log-likelihood ratio test. If the null hypothesis is not rejected, we continue the procedure by testing:

H_0 : $\boldsymbol{\Phi}_p = 0$: process $VAR(p-1)$

H_1 : $\boldsymbol{\Phi}_p \neq 0$: process $VAR(p)$

We thus have a sequence of nested tests whose goal is to determine the order p of the VAR process.

Remark 7.9 In the case of AR processes, in addition to the tests on the parameters, tests on the residuals are performed in order to validate the process. In the case of VAR processes, these tests are not very powerful, and we prefer to use a graph of the residuals. Residuals should be examined carefully especially when using VAR models for impulse response analysis, where the absence of correlation of the innovations is crucial for the interpretation.

7.4.3 Forecasting VAR Processes

Consider a process $VAR(p)$:

$$\boldsymbol{Y}_t = \boldsymbol{\Phi}_1 \boldsymbol{Y}_{t-1} + \ldots + \boldsymbol{\Phi}_p \boldsymbol{Y}_{t-p} + \boldsymbol{\varepsilon}_t \tag{7.107}$$

[12] We have assumed here that the constant c involved in the expression of the HQ criterion is equal to 1.

It is assumed that p has been chosen, that the $\boldsymbol{\Phi}_i$ have been estimated, and that the variance-covariance matrix associated with $\boldsymbol{\varepsilon}_t$ has been estimated.

Under certain conditions, the prediction in $(T+1)$ of the process is:

$$E\left[\boldsymbol{Y}_{T+1} \left| \underline{\boldsymbol{Y}_T} \right.\right] = \hat{\boldsymbol{\Phi}}_1 \boldsymbol{Y}_T + \ldots + \hat{\boldsymbol{\Phi}}_p \boldsymbol{Y}_{T-p+1} \tag{7.108}$$

where $\underline{\boldsymbol{Y}_T}$ denotes the past from $\boldsymbol{Y}$ up to and including the date T.

7.4.4 Granger Causality

The notion of **causality** plays a very important role in economics insofar as it enables us to better understand the relationships between variables. To introduce this notion, consider two variables Y_1 and Y_2. Heuristically, we say that Y_1 Granger causes Y_2 if the prediction of Y_2 based on the knowledge of the joint pasts of Y_1 and Y_2 is better than the forecast based on the knowledge of the past of Y_2 alone (see Granger, 1969). As an example, consider the following $VAR(p)$ process with two variables Y_{1t} and Y_{2t}:

$$\begin{pmatrix} Y_{1t} \\ Y_{2t} \end{pmatrix} = \begin{pmatrix} a_0 \\ b_0 \end{pmatrix} + \begin{pmatrix} a_1^1 & b_1^1 \\ a_1^2 & b_1^2 \end{pmatrix} \begin{pmatrix} Y_{1t-1} \\ Y_{2t-1} \end{pmatrix} + \ldots + \begin{pmatrix} a_p^1 & b_p^1 \\ a_p^2 & b_p^2 \end{pmatrix} \begin{pmatrix} Y_{1t-p} \\ Y_{2t-p} \end{pmatrix} + \begin{pmatrix} \varepsilon_{1t} \\ \varepsilon_{2t} \end{pmatrix} \tag{7.109}$$

Testing for the absence of causality from Y_{1t} to Y_{2t} is equivalent to performing a restriction test on the coefficients of the variables Y_{1t} of the VAR representation. Specifically:

- Y_{1t} does not cause Y_{2t} if the following null hypothesis is not rejected: $H_0 : b_1^1 = b_2^1 = \cdots = b_p^1 = 0$.
- Y_{2t} does not cause Y_{1t} if the following null hypothesis is not rejected: $H_0 : a_1^2 = a_2^2 = \cdots = a_p^2 = 0$.

These are classic Fisher tests. They are performed either equation by equation, or directly by comparison between a constrained VAR model and an unconstrained VAR model. In the latter case, we can also perform a maximum likelihood ratio test.

In the case of Fisher tests, the strategy is as follows for a test of absence of causality from Y_{1t} to Y_{2t}:

- We regress Y_{2t} on its p past values and on the p past values of Y_{1t}. This is the unconstrained model and we note RSS_{nc} the sum of squared residuals associated with this model.

- We regress Y_{2t} on its p past values and note the sum of squared residuals RSS_c. This is the constrained model in that we have imposed the nullity of the coefficients associated with the p values of Y_{1t}.
- We calculate the test statistic:

$$F = \frac{(RSS_c - RSS_{nc})/r}{RSS_{nc}/(T-k-1)} \tag{7.110}$$

where r is the number of constraints, i.e., the number of coefficients being tested for nullity, and k is the number of estimated parameters (excluding the constant) involved in the unconstrained model. Under the null hypothesis of no causality, this statistic has a Fisher distribution with $(r, T-k-1)$ degrees of freedom.

In the case of a maximum likelihood ratio test, we calculate the test statistic:

$$C = T \log \left(\frac{\det \hat{\boldsymbol{\Sigma}}_{\varepsilon}^{c}}{\det \hat{\boldsymbol{\Sigma}}_{\varepsilon}^{nc}} \right) \tag{7.111}$$

where $\hat{\boldsymbol{\Sigma}}_{\varepsilon}^{c}$ (respectively $\hat{\boldsymbol{\Sigma}}_{\varepsilon}^{nc}$) denotes the estimator of the variance-covariance matrix of the residuals of the constrained (respectively unconstrained) model, det being the determinant. Under the null hypothesis of no causality, this statistic follows a Chi-squared distribution with $2p$ degrees of freedom.

Remark 7.10 If we reject the two null hypotheses (absence of causality from Y_1 to Y_2 and absence of causality from Y_2 to Y_1), we have a bi-directional causality; we speak of **feedback loop** (**feedback effect**).

Remark 7.11 One of the practical applications of VAR models lies in the calculation of the **impulse response function**. The latter makes it possible to assess the effect of a random shock on the variables and can therefore be useful for analyzing the effects of an economic policy. This analysis is beyond the scope of this book and we refer the reader to Hamilton (1994), Lardic and Mignon (2002), or Greene (2020).

7.4.5 Empirical Application

Consider Standard and Poor's 500 (SP 500) US stock index series and the associated dividend series over the period from January 1871 to June 2021. Since the data are monthly, the number of observations is 1 806. The series are expressed in real terms, i.e., they have been deflated by the consumer price index.[13]

[13] The data come from Robert Shiller's website: http://www.econ.yale.edu/~shiller/data.htm.

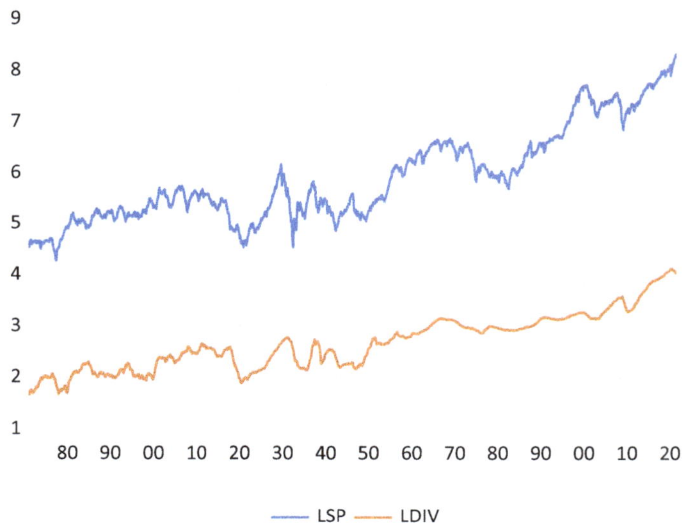

Fig. 7.15 Series LSP and LDIV, 1871.01–2021.06

Table 7.14 ADF test results

	ADF	CV at 5%	CV at 1%
LSP	1.7308	−1.95	−2.58
LDIV	−3.2391	−3.41	−3.96
DLSP	−32.3319	−1.95	−2.58
DLDIV	−13.0135	−1.95	−2.58

CV critical value

We denote LSP the logarithm of the SP 500 index and $LDIV$ the dividend series in logarithms. We are interested in the relationship between the two series, seeking to estimate a VAR-type model. Let us start by studying the characteristics of the two variables in terms of stationarity.

The two series are shown in Fig. 7.15 and appear to exhibit an overall upward trend, suggesting that they are non-stationary in the mean. In order to confirm this intuition, we perform the Dickey-Fuller unit root tests. To do this, we follow the sequential strategy presented earlier. First, we estimate the model with constant and trend. If the trend is not significant, we estimate the model with constant. Finally, if the constant is not significant, we estimate the model without constant or trend. The implementation of this strategy leads us to select:

– A model without constant or trend for the series LSP
– A model with constant and trend for the series $LDIV$

The results obtained for the value of the ADF statistic are shown in Table 7.14.

Table 7.15 Choice of p, VAR estimation

p	AIC	SIC	HQ
0	−9.230339	−9.224213	−9.228077
1	−9.664253	−9.645876	−9.657468
2	−9.700005	**−9.669376**	**−9.688696**
3	−9.702788	−9.659908	−9.686956
4	−9.709008	−9.653877	−9.688653
5	−9.712631	−9.645249	−9.687753
6	−9.712146	−9.632512	−9.682744
7	−9.711237	−9.619352	−9.677312
8	−9.710607	−9.606471	−9.672159
9	−9.709459	−9.593072	−9.666488
10	−9.708673	−9.580035	−9.661179
11	−9.706098	−9.565208	−9.65408
12	**−9.713336**	−9.560195	−9.656794

Values in bold correspond to values minimizing information criteria

It can be seen that the null hypothesis of unit root cannot be rejected for the two series considered LSP and $LDIV$. The application of the Dickey-Fuller tests on the series in first difference (denoted $DLSP$ and $DLDIV$) indicates that they are stationary. In other words, the differentiated series are integrated of order 0, implying that the series LSP and $LDIV$ are integrated of order 1.

The VAR model is then estimated on the series $DLSP$ and $DLDIV$, i.e., on the stationary series. We start by looking for the order p of the VAR process. To this end, we estimate the VAR process for values of p ranging from 1 to 12 and report the values taken by the AIC, SIC, and HQ criteria (see Table 7.15). The SIC and HQ criteria lead us to select a $VAR(2)$ process, whereas, according to the AIC criterion, we should select a $VAR(12)$ process. For the sake of parsimony, we continue the study with the $VAR(2)$ process.

The results from the estimation of the $VAR(2)$ process are shown in Table 7.16; the values in square brackets represent the t-statistics of the estimated coefficients. It can be seen that the SP returns are a function of themselves lagged by one and two periods and of dividends lagged by two periods. The logarithmic changes in dividends are a function of their one- and two-period lagged values and of the 1-month lagged values of the SP returns.

Let us now perform the Granger causality test and start by implementing Fisher tests. First, let us test the null hypothesis that the dividend growth rate does not cause the returns of the SP index. We estimate two models:

- The constrained model consisting in regressing $DLSP$ on a constant and on its values lagged by 1 and 2 months. We obtain a sum of squared residuals equal to $RSS_c = 2.7925$;
- The unconstrained model consisting in regressing $DLSP$ on a constant, its lagged values by 1 and 2 months, and the lagged values (1 and 2 months) of $DLDIV$. We obtain a sum of squared residuals equal to $RSS_{nc} = 2.7786$.

Table 7.16 Estimation of the VAR(2) process

	D(LSP)	D(LDIV)
DLSP(−1)	0.283078	−0.031786
	[11.9821]	[−4.49688]
DLSP(−2)	−0.073616	0.010414
	[−3.10058]	[1.46603]
DLDIV(−1)	−0.087335	0.459320
	[−1.11822]	[19.6562]
DLDIV(−2)	0.229337	0.165131
	[2.94504]	[7.08752]
C	0.001479	0.000546
	[1.58671]	[1.95616]
R-squared	0.078311	0.319494
Adj. R-squared	0.076260	0.317980
Sum sq. resids	2.778588	0.248729
S.E. equation	0.039311	0.011762
F-statistic	38.19147	211.0376
Log likelihood	3279.105	5454.724
Akaike AIC	−3.631841	−6.045174
Schwarz SC	−3.616596	−6.029929
Mean dependent	0.002103	0.001316
S.D. dependent	0.040902	0.014242
Determinant resid covariance (dof adj.)	2.11E-07	
Determinant resid covariance	2.10E-07	
Log likelihood	8743.940	
Akaike information criterion	−9.688231	
Schwarz criterion	−9.657742	
Number of coefficients	10	

The Fisher test statistic is:

$$F = \frac{(2.7925 - 2.7786)/2}{2.7786/(1805 - 4 - 1)} = 4.5023 \tag{7.112}$$

The number of constraints is 2 (we are testing the nullity of the two coefficients associated with the lagged dividend growth rate), the number of observations is 1 805, and the number of estimated parameters (excluding the constant) in the unconstrained model is 4. Under the null hypothesis, the F-statistic follows a Fisher distribution with (2,1800) degrees of freedom. At the 5% significance level, the critical value is 2.997. Since $4.5023 > 2.997$ we reject the null hypothesis of no causality of the dividend growth rate towards stock market returns.

Let us now consider the test of the null hypothesis that stock market returns do not cause the dividend growth rate. We estimate two models:

- The constrained model consisting in regressing $DLDIV$ on a constant and on its values lagged by one and two periods. We obtain a sum of squared residuals equal to $RSS_c = 0.2515$.
- The unconstrained model consisting in regressing $DLDIV$ on a constant, its lagged values by 1 and 2 months, and the lagged values (one and two periods) of $DLSP$. We obtain a sum of squared residuals equal to $RSS_{nc} = 0.2487$.

The Fisher test statistic is:

$$F = \frac{(0.2515 - 0.2487)/2}{0.2487/(1805 - 4 - 1)} = 10.1327 \tag{7.113}$$

If we compare this value with the critical value 2.997, we reject the null hypothesis of no causality of stock market returns towards the dividend growth rate.

We can also perform a Chi-squared test, calculating the test statistic C. The calculation of this statistic gives us:

- For the test of the null hypothesis that the dividend growth rate does not cause returns: $C = 9.0191$
- For the test of the null hypothesis that returns do not cause the dividend growth rate: $C = 20.2855$

In both cases, the statistic C is higher than the critical value of the Chi-squared distribution at the 5% significance level. The null hypothesis is rejected. There is therefore a two-way causality between stock market returns and the dividend growth rate, testifying to the presence of a feedback effect.

7.5 Cointegration and Error-Correction Models

7.5.1 The Problem of Spurious Regressions

The theory of **cointegration** was introduced by Granger (1981) to study non-stationary time series. This theory is widely used in economic and financial applications, since many macroeconomic and financial series are non-stationary. However, if we apply the usual econometric methods to non-stationary series, several problems arise, including the famous problem of **spurious regressions** addressed by Granger and Newbold (1974). Heuristically, consider two time series X_t and Y_t integrated of order 1 and without any link between them. If we run the regression $Y_t = \alpha + \beta X_t + \varepsilon_t$, we should get $\beta = 0$. Granger and Newbold (1974) show that β is significantly different from zero, meaning that X_t is an explanatory variable for Y_t, which makes no sense since, by assumption, the two series are independent. The consequence of non-stationarity is that classical inference procedures are no longer valid.

To illustrate this fundamental issue, let us give some examples of spurious regressions.[14] We also report for each estimated regression the value of the coefficient of determination R^2 and the Durbin-Watson statistic (DW). Following the usual notations, the numbers in parentheses below the estimated values of the coefficients are their t-statistics.

- *Example 1*: regression of the infant mortality rate in Egypt (MOR) on the income of US farmers (INC) and on the money supply in Honduras (M), annual data 1971–1990:

$$\widehat{MOR}_t = \underset{(16.63)}{179.9} - \underset{(-2.32)}{0.29}\, INC_t - \underset{(-4.26)}{0.04}\, M_t \tag{7.114}$$

  with $R^2 = 0.918$ and $DW = 0.47$.
- *Example 2*: regression of US exports (EXP) on Australian male life expectancy ($LIFE$), annual data 1960–1990:

$$\widehat{EXP}_t = \underset{(-16.70)}{-2943} + \underset{(17.76)}{45.79} LIFE_t \tag{7.115}$$

  with $R^2 = 0.916$ and $DW = 0.36$.
- *Example 3*: regression of South African population (POP) on US research and development expenditure (RD), annual data 1971–1990:

$$\widehat{POP}_t = \underset{(59.44)}{21698.7} + \underset{(26.40)}{111.58} RD_t \tag{7.116}$$

  with $R^2 = 0.974$ and $DW = 0.30$.

These three examples illustrate regressions that make no sense, since it is obvious that there is no link between the explanatory variables and the variable being explained in each of the three cases considered. Thus, if we take the third example, it goes without saying that finding that R&D spending in the United States has an impact on the population in South Africa makes little sense. These examples are illustrative of spurious regressions, i.e., regressions that are meaningless. This is due to the non-stationarity of the different series involved in the regressions.

Two features are common to all three regressions: firstly, the coefficient of determination is very high (above 0.9 in our examples), and, secondly, the value of the Durbin-Watson statistic is low. These two characteristics are symptomatic of spurious regressions.

[14] These examples are taken from the website of J. Gonzalo, Universidad Carlos III, Madrid.

A procedure frequently used to avoid the problem of spurious regressions is to differentiate non-stationary series in order to stationarize them and apply the usual econometric methods. However, the main limitation of this differentiation operation is that it masks the long-term properties of the series studied, since the relationships between the levels of the variables are no longer considered. The theory of cointegration alleviates this problem by offering the possibility of specifying stable long-term relationships while jointly analyzing the short-term dynamics of the variables under consideration.

7.5.2 The Concept of Cointegration

If X_t and Y_t are two series $I(d)$, then in general the linear combination z_t :

$$z_t = Y_t - \beta X_t \tag{7.117}$$

is also $I(d)$.

However, it is possible that z_t is not $I(d)$ but $I(d-b)$ where b is a positive integer ($d \geq b > 0$). In other words, z_t is integrated of an order lower than the order of integration of the two variables under consideration. In this case, the series X_t and Y_t are said to be **cointegrated**, which is noted:

$$(X_t, Y_t) \sim CI(d, b) \tag{7.118}$$

β is the **cointegration parameter** and the vector $[1, -\beta]$ is the **cointegrating vector**.

The most studied case corresponds to: $d = b = 1$. Thus, two non-stationary series ($I(1)$) are cointegrated if there exists a stationary linear combination ($I(0)$) of these two series.

The underlying idea is that, in the short term, X_t and Y_t may diverge (they are both non-stationary), but they will move in unison in the long term. There is therefore a stable long-term relationship between X_t and Y_t. This relationship is called **cointegration (or cointegrating) relationship** or the **long-term relationship**. It is given by $Y_t = \beta X_t$ (i.e., $z_t = 0$).[15] In the long term, similar movements of X_t and Y_t tend to offset each other yielding a stationary series. z_t measures the extent of the imbalance between X_t and Y_t and is called the **equilibrium error**. Examples corresponding to such a situation are numerous in economics: the relationship between consumption and income, the relationship between short- and long-term interest rates, the relationship between international stock market indices, and so on.

[15] Note that the cointegrating relationship can include a constant term, for example: $Y_t = \alpha + \beta X_t$.

Remark 7.12 For the sake of simplification, we have considered here the case of two variables. The notion of cointegration can be generalized to the case of N variables. We will not deal with this generalization in the context of this textbook and refer readers to Engle and Granger (1991), Hamilton (1994), or Lardic and Mignon (2002).

7.5.3 Error-Correction Models

One of the fundamental properties of cointegrated series is that they can be modeled as an **error-correction model**. This result was demonstrated in the **Granger representation theorem** (Granger, 1981), valid for series $CI(1, 1)$. Such models allow us to model the adjustments that lead to a long-term equilibrium situation. They are dynamic models, incorporating both short-term and long-term changes in variables.

Let X_t and Y_t be two $CI(1, 1)$ variables. Assuming that Y_t is the endogenous variable and X_t is the explanatory variable, the error-correction model is written:

$$\Delta Y_t = \gamma \hat{z}_{t-1} + \sum_i \beta_i \Delta X_{t-i} + \sum_j \delta_j \Delta Y_{t-j} + d(L)\,\varepsilon_t \tag{7.119}$$

where ε_t is white noise. $\hat{z}_t = Y_t - \hat{\beta} X_t$ is the residual from the estimation of the cointegration relationship between X_t and Y_t. $d(L)$ is a finite polynomial in L. In practice, we frequently have $d(L) = 1$ and the error-correction model is written more simply:

$$\Delta Y_t = \gamma \hat{z}_{t-1} + \sum_i \beta_i \Delta X_{t-i} + \sum_j \delta_j \Delta Y_{t-j} + \varepsilon_t \tag{7.120}$$

The coefficient γ associated with $\hat{z}_{t-1}$ is the error-correction coefficient. It provides a measure of the speed of adjustment towards the long-term target, given by the cointegration relationship. The coefficient γ must be significantly non-zero and negative for the error-correction mechanism to be present. Otherwise, there is no return-to-equilibrium phenomenon. The error-correction model allows short-term fluctuations to be accounted for around the long-term equilibrium (given by the cointegration relationship). It thus describes an adjustment process and combines two types of variables:

- First-difference (stationary) variables representing short-term fluctuations
- Variables in levels, here a variable $\hat{z}_t$ which is a stationary linear combination of non-stationary variables, which ensures that the long term is taken into account

7.5.4 Estimation of Error-Correction Models and Cointegration Tests: The Engle and Granger (1987) Approach

Two-Step Estimation Method

The two-step estimation method, valid for $CI(1, 1)$ series, was proposed by Engle and Granger (1987).

First step: Estimation of the long-term relationship.

The long-term relationship is estimated by OLS:[16]

$$Y_t = \alpha + \beta X_t + z_t \tag{7.121}$$

where z_t is the error term. If the variables are cointegrated, we proceed to the second step.

Second step: Estimation of the error-correction model.

The error-correction model is estimated by OLS:

$$\Delta Y_t = \gamma \widehat{z}_{t-1} + \sum_i \beta_i \Delta X_{t-i} + \sum_j \delta_j \Delta Y_{t-j} + \varepsilon_t \tag{7.122}$$

where $\varepsilon_t \sim WN$ and $\widehat{z}_{t-1}$ is the residual from the estimation of the one-period-lagged long-term relationship: $\widehat{z}_{t-1} = Y_{t-1} - \hat{\alpha} - \hat{\beta} X_{t-1}$.

In the first step of the Engle and Granger (1987) estimation method, it is necessary to check that the series X_t and Y_t are cointegrated, i.e., that the residuals of the long-term relationship are stationary ($I(0)$). It is important to remember that if $\hat{z}_t$ is not stationary, i.e., if the variables X_t and Y_t are not cointegrated, then the relationship (7.121) is a spurious regression. On the other hand, if $\hat{z}_t$ is stationary, the relationship (7.121) is a cointegrating relationship. To test whether the residual term of the long-term relationship is stationary or not, cointegration tests are performed. There are several such tests (see in particular Engle and Granger, 1987; Johansen, 1988 and 1991); here we propose the Dickey-Fuller test.

Dickey-Fuller Test of No Cointegration

The Dickey-Fuller (DF) and augmented Dickey-Fuller (ADF) tests allow us to test the null hypothesis of no cointegration against the alternative hypothesis that the series under consideration are cointegrated. Their purpose is thus to test the existence of a unit root in the residuals $\widehat{z}_t$ derived from the estimation of the long-term relationship:

$$\widehat{z}_t = Y_t - \hat{\alpha} - \hat{\beta} X_t \tag{7.123}$$

[16] We assume here that the long-term relationship includes a constant term.

Table 7.17 Engle and Yoo's (1987) critical values for the DF test of no cointegration ($p = 0$)

	T	*1%*	*5%*	*10%*
$N = 2$	50	−4.32	−3.67	−3.28
	100	−4.07	−3.37	−3.03
	200	−4.00	−3.37	−3.02
$N = 3$	50	−4.84	−4.11	−3.73
	100	−4.45	−3.93	−3.59
	200	−4.35	−3.78	−3.47
$N = 4$	50	−4.94	−4.35	−4.02
	100	−4.75	−4.22	−3.89
	200	−4.70	−4.18	−3.89
$N = 5$	50	−5.41	−4.76	−4.42
	100	−5.18	−4.58	−4.26
	200	−5.02	−4.48	−4.18

- In the case of the DF test, we estimate the relationship:

$$\Delta\widehat{z}_t = \phi\widehat{z}_{t-1} + u_t \tag{7.124}$$

- In the case of the ADF test, we estimate the relationship:

$$\Delta\widehat{z}_t = \phi\widehat{z}_{t-1} + \sum_{i=1}^{p}\phi_i\Delta\widehat{z}_{t-i} + u_t \tag{7.125}$$

with, in both cases, $u_t \sim WB$.

We test the null hypothesis $H_0 : \widehat{z}_t$ non-stationary ($\phi = 0$) reflecting the fact that the variables X_t and Y_t are not cointegrated, against the alternative hypothesis H_1: $\widehat{z}_t$ stationary ($\phi < 0$), indicating that the series X_t and Y_t are cointegrated.

It is important to stress that this test of no cointegration is based on the estimated residuals $\hat{z}_t$ and not on the true values z_t. The consequence is that the critical values tabulated by Dickey and Fuller are no longer valid. It is therefore appropriate to use the critical values tabulated by Engle and Yoo (1987) (Tables 7.17 and 7.18) or by MacKinnon (1991) (Table 7.19).[17] In these tables, N designates the number of variables considered and T the number of observations.

Since the critical values are negative, the decision rule is as follows (noting $t_{\hat{\phi}}$ the value of the t-statistic associated with the estimated coefficient $\hat{\phi}$):

- If $t_{\hat{\phi}}$ is lower than the critical value, we reject H_0: the series X_t and Y_t are cointegrated.

[17] In the MacKinnon table, critical values are distinguished according to whether or not a trend is included in the cointegration relationship.

Table 7.18 Engle and Yoo's (1987) critical values for the ADF test of no cointegration with $p = 4$

	T	1%	5%	10%
$N = 2$	50	−4.12	−3.29	−2.90
	100	−3.73	−3.17	−2.91
	200	−3.78	−3.25	−2.98
$N = 3$	50	−4.45	−3.75	−3.36
	100	−4.22	−3.62	−3.32
	200	−4.34	−3.78	−3.51
$N = 4$	50	−4.61	−3.98	−3.67
	100	−4.61	−4.02	−3.71
	200	−4.72	−4.13	−3.83
$N = 5$	50	−4.80	−4.15	−3.85
	100	−4.98	−4.36	−4.06
	200	−4.97	−4.43	−4.14

Table 7.19 MacKinnon's (1991) critical values for the ADF test of no cointegration

		1%	5%	10%
$N = 2$	Without trend	−3.90	−3.34	−3.04
	With trend	−4.32	−3.78	−3.50
$N = 3$	Without trend	−4.30	−3.74	−3.45
	With trend	−4.67	−4.12	−3.84
$N = 4$	Without trend	−4.65	−4.10	−3.81
	With trend	−4.97	−4.43	−4.15
$N = 5$	Without trend	−4.96	−4.41	−4.13
	With trend	−5.25	−4.72	−4.44
$N = 6$	Without trend	−5.24	−4.71	−4.42
	With trend	−5.51	−4.98	−4.70

– If $t_{\hat{\phi}}$ is higher than the critical value, we do not reject H_0: the variables X_t and Y_t are not cointegrated.

Remark 7.13 The method of Engle and Granger (1987) provides us with a simple way to test the hypothesis of no cointegration and to estimate an error-correction model in two steps. The disadvantage of this approach is that it does not allow multiple cointegration vectors to be distinguished. This is problematic when we study N variables simultaneously, with $N > 2$, or, if preferred, when we have more than one explanatory variable ($k > 1$). Indeed, we know that if we analyze the behavior of N variables (with $N > 2$), we can have up to $(N - 1)$ cointegration relationships, the Engle-Granger approach allowing us to obtain only one cointegration relationship. To overcome this difficulty, Johansen (1988) proposed a multivariate approach to cointegration based on the maximum likelihood method (see also Johansen and Juselius, 1990 and Johansen, 1991). A presentation of this approach is beyond the scope of this book, and readers can consult Engle and Granger (1991), Hamilton (1994), or Lardic and Mignon (2002).

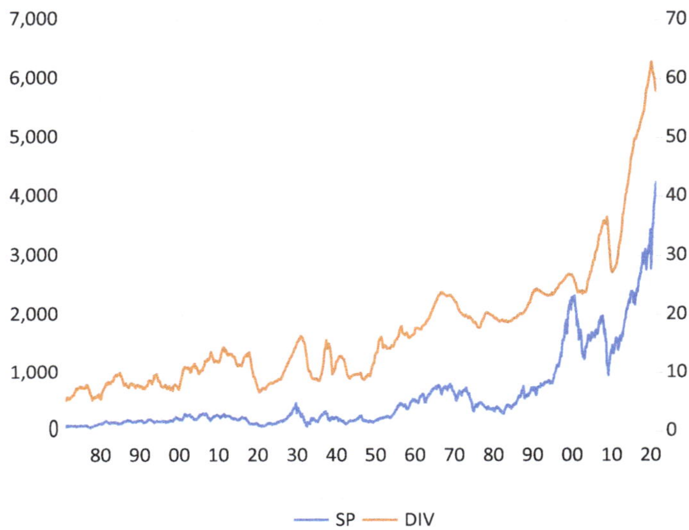

Fig. 7.16 Evolution of stock prices and dividends, United States, 1871.01–2021.06

Example: The Relationship Between Prices and Dividends

The efficient capital market theory forms the core of modern financial theory. This theory assumes that every asset has a "fundamental value," reflecting the underlying economic fundamentals. More precisely, in line with the dividend discount model, the fundamental value of a stock or stock index is defined as the discounted sum of future dividends rationally anticipated by agents. We deduce that, based on this approach, prices and dividends are linked through a stable long-term relationship: prices and dividends must be cointegrated. Indeed, if prices and dividends are not cointegrated, i.e., if the residuals of the relationship between prices and dividends are non-stationary, then there is a long-lasting deviation between the price and the fundamental value under the dividend discount model. The price does not return to the fundamental value, which can be interpreted as evidence of informational inefficiency. Conversely, if prices and dividends are cointegrated, the residuals are stationary, and there is no lasting deviation between the price and the fundamental value, which is consistent with the discount model and therefore with the informational efficiency of the market under consideration.

In order to grasp this issue, consider the Shiller data set[18] relating to the US market over the period January 1871 to June 2021 (monthly data). Figure 7.16 plots the dynamics of the SP 500 index (SP) of the New York Stock Exchange as well as the corresponding dividends (DIV), both variables being expressed in real terms (i.e., deflated by the consumer price index). Looking at this figure, we can see that

[18] www.econ.yale.edu/~shiller.

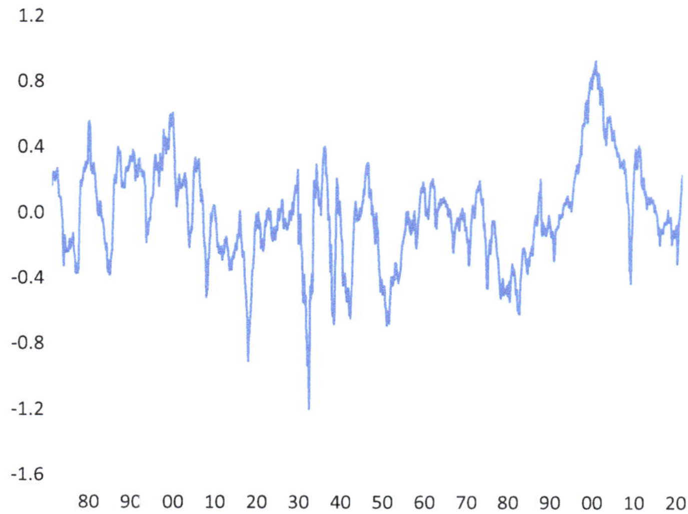

Fig. 7.17 Residuals of the long-term relationship between prices and dividends

prices and dividends follow a common trend, even though prices vary much more than dividends. This is representative of the well-known phenomenon of excessive stock price volatility.

In any case, and having confirmed that the two series under consideration are indeed non-stationary and integrated of the same order (order 1), it is legitimate to address the question of cointegration between the two variables. To this end, we regress prices on dividends and study the stationarity of the residuals resulting from the estimation of this relationship.

Figure 7.17 plots the pattern of this residual series. No particular structure emerges, suggesting that the residuals appear stationary. Let us check this intuition by applying the augmented Dickey-Fuller test to the residual series. We select a number of lags equal to 1 and obtain a calculated value of the ADF statistic equal to -4.5805. The 5% critical value for 2 variables, more than 200 observations, and zero lags is equal to -3.37. Since $-4.5805 < -3.37$, the null hypothesis of non-stationarity of the residual series is rejected. It follows that the null hypothesis of no cointegration between prices and dividends is rejected. Prices and dividends are therefore cointegrated: there is a stable long-term relationship between the two series, which is consistent with the efficient capital market hypothesis for the United States over the period 1871–2021.

7.5.5 Empirical Application

We consider the long-term (10-year) interest rate series for Germany (GER) and Austria (AUT) at daily frequency over the period from January 2, 1986, to July

Table 7.20 ADF test results

	ADF	CV at 5%
GER	−1.4007	−1.95
AUT	−1.8859	−1.95
DGER	−74.7409	−1.95
DAUT	−98.4491	−1.95

CV critical value

13, 2021, i.e., a total of 9 269 observations. These series are extracted from the Macrobond database. Since the Engle-Granger approach applies for $CI(1, 1)$ series, we first implement the ADF unit root test on the series GER and AUT. To this end, we follow the previously presented strategy, consisting in starting from the estimation of a model with trend and constant, then estimating a model with constant without trend if the latter is not significant, and finally a model without constant or trend if neither of them proves to be significant. The application of this strategy leads to the results shown in Table 7.20. We have chosen a model without constant or trend for both series. Since the calculated value of the ADF statistic for the series GER and AUT is higher than the critical value, we do not reject the null hypothesis of unit root at the 5% significance level. To determine the order of integration of the two series, we differentiate them and apply the test procedure on the series in first-difference $DGER$ and $DAUT$. In both cases, a model without constant or trend is used. It appears that the calculated value of the ADF statistic is lower than the critical value at the 5% significance level: the null hypothesis of unit root is therefore rejected. In other words, $DGER$ and $DAUT$ are integrated of order zero, implying that GER and AUT are integrated of order 1.

The two series are integrated of the same order (order 1), which is a necessary condition for implementing the Engle-Granger method.

Figure 7.18 representing the joint evolution of the two series further indicates that GER and AUT are characterized by a common trend over the entire period. Thus, since GER and AUT are non-stationary and integrated of the same order, and follow a similar pattern, it is legitimate ask whether the two variables are cointegrated.

We begin by estimating the static relationship between GER and AUT, i.e.:

$$AUT_t = \alpha + \beta GER_t + z_t \tag{7.126}$$

The results from estimating this relationship allow us to deduce the residual series:

$$\hat{z}_t = AUT_t - 0.2237 - 1.0031 GER_t \tag{7.127}$$

Recall that:

- If the residuals are non-stationary, the estimated static relationship is a spurious regression.

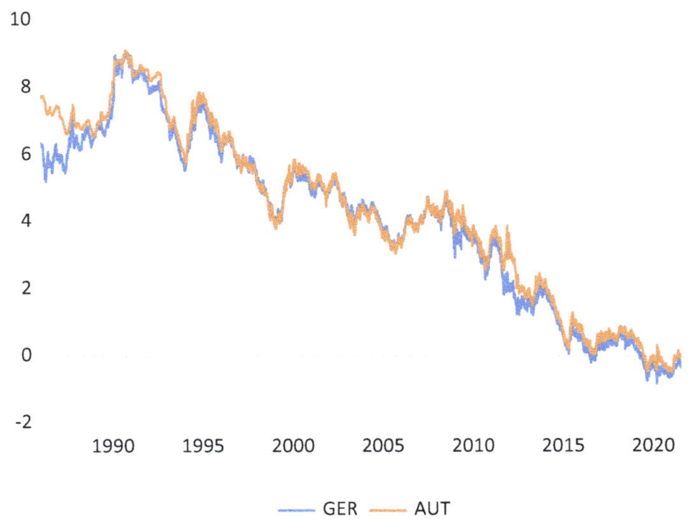

Fig. 7.18 10-year interest rates, Germany (GER) and Austria (AUT), January 2, 1986–July 13, 2021

– If the residuals are stationary, the estimated static relationship is a cointegrating relationship.

To discriminate between these two possibilities, we apply the ADF test of no cointegration. Table 7.21 shows the results of the ADF test on the residual series $\hat{z}_t$ (noted $RESIDS$ in the table).

The calculated value of the ADF statistic should be compared with the critical values of Engle and Yoo (1987) or MacKinnon (1991) (see Tables 7.17, 7.18, and 7.19). A number of lags $p = 4$ in the implementation of the ADF test have been selected here. The critical value for $p = 4$ is equal to -3.25 at the 5% significance level. The calculated value -4.9615 being lower than the 5% critical value, the null hypothesis of non-stationarity of the residuals is rejected. It follows that the series GER and AUT are cointegrated and the estimated static relationship is indeed a cointegrating relationship. It is then possible to estimate an error-correction model; the results are shown in Table 7.22.

The results in Table 7.22 show that the coefficient associated with the one-period lagged residual term is negative (-0.0039) and significantly different from zero (its t-statistic is higher than 1.96 in absolute value). There is thus an error-correction mechanism: in the long term, the differences (or imbalances) between the two series tend to offset each other, leading the variables to evolve in a similar way. We also note that, in the short term, the change in the Austrian interest rate is a function of itself, lagged by one period, and of the variation in the German interest rate, also lagged by one period.

Table 7.21 ADF test on residuals

Null hypothesis: RESIDS has a unit root				
Exogenous: none				
Lag length: 4 (automatic – based on SIC, maxlag = 37)				
			t-Statistic	Prob.*
Augmented Dickey-Fuller test statistic			−4.961462	0.0000
Test critical values:	1%level		−2.565212	
	5% level		−1.940858	
	10% level		−1.616677	
*MacKinnon (1996) one-sided p-values				
Augmented Dickey-Fuller test equation				
Dependent variable: D(RESIDS)				
Method: least squares				
Sample (adjusted): 1/09/1986 7/13/2021				
Included observations: 9264 after adjustments				
Variable	Coefficient	Std. error	t-Statistic	Prob.
RESIDS(−1)	−0.007200	0.001451	−4.961462	0.0000
D(RESIDS(−1))	−0.311232	0.010406	−29.90853	0.0000
D(RESIDS(−2))	−0.157829	0.010850	−14.54651	0.0000
D(RESIDS(−3))	−0.092181	0.010841	−8.502916	0.0000
D(RESIDS(−4))	−0.043617	0.010374	−4.204419	0.0000
R-squared	0.097161	Mean dependent var	−0.000137	
Adjusted R-squared	0.096771	S.D. dependent var	0.049113	
S.E. of regression	0.046676	Akaike info criterion	−3.290644	
Sum squared resid	20.17191	Schwarz criterion	−3.286793	
Log likelihood	15247.26	Hannan-Quinn criterion	−3.289335	
Durbin-Watson stat	2.000780			

Table 7.22 Estimation of the error-correction model

Dependent variable: D(AUT)				
Variable	Coefficient	Std. error	t-Statistic	Prob.
C	−0.000848	0.000421	−2.015442	0.0439
RESIDS(−1)	−0.003948	0.001252	−3.152072	0.0016
D(AUT(−1))	−0.074325	0.011648	−6.380673	0.0000
D(GER(−1))	0.088085	0.009155	9.621747	0.0000
R-squared	0.011960	Mean dependent var	−0.000847	
Adjusted R-squared	0.011640	S.D. dependent var	0.040712	
S.E. of regression	0.040474	Akaike info criterion	−3.575884	
Sum squared resid	15.17411	Schwarz criterion	−3.572805	
Log likelihood	16572.86	Hannan-Quinn criterion	−3.574838	
F-statistic	37.37579	Durbin-Watson stat	2.002893	
Prob(F-statistic)	0.000000			

Conclusion

This chapter has introduced the basics of time series econometrics, a branch of econometrics that is still undergoing many developments. In addition to univariate time series models, we have dealt with multivariate analysis through VAR processes. In these processes, all variables have the same status, in the sense that no distinction is made between endogenous and exogenous variables. An alternative to VAR processes are the simultaneous equations models which are discussed in the next chapter. Unlike VAR models, which have no theoretical content, simultaneous equations models are structural macroeconomic models.

The Gist of the Chapter

Stationarity	
Definition	$E\left(Y_t^2\right) < \infty \forall t \in Z$ $E\left(Y_t\right) = m\ \forall t \in Z$ $Cov\left(Y_t, Y_{t+h}\right) = \gamma_h,\ \forall t, h \in Z,$ γ : autocovariance function
Unit root test	Dickey-Fuller tests
Functions	
Autocovariance	$\gamma_h = Cov\left(Y_t, Y_{t+h}\right), h \in Z$
Autocorrelation	$\rho_h = \frac{\gamma_h}{\gamma_0}, h \in Z$
Partial autocorrelation	ϕ_{hh} : calculation using the Durbin algorithm
Process	
$AR(p)$	$Y_t - \phi_1 Y_{t-1} - \cdots - \phi_p Y_{t-p} = \varepsilon_t$ $\phi_{hh} = 0\ \forall h > p$
$MA(q)$	$Y_t = \varepsilon_t - \theta_1 \varepsilon_{t-1} - \cdots - \theta_q \varepsilon_{t-q}$ $\rho_h = 0\ \forall h > q$
$ARMA(p, q)$	$Y_t - \phi_1 Y_{t-1} - \cdots - \phi_p Y_{t-p} =$ $\varepsilon_t - \theta_1 \varepsilon_{t-1} \cdots \theta_q \varepsilon_{t-q}$
Information criteria	
Akaike	$AIC = \log \widehat{\sigma}_\varepsilon^2 + \frac{2(p+q)}{T}$
Schwarz	$SIC = \log \widehat{\sigma}_\varepsilon^2 + (p+q) \frac{\log T}{T}$
Hannan-Quinn ($c = 1$)	$HQ = \log \widehat{\sigma}_\varepsilon^2 + 2(p+q) \frac{\log(\log T)}{T}$
$VAR(p)$	$\boldsymbol{Y}_t - \sum_{i=1}^{p} \boldsymbol{\Phi}_i \boldsymbol{Y}_{t-i} = \boldsymbol{\Phi}_0 + \boldsymbol{\varepsilon}_t$
Cointegration	$(X_t, Y_t) \sim CI\ (d, b)$ if $z_t = Y_t - \beta X_t \sim I\ (d-b)$ with X_t and $Y_t \sim I\ (d), d \geq b > 0$
Error-correction model	$\Delta Y_t = \gamma \hat{z}_{t-1} + \sum_i \beta_i \Delta X_{t-i} + \sum_j \delta_j \Delta Y_{t-j} + \varepsilon_t,$ γ: speed of adjustment to the long-term target

Further Reading

There are many textbooks on time series econometrics. In addition to the pioneering work by Box and Jenkins (1970), let us mention the manuals by Harvey (1990), Mills (1990), Hamilton (1994), Gouriéroux and Monfort (1996), or Brockwell and Davis (1998); Hamilton's (1994) work in particular includes numerous developments on multivariate models.

On the econometrics of non-stationary time series, in addition to the textbooks cited above and the many references included in this chapter, readers may usefully consult Engle and Granger (1991), Banerjee et al. (1993), Johansen (1995), as well as Maddala and Kim (1998).

As mentioned in this chapter, time series econometrics has undergone, and continues to undergo, many developments. There are therefore references specific to certain fields:

- For developments relating to nonlinear time series econometrics, readers can refer to Granger and Teräsvirta (1993), Lardic and Mignon (2002), or Teräsvirta et al. (2010). A particular category of these processes concerns processes with nonlinearities in variance (ARCH-type models), which are widely used in finance. For a presentation of these models, in addition to the previously cited references in nonlinear time series econometrics, see also the literature reviews by Bollerslev et al. (1992, 1994), Palm (1996), Gouriéroux (1997), Bollerslev (2008), and Bauwens et al. (2012).
- Readers interested in the econometrics of long-memory processes may refer to Beran (1994) and Lardic and Mignon (1999, 2002).
- Concerning extensions of the notion of cointegration, let us mention the work of Dufrénot and Mignon (2002a,b) on nonlinear cointegration and that of Lardic and Mignon (2002) and Lardic et al. (2005) on fractional cointegration.
- Finally, let us mention a field whose development has been particularly notable in recent years: non-stationary panel data econometrics. For pedagogical presentations in French, interested readers may refer to Hurlin and Mignon (2005, 2007).

Simultaneous Equations Models 8

So far, with the exception of the VAR models presented in the previous chapter, we have considered models with only one equation. However, many economic theories are based on models with several equations, i.e., on **systems of equations**. Since these equations are not independent of each other, the interaction of the different variables has important consequences for the estimation of each equation and for the system as a whole.

We start by outlining the analytical framework before turning to the possibility or not of estimating the parameters of the model, known as **identification**. We then present the estimation methods relating to **simultaneous equations models**, as well as the specification test proposed by Hausman (1978). We conclude with an empirical application.

8.1 The Analytical Framework

In the single-equation models we have studied so far, there is only one endogenous variable, the latter being explained by one or more exogenous variables. If a causal relationship exists, it runs from the exogenous variables to the endogenous variable.

In a simultaneous equations model, each equation is relative to an endogenous variable, and it is very common for an explained variable in one equation to become an explanatory variable in another equation of the model. The distinction between endogenous and exogenous variables is therefore no longer as marked as in the case of single-equation models and, in a simultaneous equations model, the variables are determined simultaneously. This dual status of the variables appearing in a simultaneous equations model means that it is impossible to estimate the parameters of one equation without taking into account the information provided by the other equations in the system. In particular, the OLS estimators are biased and non-consistent, in the sense that they do not converge to their true values when the sample

V. Mignon, *Principles of Econometrics*, Classroom Companion: Economics,
https://doi.org/10.1007/978-3-031-52535-3_8

size increases. As an example, consider the following system:

$$Y_t = \alpha + \beta X_t + \varepsilon_t \tag{8.1}$$

$$X_t = Y_t + Z_t \tag{8.2}$$

where ε_t is an error term. In Eq. (8.1), the variable Y is explained by X; Y is therefore an endogenous variable. Equation (8.2) shows that the variable X is in turn explained by Y and Z. All in all, in this system, Y and X are endogenous variables and Z is an exogenous variable.

Suppose that ε_t follows a normal distribution of zero mean and constant variance σ_ε^2, and that ε_t and Z_t are independent. We can rewrite the system in the following form:

$$Y_t = \alpha + \beta X_t + \varepsilon_t = \alpha + \beta \left(Y_t + Z_t\right) + \varepsilon_t \tag{8.3}$$

hence:

$$Y_t = \frac{\alpha}{1-\beta} + \frac{\beta}{1-\beta} Z_t + \frac{1}{1-\beta} \varepsilon_t \tag{8.4}$$

We deduce:

$$X_t = \frac{\alpha}{1-\beta} + \frac{\beta}{1-\beta} Z_t + \frac{1}{1-\beta} \varepsilon_t + Z_t \tag{8.5}$$

hence:

$$X_t = \frac{\alpha}{1-\beta} + \frac{1}{1-\beta} Z_t + \frac{1}{1-\beta} \varepsilon_t \tag{8.6}$$

The system is finally written:

$$Y_t = \frac{\alpha}{1-\beta} + \frac{\beta}{1-\beta} Z_t + \mu_t \tag{8.7}$$

$$X_t = \frac{\alpha}{1-\beta} + \frac{1}{1-\beta} Z_t + \mu_t \tag{8.8}$$

with $\mu_t = \frac{1}{1-\beta}\varepsilon_t$.

Equation (8.8) shows that X_t is influenced by μ_t and, consequently, by ε_t. It follows that $Cov\left(X_t, \varepsilon_t\right) \neq 0$, implying that the OLS estimator is not consistent.

In order to introduce some concepts relating to simultaneous equations models, let us start with an introductory example.

8.1.1 Introductory Example

Consider the following three-equation system composed of centered variables:

$$q_t^d = \alpha_1 p_t + \alpha_2 y_t + \varepsilon_t^d \tag{8.9}$$

$$q_t^s = \beta_1 p_t + \varepsilon_t^s \tag{8.10}$$

$$q_t^d = q_t^s = q_t \tag{8.11}$$

where Eq. (8.9) is the demand equation, q_t^d denoting the quantity demanded of any good, p_t the price of that good, and y_t income. Equation (8.10) is the supply equation, q_t^s denoting the quantity offered of the good under consideration. ε_t^d and ε_t^s are error terms, also known as **disturbances**. The demand and supply equations are **behavioral equations**. Finally, Eq. (8.11) is called the **equilibrium equation**: it is the equilibrium condition represented by the equality between demand and supply. Equilibrium equations contain no error term.

The equations of this system, derived from economic theory, are called **structural equations**. This is referred to as a model expressed in **structural form**. In this system, price and quantity variables are interdependent, so they are mutually dependent or endogenous. Income y_t is an exogenous variable, in the sense that it is determined outside the system.

Since the system incorporates a demand equation, a supply equation, and an equilibrium condition, it is referred to as a **complete system** in the sense that it has as many equations as there are endogenous variables.

Let us express each of the endogenous variables in terms of the exogenous variable and the error terms ε_t^d and ε_t^s. From Eq. (8.10), we can write:

$$p_t = \frac{1}{\beta_1} q_t - \frac{1}{\beta_1} \varepsilon_t^s \tag{8.12}$$

We transfer this expression into (8.9), which gives:

$$q_t = \alpha_1 \left(\frac{1}{\beta_1} q_t - \frac{1}{\beta_1} \varepsilon_t^s \right) + \alpha_2 y_t + \varepsilon_t^d \tag{8.13}$$

Hence:

$$q_t = \frac{\alpha_2 \beta_1}{\beta_1 - \alpha_1} y_t + \frac{1}{\beta_1 - \alpha_1} \left(\beta_1 \varepsilon_t^d - \alpha_1 \varepsilon_t^s \right) \tag{8.14}$$

Positing:

$$\gamma_1 = \frac{\alpha_2 \beta_1}{\beta_1 - \alpha_1} \tag{8.15}$$

and:

$$\mu_{1t} = \frac{1}{\beta_1 - \alpha_1}\left(\beta_1 \varepsilon_t^d - \alpha_1 \varepsilon_t^s\right) \tag{8.16}$$

we can rewrite Eq. (8.14) as follows:

$$q_t = \gamma_1 y_t + \mu_{1t} \tag{8.17}$$

Now we transfer the expression (8.14) into (8.12), and we obtain:

$$p_t = \frac{1}{\beta_1}\left(\frac{\alpha_2 \beta_1}{\beta_1 - \alpha_1} y_t + \frac{1}{\beta_1 - \alpha_1}\left(\beta_1 \varepsilon_t^d - \alpha_1 \varepsilon_t^s\right)\right) - \frac{1}{\beta_1}\varepsilon_t^s \tag{8.18}$$

that is:

$$p_t = \frac{\alpha_2}{\beta_1 - \alpha_1} y_t + \frac{1}{\beta_1 - \alpha_1}\left(\varepsilon_t^d - \varepsilon_t^s\right) \tag{8.19}$$

By positing:

$$\gamma_2 = \frac{\alpha_2}{\beta_1 - \alpha_1} \tag{8.20}$$

and:

$$\mu_{2t} = \frac{1}{\beta_1 - \alpha_1}\left(\varepsilon_t^d - \varepsilon_t^s\right) \tag{8.21}$$

this last equation can be rewritten as:

$$p_t = \gamma_2 y_t + \mu_{2t} \tag{8.22}$$

Putting together Eqs. (8.17) and (8.22), the system of equations is finally written as:

$$q_t = \gamma_1 y_t + \mu_{1t} \tag{8.23}$$

$$p_t = \gamma_2 y_t + \mu_{2t} \tag{8.24}$$

Each of the endogenous variables is expressed as a function of the exogenous variable and a random error term. This is known as the **reduced form** of the model (none of the endogenous variables is any longer expressed as a function of the other endogenous variables). Equations (8.17) and (8.22) are called **reduced-form equations**.

In this system, the endogenous variables are correlated with the error terms, with the result that the OLS estimators are no longer consistent. As we will see later, it is possible to use an instrumental variables estimator or a two-stage least squares estimator.

Remark 8.1 When a model includes lagged endogenous variables, these are referred to as **predetermined variables**. As an example, consider the following model:

$$C_t = \alpha_0 + \alpha_1 Y_t + \alpha_2 C_{t-1} + \varepsilon_{1t} \tag{8.25}$$

$$I_t = \beta_0 + \beta_1 R_t + \beta_2 \left(Y_t - Y_{t-1} \right) + \varepsilon_{2t} \tag{8.26}$$

$$Y_t = C_t + I_t + G_t \tag{8.27}$$

Equation (8.25) is the consumption equation, (8.26) is the investment equation, and (8.27) is the equilibrium condition. This model has three endogenous variables (C_t, I_t, and Y_t), two exogenous variables (R_t and G_t), and two lagged endogenous variables (C_{t-1} and Y_{t-1}). The latter two variables are said to be predetermined in the sense that they are considered to be already determined with respect to the current values of the endogenous variables.

More generally, variables that are independent of all future error terms of the structural form are called predetermined variables.

8.1.2 General Form of Simultaneous Equations Models

In the general case, the structural form of the simultaneous equations model is written:

$$\begin{aligned}
&\beta_{11} Y_{1t} + \beta_{12} Y_{2t} + \ldots + \beta_{1M} Y_{Mt} + \gamma_{11} X_{1t} + \gamma_{12} X_{2t} + \ldots + \gamma_{1k} X_{kt} = \varepsilon_{1t} \\
&\beta_{21} Y_{1t} + \beta_{22} Y_{2t} + \ldots + \beta_{2M} Y_{Mt} + \gamma_{21} X_{1t} + \gamma_{22} X_{2t} + \ldots + \gamma_{2k} X_{kt} = \varepsilon_{2t} \\
&\ldots \\
&\beta_{M1} Y_{1t} + \beta_{M2} Y_{2t} + \ldots + \beta_{MM} Y_{Mt} + \gamma_{M1} X_{1t} + \gamma_{M2} X_{2t} + \ldots + \gamma_{Mk} X_{kt} = \varepsilon_{Mt}
\end{aligned} \tag{8.28}$$

This model includes M equations and M endogenous variables ($Y_{1t}, Y_{2t}, \ldots, Y_{Mt}$). It comprises k exogenous variables ($X_{1t}, X_{2t}, \ldots, X_{kt}$) which may also contain predetermined values of the endogenous variables.[1] One of the variables may consist of 1 in order to account for the constant term in each of the equations. The error terms ($\varepsilon_{1t}, \varepsilon_{2t}, \ldots, \varepsilon_{kt}$) are called **structural disturbances**.

[1] The predetermined variables can thus be divided into two categories: exogenous variables and lagged endogenous variables.

This model can also be written in matrix form:

$$\underset{(M,M)}{\boldsymbol{B}}\underset{(M,1)}{\boldsymbol{Y}} + \underset{(M,k)}{\boldsymbol{\Gamma}}\underset{(k,1)}{\boldsymbol{X}} = \underset{(M,1)}{\boldsymbol{\varepsilon}} \tag{8.29}$$

with:

$$\boldsymbol{B} = \begin{pmatrix} \beta_{11} & \beta_{12} & \cdots & \beta_{1M} \\ \beta_{21} & \beta_{22} & \cdots & \beta_{2M} \\ & & \vdots & \\ \beta_{M1} & \beta_{M2} & \cdots & \beta_{MM} \end{pmatrix} \tag{8.30}$$

$$\boldsymbol{Y} = \begin{pmatrix} Y_{1t} \\ Y_{2t} \\ \vdots \\ Y_{Mt} \end{pmatrix} \tag{8.31}$$

$$\boldsymbol{\Gamma} = \begin{pmatrix} \gamma_{11} & \gamma_{12} & \cdots & \gamma_{1k} \\ \gamma_{21} & \gamma_{22} & \cdots & \gamma_{2k} \\ & & \vdots & \\ \gamma_{M1} & \gamma_{M2} & \cdots & \gamma_{Mk} \end{pmatrix} \tag{8.32}$$

$$\boldsymbol{X} = \begin{pmatrix} X_{1t} \\ X_{2t} \\ \vdots \\ X_{kt} \end{pmatrix} \tag{8.33}$$

and:

$$\boldsymbol{\varepsilon} = \begin{pmatrix} \varepsilon_{1t} \\ \varepsilon_{2t} \\ \vdots \\ \varepsilon_{Mt} \end{pmatrix} \tag{8.34}$$

In each equation, one of the endogenous variables has its coefficient equal to 1: this is the dependent variable. There is therefore one dependent variable per equation. In other words, in the matrix $\boldsymbol{B}$, each column has at least one value equal to 1. This is known as **normalization**. On the other hand, equations in which all coefficients are equal to 1 and involve no disturbance are equilibrium equations.

If the matrix $\boldsymbol{B}$ is non-singular, it is invertible and it is possible to derive the reduced form of the model allowing the matrix $\boldsymbol{Y}$ to be expressed in terms of the matrix $\boldsymbol{X}$:

$$\boldsymbol{Y} = -\boldsymbol{B}^{-1}\boldsymbol{\Gamma}\boldsymbol{X} + \boldsymbol{B}^{-1}\boldsymbol{\varepsilon} \tag{8.35}$$

The condition that the matrix $\boldsymbol{B}$ is non-singular is called the **completeness condition**. The reduced form allows each endogenous variable to be expressed in terms of the exogenous or predetermined variables and the disturbances.

The reduced-form equations can be estimated by OLS. In these equations, the endogenous variables are expressed as a function of the exogenous or predetermined variables, assumed to be uncorrelated with the error terms. After estimating the parameters of the reduced-form equations, it is possible to determine the parameters of the structural equations by applying the **indirect least squares method** (see below).

While the transition from the structural form to the reduced form seems easy in theory, it is not the same in practice. In the reduced form, knowing the elements of the matrix $\left(\boldsymbol{B}^{-1}\boldsymbol{\Gamma}\right)$ does not allow us to determine, i.e., to identify, the matrices $\boldsymbol{B}$ and $\boldsymbol{\Gamma}$ separately. This is known as the **identification problem**: we have a system of $(M \times k)$ equations with $(M \times M) + (M \times k)$ unknowns, which therefore cannot be solved without imposing certain restrictions.

Remark 8.2 If the matrix $\boldsymbol{B}$ is an upper triangular matrix, the system is described as **triangular** or **recursive**. Its form is as follows:

$$\begin{aligned}
Y_{1t} &= f_1\left(X_{1t}, X_{2t}, \ldots, X_{kt}\right) + \varepsilon_{1t} \\
Y_{2t} &= f_2\left(Y_{1t}, X_{1t}, X_{2t}, \ldots, X_{kt}\right) + \varepsilon_{2t} \\
\ldots & \\
Y_{Mt} &= f_M\left(Y_{1t}, Y_{2t}, \ldots, Y_{M-1t}, X_{1t}, X_{2t}, \ldots, X_{kt}\right) + \varepsilon_{Mt}
\end{aligned} \tag{8.36}$$

Each endogenous variable is determined sequentially or recursively. The first equation contains no endogenous variables and is entirely determined by the exogenous variables. In the second equation, the explanatory variables include the endogenous variable from the first equation, and so on. In a triangular system of the kind, the OLS method can be applied equation by equation, since the endogenous variables do not depend on the disturbances.

8.2 The Identification Problem

8.2.1 Problem Description

The question posed here is whether it is possible to derive estimators of the structural form parameters from estimators of the reduced-form parameters. The problem arises from the fact that several structural coefficient estimates can be compatible with the same data sets. In other words, one reduced-form equation may correspond to several structural equations.

The identification conditions are determined equation by equation. Several cases may arise:

- If it is impossible to obtain the estimators of the structural form parameters from the estimators of the reduced form, the model is said to be unidentified or **underidentified**. Thus, a model is underidentified if one equation of the model is underidentifiable. This means that the number of equations is smaller than the number of parameters to be identified in the structural form, and it is then impossible to solve the system.
- If it is possible to obtain the estimators of the parameters of the structural form from the estimators of the reduced form, the model is said to be **identified**. There are two possible scenarios here:
  - The model is **exactly** (or fully or strictly) **identified** if all its equations are strictly identifiable, i.e., if unique values of the structural parameters can be obtained.
  - The model is **overidentified** if the equations are overidentifiable, i.e., if several values correspond to the structural parameters.

We will come back to these various cases later.

8.2.2 Rank and Order Conditions for Identification

Recall that the structural form is given by:

$$\boldsymbol{B}\boldsymbol{Y} + \boldsymbol{\Gamma}\boldsymbol{X} = \boldsymbol{\varepsilon} \tag{8.37}$$

and the reduced form by:

$$\boldsymbol{Y} = -\boldsymbol{B}^{-1}\boldsymbol{\Gamma}\boldsymbol{X} + \boldsymbol{B}^{-1}\boldsymbol{\varepsilon} \tag{8.38}$$

or:

$$\boldsymbol{Y} = \boldsymbol{\Pi}\boldsymbol{X} + \boldsymbol{\upsilon} \tag{8.39}$$

with $\boldsymbol{\Pi} = -\boldsymbol{B}^{-1}\boldsymbol{\Gamma}$ and $\boldsymbol{\upsilon} = \boldsymbol{B}^{-1}\boldsymbol{\varepsilon}$.

Thus, three types of structural parameters are unknown:

- The matrix $\boldsymbol{B}$ which is a non-singular matrix of size $(M \times M)$
- The parameter matrix $\boldsymbol{\Gamma}$ of size $(M \times k)$
- The variance-covariance matrix of structural disturbances, denoted $\boldsymbol{\Omega}_{\varepsilon}$

The reduced form includes the following known parameters:

- The matrix of coefficients of the reduced form $\mathbf{\Pi}$ of size $(M \times k)$
- The variance-covariance matrix of the disturbances of the reduced form, noted $\mathbf{\Omega}_{\upsilon}$

In other words, the number of structural parameters is equal to $M^2 + Mk + \frac{M(M+1)}{2}$ and the number of parameters of the reduced form is given by: $Mk + \frac{M(M+1)}{2}$. The difference between the number of structural parameters and that of the reduced form is therefore equal to M^2, which corresponds to the number of unknown elements in the matrix $\boldsymbol{B}$. Consequently, if no additional information is available, identification is impossible. The additional information can be of several types, depending on the nature of the restrictions or constraints imposed on the coefficients of the structural form: normalization, identities, exclusion relations, linear restrictions, or even restrictions on the variance-covariance matrix of disturbances. Let us consider each of these five points in turn.

Restrictions

- *Normalization*. As previously mentioned, in each equation, one of the endogenous variables has its coefficient equal to 1: this is the dependent variable. There is one such dependent variable per equation. Imposing a value of 1 on a coefficient is called **normalization**. This operation reduces the number of unknown elements in the matrix $\boldsymbol{B}$, since we then have $M(M-1)$ and no longer M^2 undetermined elements.
- *Identities*. We know that a model can contain behavioral relations and equilibrium relations or accounting identities. These equilibrium relations and accounting identities do not have to be identified: the coefficients associated with the variables in these relations are in fact known and are frequently equal to 1. In the introductory example we studied earlier, Eq. (8.11) is the equilibrium condition and does not have to be identified.
- *Exclusion relations*. Not introducing a variable into one of the equations of the system is considered as an exclusion relation. In effect, this amounts to assigning a zero coefficient to the variable in question. In other words, it consists in placing zeros in the elements of the matrices $\boldsymbol{B}$ and/or $\mathbf{\Gamma}$. Such a procedure obviously reduces the number of unknown parameters and thus provides an aid to identification.
- *Linear restrictions*. In line with economic theory, some models contain variables with identical coefficients. Imposing such restrictions on parameters facilitates the estimation procedure by reducing the number of unknown parameters.
- *Restrictions on the variance-covariance matrix of the disturbances*. Such restrictions are similar to those imposed on the model parameters. They consist, for example, in introducing zeros for certain elements of the variance-covariance matrix when imposing the absence of correlation between the structural disturbances of several equations.

Conditions for Identification

Let us first introduce some notations. Consider a particular equation j of the model with M simultaneous equations. The coefficients associated with this equation appear accordingly in the j-th columns of the matrices $\boldsymbol{B}$ and/or $\boldsymbol{\Gamma}$. It is further assumed that:

- In this equation, one of the elements of the matrix $\boldsymbol{B}$ is equal to 1 (normalization).
- Some variables appearing in other equations are excluded from this equation (exclusion relations).

Note:

- M the number of endogenous variables in the model, i.e., the number of equations in the model,
- k the number of exogenous variables introduced into the model,
- M_j the number of endogenous variables included in the equation j considered, M_j^* designating the number of endogenous variables excluded from the equation j,
- k_j the number of exogenous variables present in the equation j under consideration, k_j^* denoting the number of exogenous variables excluded from the equation j.

The number of equations in the model M is therefore given by:

$$M = M_j + M_j^* + 1 \tag{8.40}$$

and the number of exogenous variables k is equal to:

$$k = k_j + k_j^* \tag{8.41}$$

Since the number of equations must be at least equal to the number of unknowns, we deduce the **order condition** for the identification of the equation j:

$$k_j^* \geq M_j \tag{8.42}$$

According to this condition, the number of variables excluded from the equation j must be at least equal to the number of endogenous variables included in this same equation j. The order condition is a necessary condition for identification, but not a sufficient one. In other words, it ensures that the j-th equation of the reduced form admits a solution, but we do not know whether or not it is unique. In order to guarantee the uniqueness of the solution, the **rank condition** is necessary. This condition (see Greene, 2020) imposes a restriction on the submatrix of the reduced-form coefficient matrix and ensures that there is a unique solution for the structural parameters given the parameters of the reduced form. This rank condition can be expressed as follows: the equation j is identified if and only if it is possible to obtain at least one non-zero determinant of order $(M - 1, M - 1)$ from the coefficients of the variables excluded from the equation j, but included in the other equations of the model. In large models, only the order condition is used, as it is very difficult, if not impossible, to apply the rank condition.

Three cases are then possible, as discussed at the beginning of this section:

- If $k_j^* < M_j$, or if the rank condition is not verified, the model is underidentified.
- If $k_j^* = M_j$, and the rank condition is verified, the model is exactly identified.
- If $k_j^* > M_j$, and the rank condition is verified, the model is overidentified (there are more restrictions than those necessary for identification).

To establish these identification conditions, we have considered only exclusion relations. If we also have linear restrictions on the parameters, the order condition becomes:

$$r_j + k_j^* \geq M_j \tag{8.43}$$

where r_j denotes the number of restrictions other than the exclusion restrictions. It is possible to reformulate this order condition by taking into account both the exclusion relations and the linear restrictions. By noting s_j the total number of restrictions, i.e.:

$$s_j = r_j + k_j^* + M_j^* \tag{8.44}$$

we can write the order condition as:

$$s_j \geq M - 1 \tag{8.45}$$

Knowing that $M - 1 = M_j + M_j^*$, we have, by transferring (8.44) into (8.45): $r_j + k_j^* + M_j^* \geq M_j + M_j^*$ and we find Eq. (8.43). We then obtain the three cases previously presented:

- If $r_j + k_j^* < M_j$, or if the rank condition does not hold, the model is underidentified.
- If $r_j + k_j^* = M_j$, and the rank condition does not hold, the model is exactly identified.
- If $r_j + k_j^* > M_j$, and the rank condition does not hold, the model is overidentified.

Remark 8.3 It is also possible to use restrictions on the variance-covariance matrix of the disturbances. We know that $\boldsymbol{\Omega}_{\upsilon} = \boldsymbol{B}^{-1}\boldsymbol{\Omega}_{\varepsilon}\left(\boldsymbol{B}^{-1}\right)'$. If restrictions are imposed on $\boldsymbol{\Omega}_{\upsilon}$, more information than necessary will be available for estimating $\boldsymbol{\Omega}_{\upsilon}$. As a result, it is possible to use the additional information to identify the elements of $\boldsymbol{B}$. Thus, imposing zero covariances between disturbances can help in the identification.

On this point, reference can be made to Johnston and Dinardo (1996) and Greene (2020).

8.3 Estimation Methods

Identification is a prerequisite for estimating a simultaneous equations model, since if the model is underidentified, it cannot be estimated. Only exactly identified or overidentified models are estimable.

We have seen that while the reduced form can be estimated by OLS, this is not the case for the structural form. The OLS estimators of the structural parameters are not consistent insofar as the endogenous variables in each of the equations are correlated with the disturbances.[2]

Methods for estimating simultaneous equations models are for the most part **instrumental variables methods** (see Chap. 5) and can be classified into two broad categories:

- **Limited-information estimation methods**: each equation is estimated separately.
- **Full-information estimation methods**: the system as a whole is estimated, i.e., the M equations of the model are estimated simultaneously.

In limited-information estimation methods, the information contained in the other equations is ignored, hence the name given to these techniques. This category includes the methods of indirect least squares, two-stage least squares, generalized moments, limited-information maximum likelihood, and K-class estimators. On the contrary, in the full-information methods, all the information contained in the set of M equations is used, hence their name. This category includes the three-stage least squares method, the full-information maximum likelihood method, or the system generalized method of moments. Logically, full-information methods are expected to perform better than limited-information methods, as the joint estimation should lead to efficiency gains. Despite this advantage, these methods tend to be less widely used in practice than limited-information methods, for essentially three reasons: computational complexity, existence of nonlinear solutions on the parameters, and sensitivity to specification errors.

We focus here essentially on two limited-information estimation methods: indirect least squares and two-stage least squares.

[2] However, the OLS method can be applied in the case of triangular (or recursive) systems.

8.3.1 Indirect Least Squares

This estimation method applies only to equations that are exactly identified. Generally speaking, the principle of **indirect least squares** (ILS) consists in estimating the parameters of the reduced form by OLS and deducing the structural coefficients by an appropriate transformation of the reduced form coefficients. This technique can be described in three steps:

- The first step is to write the model in reduced form. This involves expressing the dependent variable of each equation as a function of the predetermined variables (exogenous and lagged endogenous variables) and the disturbances.
- The second step aims to estimate the parameters of each of the reduced-form equations by OLS. The application of OLS is made possible by the fact that the explanatory variables (predetermined variables) of the reduced-form equations are no longer correlated with the disturbances.
- The purpose of the third step is to deduce the parameters of the structural form from the estimated parameters of the reduced form. This determination is made using the algebraic relations linking the structural and the reduced form coefficients. The solution is unique since the model is exactly identifiable: there is thus a one-to-one correspondence between the structural coefficients and those of the reduced form.

The ILS estimator of the reduced form—which is therefore the OLS estimator—is a BLUE estimator. In contrast, the ILS estimator of the structural form coefficients is a biased estimator in the case of small samples. In addition, since the reduced form of a model is not always easy to establish—especially when the model comprises a large number of equations—and the existence of an exactly identified relationship is quite rare, the ILS method is not often used in practice. The two-stage least squares method is used more frequently.

8.3.2 Two-Stage Least Squares

The **two-stage least squares (2SLS) method** is the most widely used estimation method for simultaneous equations models. This estimation procedure was introduced by Theil (1953) and Basmann (1957) and applies to models that are exactly identifiable or overidentifiable.

As the name suggests, this technique involves applying the OLS method twice. Consider the simultaneous equations model with M endogenous variables and k predetermined variables:

$$\begin{aligned}
Y_{1t} &= \beta_{12}Y_{2t} + \ldots + \beta_{1M}Y_{Mt} + \gamma_{11}X_{1t} + \gamma_{12}X_{2t} + \ldots + \gamma_{1k}X_{kt} + \varepsilon_{1t} \\
Y_{2t} &= \beta_{21}Y_{1t} + \ldots + \beta_{2M}Y_{Mt} + \gamma_{21}X_{1t} + \gamma_{22}X_{2t} + \ldots + \gamma_{2k}X_{kt} + \varepsilon_{2t} \\
&\ldots \\
Y_{Mt} &= \beta_{M1}Y_{1t} + \ldots + \beta_{MM}Y_{Mt} + \gamma_{M1}X_{1t} + \gamma_{M2}X_{2t} + \ldots + \gamma_{Mk}X_{kt} + \varepsilon_{Mt}
\end{aligned} \tag{8.46}$$

The first step consists of regressing each of the endogenous variables $(Y_{1t}, Y_{2t}, \ldots, Y_{Mt})$ on the set of predetermined variables $(X_{1t}, X_{2t}, \ldots, X_{kt})$—the aim being to remove the correlation between endogenous variables and disturbances. We thus have the following system:

$$\begin{aligned} Y_{1t} &= \alpha_{11}X_{1t} + \alpha_{12}X_{2t} + \ldots + \alpha_{1k}X_{kt} + u_{1t} \\ Y_{2t} &= \alpha_{21}X_{1t} + \alpha_{22}X_{2t} + \ldots + \alpha_{2k}X_{kt} + u_{2t} \\ &\ldots \\ Y_{Mt} &= \alpha_{M1}X_{1t} + \alpha_{M2}X_{2t} + \ldots + \alpha_{Mk}X_{kt} + u_{Mt} \end{aligned} \tag{8.47}$$

The terms $(u_{1t}, u_{2t}, \ldots, u_{Mt})$ denote the error terms associated with each of the equations in this system. This system corresponds to a reduced form system insofar as no endogenous variables appear on the right-hand side of the various equations. We deduce from the estimation of these equations the estimated values $\left(\hat{Y}_{1t}, \hat{Y}_{2t}, \ldots, \hat{Y}_{Mt}\right)$:

$$\begin{aligned} \hat{Y}_{1t} &= \hat{\alpha}_{11}X_{1t} + \hat{\alpha}_{12}X_{2t} + \ldots + \hat{\alpha}_{1k}X_{kt} \\ \hat{Y}_{2t} &= \hat{\alpha}_{21}X_{1t} + \hat{\alpha}_{22}X_{2t} + \ldots + \hat{\alpha}_{2k}X_{kt} \\ &\ldots \\ \hat{Y}_{Mt} &= \hat{\alpha}_{M1}X_{1t} + \hat{\alpha}_{M2}X_{2t} + \ldots + \hat{\alpha}_{Mk}X_{kt} \end{aligned} \tag{8.48}$$

The second step consists in replacing the endogenous variables appearing on the right-hand side of the structural equations with their values estimated in the first step, i.e.:

$$\begin{aligned} Y_{1t} &= \beta_{12}\hat{Y}_{2t} + \ldots + \beta_{1M}\hat{Y}_{Mt} + \gamma_{11}X_{1t} + \gamma_{12}X_{2t} + \ldots + \gamma_{1k}X_{kt} + v_{1t} \\ Y_{2t} &= \beta_{21}\hat{Y}_{1t} + \ldots + \beta_{2M}\hat{Y}_{Mt} + \gamma_{21}X_{1t} + \gamma_{22}X_{2t} + \ldots + \gamma_{2k}X_{kt} + v_{2t} \\ &\ldots \\ Y_{Mt} &= \beta_{M1}\hat{Y}_{1t} + \ldots + \beta_{MM}\hat{Y}_{Mt} + \gamma_{M1}X_{1t} + \gamma_{M2}X_{2t} + \ldots + \gamma_{Mk}X_{kt} + v_{Mt} \end{aligned} \tag{8.49}$$

where the terms $(v_{1t}, v_{2t}, \ldots, v_{Mt})$ designate the disturbances associated with the equations of the latter system.

The two-stage least squares estimator can be interpreted as an instrumental variables estimator where the instruments used are the estimated values of the endogenous variables (for an in-depth description, see in particular Johnston and Dinardo, 1996). It can be shown that in the absence of autocorrelation and heteroskedasticity, the two-stage least squares estimator is the most efficient instrumental variables estimator.

Remark 8.4 If the coefficients of determination associated with the reduced-form equations of the first stage are very high, the OLS and two-stage least squares estimators will be similar. Indeed, if the coefficient of determination is large, the estimated values of the endogenous variables $\left(\hat{Y}_{1t}, \hat{Y}_{2t}, \ldots, \hat{Y}_{Mt}\right)$ are close to the true values $(Y_{1t}, Y_{2t}, \ldots, Y_{Mt})$. As a result, the estimators in the second step will be very close to those in the first step. Conversely, if the coefficients of determination associated with the reduced-form equations of the first stage are low, the regressions are poorly explanatory, and the estimated values $\left(\hat{Y}_{1t}, \hat{Y}_{2t}, \ldots, \hat{Y}_{Mt}\right)$ used in the second stage will be largely composed of the errors of the first-stage regressions. The significance of the two-stage least squares estimators is then greatly reduced.

Remark 8.5 When the model is exactly identified, the indirect least squares and two-stage least squares methods lead to identical results.

Remark 8.6 There are other limited-information methods for estimating simultaneous equations models: the generalized moments estimator (used when there is a presumption of heteroskedasticity), the limited-information maximum likelihood estimator, or K-class estimators. For a presentation of these various techniques, see Theil (1971), Davidson and MacKinnon (1993), Florens et al. (2007), or Greene (2020).

8.3.3 Full-Information Methods

We will not develop these techniques in this book but refer readers instead to Zellner and Theil (1962), Theil (1971), Johnston and Dinardo (1996), or Greene (2020). Let us simply mention that these procedures consist in estimating the M equations of the system simultaneously. Thus, all the information about the set of the structural equations is taken into account during the estimation. In this framework, the most commonly used methods are:

- the **Three-stage least squares method**, due to Zellner and Theil (1962). Heuristically, this technique involves (i) estimating the reduced form coefficients by OLS, (ii) determining the two-stage least squares estimators for each equation, and (iii) calculating the GLS estimator. The three-stage least squares estimator is an asymptotically efficient instrumental variables estimator. It is particularly appropriate when the disturbances are heteroskedastic and correlated with each other.
- The **full-information maximum likelihood method**. Like the previous one, this technique considers all the equations and all the model parameters jointly. It is based on the assumption that the disturbances are normally distributed and consists in maximizing the log likelihood associated with the model. In addition to Theil (1971), Dhrymes (1973) and Hausman (1975, 1983) can also be consulted on this technique.
- The **system generalized method of moments**. This method is mainly used in the presence of heteroskedasticity. If the disturbances are homoskedastic, this leads to results asymptotically equivalent to those derived from the three-stage least squares method.

Remark 8.7 The three-stage least squares method can be seen as a two-stage least squares version of the **SUR** (seemingly unrelated regressions) method of Zellner (1962). A **SUR model** is a system composed of M equations and T observations of the type:

$$\begin{cases} \boldsymbol{Y}_1 = \boldsymbol{X}_1\boldsymbol{\beta}_1 + \boldsymbol{\varepsilon}_1 \\ \boldsymbol{Y}_2 = \boldsymbol{X}_2\boldsymbol{\beta}_2 + \boldsymbol{\varepsilon}_2 \\ \dots \\ \boldsymbol{Y}_M = \boldsymbol{X}_M\boldsymbol{\beta}_M + \boldsymbol{\varepsilon}_M \end{cases} \tag{8.50}$$

which can be written as:

$$\boldsymbol{Y}_i = \boldsymbol{X}_i\boldsymbol{\beta}_i + \boldsymbol{\varepsilon}_i \tag{8.51}$$

with $i = 1, \dots, M$, $\boldsymbol{\varepsilon} = [\boldsymbol{\varepsilon}_1, \boldsymbol{\varepsilon}_2, \dots, \boldsymbol{\varepsilon}_M]'$, $E[\boldsymbol{\varepsilon}\,|\boldsymbol{X}_1, \dots, \boldsymbol{X}_M] = 0$, and $E\left[\boldsymbol{\varepsilon}\boldsymbol{\varepsilon}'\,|\boldsymbol{X}_1, \dots, \boldsymbol{X}_M\right] = \boldsymbol{\Omega}_{\varepsilon}$.

This model is called a seemingly unrelated regressions model because the equations are linked only by their disturbances. The SUR method, which consists in applying the GLS technique to the system of M equations, is appropriate if all the variables on the right-hand side of the equations are exogenous (which is not the case in the structural form of simultaneous equations models) and enables the parameters of a system to be estimated, taking into account heteroskedasticity and correlation between the error terms of the different equations.

Remark 8.8 The three-stage least squares and full-information maximum likelihood estimators are instrumental variables estimators. Both estimators have the same asymptotic variance-covariance matrix. Therefore, under the assumption of normality of the disturbances, the two estimators have the same asymptotic distribution. The three-stage least squares estimator is, however, easier to calculate than the full-information maximum likelihood estimator.

Remark 8.9 One may ask under what conditions the three-stage least squares method is more efficient than the two-stage least squares method. Generally speaking, a full-information method is more efficient than a limited-information method if the model specification is correct. This is a very strong condition, especially in the case of large models. A misspecification in the model structure will affect the whole system with full-information three-stage least squares and maximum likelihood methods, whereas limited-information methods generally restrict the problem to the equation affected by the misspecification. Furthermore, if the disturbances of the structural equations are not correlated with each other, the two-stage and three-stage least squares methods yield identical results. Similarly, both techniques lead to identical results if the model equations are exactly identified.

8.4 Specification Test

We have seen that OLS estimators are not consistent in the case of simultaneous equations. In the presence of simultaneity, it is appropriate to use other estimation techniques that we presented in the previous section (instrumental variables methods). However, if simultaneity does not exist, the instrumental variables techniques lead to efficient, but non-consistent estimators. The question of simultaneity is therefore crucial. It arises insofar as the endogenous variables appear among the regressors of a simultaneous equations model and insofar as such variables are likely to be correlated with the disturbances. Testing simultaneity therefore amounts to testing the correlation between an endogenous regressor and the error term. If the test concludes that simultaneity is present, it is appropriate to use the techniques presented in the previous section, i.e., the instrumental variables methods. On the other hand, in the absence of simultaneity, OLS should be used.

The test proposed by Hausman (1978) provides a way of dealing with the simultaneity problem. The general principle of the test is to compare two sets of estimators: (i) a set of estimators assumed to be consistent under the null hypothesis (absence of simultaneity) and under the alternative hypothesis (presence of simultaneity) and (ii) a set of estimators assumed to be consistent only under the null hypothesis. To illustrate this test, consider the following example, inspired by Pindyck and Rubinfeld (1991). The model consists of a demand equation:

$$Q_t = \alpha_0 + \alpha_1 P_t + \alpha_2 Y_t + \alpha_3 R_t + \varepsilon_{1t} \tag{8.52}$$

and a supply equation:

$$Q_t = \beta_0 + \beta_1 P_t + \varepsilon_{2t} \tag{8.53}$$

where Q denotes quantity, P price, Y income, and R wealth. It is assumed that Y and R are exogenous, with P and Q being endogenous. To determine whether there is a simultaneity problem between P and Q, we proceed as follows.

In the first step, from the structural model formed by Eqs. (8.52) and (8.53), we deduce the reduced form, which can be expressed in a general way as follows:

$$Q_t = a_0 + a_1 Y_t + \alpha_2 R_t + u_{1t} \tag{8.54}$$

$$P_t = b_0 + b_1 Y_t + b_2 R_t + u_{2t} \tag{8.55}$$

We estimate (8.55) by OLS, which gives:

$$\hat{P}_t = \hat{b}_0 + \hat{b}_1 Y_t + \hat{b}_2 R_t \tag{8.56}$$

from which we derive the residuals:

$$\hat{u}_{2t} = P_t - \hat{P}_t \tag{8.57}$$

Replacing P_t by $\hat{P}_t + \hat{u}_{2t}$ in (8.53), we obtain:

$$Q_t = \beta_0 + \beta_1 \hat{P}_t + \beta_1 \hat{u}_{2t} + \varepsilon_{2t} \tag{8.58}$$

Under the null hypothesis of no simultaneity, the correlation between $\hat{u}_{2t}$ and ε_{2t} is zero.

The second step is to estimate the relationship (8.58) and perform a significance test (usual t-test) of the coefficient assigned to $\hat{u}_{2t}$. If this coefficient is not significantly different from zero, the null hypothesis is not rejected and there is no simultaneity problem: the OLS method can be applied. On the other hand, if it is significantly different from zero, the instrumental variables methods presented in the previous section should be preferred.

Remark 8.10 In Eq. (8.58), Pindyck and Rubinfeld (1991) suggest regressing Q_t on P_t (instead of $\hat{P}_t$) and $\hat{u}_{2t}$.

8.5 Empirical Application

To illustrate the various concepts presented in this chapter, let us consider Klein's (1950) model of the US economy over the period 1920–1941.

8.5.1 Writing the Model

This model is composed of the following six equations:

$$C_t = \alpha_0 + \alpha_1 \pi_t + \alpha_2 \pi_{t-1} + \alpha_3 (W_{1t} + W_{2t}) + \varepsilon_{1t} \tag{8.59}$$

$$I_t = \beta_0 + \beta_1 \pi_t + \beta_2 \pi_{t-1} + \beta_3 K_{t-1} + \varepsilon_{2t} \tag{8.60}$$

$$W_{1t} = \gamma_0 + \gamma_1 Y_t + \gamma_2 Y_{t-1} + \gamma_3 t + \varepsilon_{3t} \tag{8.61}$$

$$C_t + I_t + G_t = Y_t \tag{8.62}$$

$$\pi_t = Y_t - W_{1t} - W_{2t} - T_t \tag{8.63}$$

$$K_t - K_{t-1} = I_t \tag{8.64}$$

where C denotes consumption (in constant dollars), π profits (in constant dollars), W_1 private sector wages, W_2 government wage payments (public sector wages), I net investment (in constant dollars), K_{t-1} is the capital stock at the beginning of the year, Y output (in constant dollars), G government expenditures, T taxes on profits, and t a time trend.

Equation (8.59) is the consumption equation, Eq. (8.60) is the investment equation, and Eq. (8.61) describes the demand for labor. The last three equations are identities. Equation (8.62) stresses that output is equal to the sum of consumer demand for goods, firm investment, and government spending. According to Eq. (8.63), output, i.e., income, is equal to the sum of profits, taxes on profits, and wages. Finally, Eq. (8.64) defines investment as the change in the capital stock.

The endogenous variables of the system are consumption, investment, private sector wages, output, profits, and the capital stock. With our notations, we therefore have $M = 6$. For the predetermined variables, we distinguish:

- Lagged variables: π_{t-1}, K_{t-1}, Y_{t-1}
- Exogenous variables: W_{2t}, T_t, G_t, and the trend t

If we add the constant term present in each of the first three equations, the number of exogenous variables k is equal to 8.

8.5.2 Conditions for Identification

Prior to estimation, it is necessary to check that the model is not underidentified, in which case estimation is impossible. The identification condition (order condition) established previously is written:

$$k_j^* \geq M_j \tag{8.65}$$

or, where there are linear restrictions on the parameters:

$$r_j + k_j^* \geq M_j \tag{8.66}$$

where M_j is the number of endogenous variables included in the equation j considered, k_j^* is the number of exogenous variables excluded from the equation j, and r_j designates the number of restrictions other than those of exclusion. Recall further that we have:

$$k = k_j + k_j^* \tag{8.67}$$

where k_j is the number of exogenous variables in the equation j under consideration, with k denoting the total number of exogenous variables in the model.

With these points in mind, let us study the identification conditions equation by equation:

- For Eq. (8.59), we have: $M_j = 3$ (three endogenous variables) and $k_j = 3$ (two exogenous variables plus the constant term). A linear restriction is also imposed on the parameters, since the coefficients associated with W_1 and W_2 are assumed to be identical. We thus have $r_j = 1$. We therefore use the order condition (8.66) with $k_j^* = k - k_j = 8 - 3 = 5$. We have: $r_j + k_j^* = 1 + 5 = 6$ which is greater than $M_j = 3$. We deduce that Eq. (8.59) is overidentified.

- In Eq. (8.60), we have: $M_j = 2$ and $k_j = 3$ (two exogenous variables plus the constant term). No restriction is imposed on the parameters $(r_j = 0)$ and we then use the order condition (8.65). $k_j^* = 8 - 3 = 5$ is greater than $M_j = 2$, implying that Eq. (8.60) is also overidentified.
- Finally, Eq. (8.61) is such that $M_j = 2$ and $k_j = 3$ (two exogenous variables plus the constant term). Due to the absence of restrictions on the parameters $(r_j = 0)$, using the order condition (8.65) gives us: $k_j^* > M_j$. Consequently, Eq. (8.61) is also overidentified.

All three equations of the Klein model are overidentified. The model can then be estimated.

8.5.3 Data

The data concern the United States over the period 1920–1941 and are annual. Table 8.1 gives the values taken by the various variables used in the model.

8.5.4 Model Estimation

In order to estimate the Klein model, instrumental variables methods must be used. We propose to apply one limited-information method (two-stage least squares) and two full-information methods (three-stage least squares and full-information maximum likelihood). First, we estimate each of the equations using OLS.

OLS Estimation Equation by Equation

As previously mentioned, OLS estimators are not consistent when there is interdependence between endogenous variables, which is the case here. However, we apply this estimation procedure for comparison with the results obtained by instrumental variables methods. The results of the OLS estimation of each equation are reported in Tables 8.2 (consumption equation), 8.3 (investment equation), and 8.4 (labor-demand equation).

Two-Stage Least Squares Estimation

We now propose to apply the two-stage least squares method to the three equations of the Klein model. This method is a priori appropriate insofar as the model is overidentified (we cannot apply the indirect least squares method, which requires the model to be exactly identified).

Applying this procedure involves selecting a certain number of instruments. We have chosen the same instruments for each of the equations, i.e., the set of exogenous variables: the constant term, one-period lagged profits, one-period lagged capital

Table 8.1 Data from Klein's model

	C_t	π_t	W_{1t}	W_{2t}	K_{t-1}	Y_t	G_t	I_t	T_t
1920	39.8	12.7	28.8	2.2	180.1	44.9	2.4	2.7	3.4
1921	41.9	12.4	25.5	2.7	182.8	45.6	3.9	−0.2	7.7
1922	45	16.9	29.3	2.9	182.6	50.1	3.2	1.9	3.9
1923	49.2	18.4	34.1	2.9	184.5	57.2	2.8	5.2	4.7
1924	50.6	19.4	33.9	3.1	189.7	57.1	3.5	3	3.8
1925	52.6	20.1	35.4	3.2	192.7	61	3.3	5.1	5.5
1926	55.1	19.6	37.4	3.3	197.8	64	3.3	5.6	7
1927	56.2	19.8	37.9	3.6	203.4	64.4	4	4.2	6.7
1928	57.3	21.1	39.2	3.7	207.6	64.5	4.2	3	4.2
1929	57.8	21.7	41.3	4	210.6	67	4.1	5.1	4
1930	55	15.6	37.9	4.2	215.7	61.2	5.2	1	7.7
1931	50.9	11.4	34.5	4.8	216.7	53.4	5.9	−3.4	7.5
1932	45.6	7	29	5.3	213.3	44.3	4.9	−6.2	8.3
1933	46.5	11.2	28.5	5.6	207.1	45.1	3.7	−5.1	5.4
1934	48.7	12.3	30.6	6	202	49.7	4	−3	6.8
1935	51.3	14	33.2	6.1	199	54.4	4.4	−1.3	7.2
1936	57.7	17.6	36.8	7.4	197.7	62.7	2.9	2.1	8.3
1937	58.7	17.3	41	6.7	199.8	65	4.3	2	6.7
1938	57.5	15.3	38.2	7.7	201.8	60.9	5.3	−1.9	7.4
1939	61.6	19	41.6	7.8	199.9	69.5	6.6	1.3	8.9
1940	65	21.1	45	8	201.2	75.7	7.4	3.3	9.6
1941	69.7	23.5	53.3	8.5	204.5	88.4	13.8	4.9	11.6

Source: Klein (1950)

Table 8.2 OLS estimation of the consumption equation

Dependent variable: C				
Method: least squares				
Variable	Coefficient	Std. error	t-Statistic	Prob.
Constant	16.23660	1.302698	12.46382	0.0000
π	0.192934	0.091210	2.115273	0.0495
$\pi(-1)$	0.089885	0.090648	0.991582	0.3353
$(W_1 + W_2)$	0.796219	0.039944	19.93342	0.0000
R-squared	0.981008	Mean dependent var	53.99524	
Adjusted R-squared	0.977657	S.D. dependent var	6.860866	
S.E. of regression	1.025540	Akaike info criterion	3.057959	
Sum squared resid	17.87945	Schwarz criterion	3.256916	
Log likelihood	−28.10857	F-statistic	292.7076	
Durbin-Watson stat	1.367474	Prob(F-statistic)	0.000000	

Table 8.3 OLS estimation of the investment equation

Dependent variable: I				
Method: least squares				
Variable	Coefficient	Std. error	t-Statistic	Prob.
Constant	10.12579	5.465547	1.852658	0.0814
π	0.479636	0.097115	4.938864	0.0001
$\pi(-1)$	0.333039	0.100859	3.302015	0.0042
$K(-1)$	−0.111795	0.026728	−4.182749	0.0006
R-squared	0.931348	Mean dependent var	1.266667	
Adjusted R-squared	0.919233	S.D. dependent var	3.551948	
S.E. of regression	1.009447	Akaike info criterion	3.026325	
Sum squared resid	17.32270	Schwarz criterion	3.225282	
Log likelihood	−27.77641	F-statistic	76.87537	
Durbin-Watson stat	1.810184	Prob(F-statistic)	0.000000	

Table 8.4 OLS estimation of the labor-demand equation

Dependent variable: W_1				
Method: least squares				
Variable	Coefficient	Std. error	t-Statistic	Prob.
Constant	0.064346	1.151797	0.055866	0.9561
Y	0.439477	0.032408	13.56093	0.0000
$Y(-1)$	0.146090	0.037423	3.903734	0.0011
$@TREND$	0.130245	0.031910	4.081604	0.0008
R-squared	0.987414	Mean dependent var	36.36190	
Adjusted R-squared	0.985193	S.D. dependent var	6.304401	
S.E. of regression	0.767147	Akaike info criterion	2.477367	
Sum squared resid	10.00475	Schwarz criterion	2.676324	
Log likelihood	−22.01235	F-statistic	444.5682	
Durbin-Watson stat	1.958434	Prob(F-statistic)	0.000000	

stock, one-period lagged output, public sector wages, government expenditure, income taxes, and the trend. The results from applying the two-stage least squares method to each of the equations are given in Tables 8.5 (consumption equation), 8.6 (investment equation), and 8.7 (labor-demand equation).

If we compare the results obtained with the two-stage least squares technique with those obtained with the OLS method, we see that the coefficients still have the same signs, but their orders of magnitude are different. This is particularly noticeable for profits and one-period lagged profits in the consumption and investment equations. In particular, the OLS method gives more weight to current profits, unlike the two-stage least squares procedure, which gives greater weight to one-period lagged profits. However, the sum of the coefficients associated with π_t and π_{t-1} is similar for both methods. With regard to the labor-demand equation, the results obtained by the two methods are very similar.

Table 8.5 Two-stage least squares estimation of the consumption equation

Dependent variable: C				
Method: two-stage least squares				
Instrument list: constant $\pi(-1)$ $K(-1)$ $Y(-1)$ W_2 $@TREND$ G T				
Variable	Coefficient	Std. error	t-Statistic	Prob.
Constant	16.55476	1.467979	11.27725	0.0000
π	0.017302	0.131205	0.131872	0.8966
$\pi(-1)$	0.216234	0.119222	1.813714	0.0874
$(W_1 + W_2)$	0.810183	0.044735	18.11069	0.0000
R-squared	0.976711	Mean dependent var	53.99524	
Adjusted R-squared	0.972601	S.D. dependent var	6.860866	
S.E. of regression	1.135659	F-statistic	225.9334	
Sum squared resid	21.92525	Prob(F-statistic)	0.0000	
Durbin-Watson stat	1.485072			

Table 8.6 Two-stage least squares estimation of the investment equation

Dependent variable: I				
Method: two-stage least squares				
Instrument list: constant $\pi(-1)$ $K(-1)$ $Y(-1)$ W_2 $@TREND$ G T				
Variable	Coefficient	Std. error	t-Statistic	Prob.
Constant	20.27821	8.383249	2.418896	0.0271
π	0.150222	0.192534	0.780237	0.4460
$\pi(-1)$	0.615944	0.180926	3.404398	0.0034
$K(-1)$	−0.157788	0.040152	−3.929751	0.0011
R-squared	0.884884	Mean dependent var	1.266667	
Adjusted R-squared	0.864569	S.D. dependent var	3.551948	
S.E. of regression	1.307149	F-statistic	41.20019	
Sum squared resid	29.04686	Prob(F-statistic)	0.000000	
Durbin-Watson stat	1.810184			

Three-Stage Least Squares Estimation

The three-stage least squares method is a full-information estimation method, since we estimate the model as a whole, i.e., all the equations simultaneously. All the information contained in the system is thus taken into account. Such a technique is particularly appropriate in the presence of heteroskedasticity and cross-correlation between the disturbances.

The list of instruments used for the estimation is identical to the previous one, namely: the constant, one-period lagged profits, one-period lagged capital stock, one-period lagged output, public sector wages, government expenditure, taxes on profits, and the trend. The results from the estimation are shown in Table 8.8. Table 8.9 sets out the estimation statistics for each of the three equations.

Table 8.8 shows that the results obtained by three-stage least squares are similar to those obtained by applying the two-stage least squares method. The coefficients

Table 8.7 Two-stage least squares estimation of the labor-demand equation

Dependent variable: W_1				
Method: two-stage least squares				
Instrument list: constant $\pi(-1)$ $K(-1)$ $Y(-1)$ W_2 $@TREND$ G T				
Variable	Coefficient	Std. error	t-Statistic	Prob.
Constant	0.065944	1.153313	0.057178	0.9551
Y	0.438859	0.039603	11.08155	0.0000
$Y(-1)$	0.146674	0.043164	3.398063	0.0034
$@TREND$	0.130396	0.032388	4.026001	0.0009
R-squared	0.987414	Mean dependent var	36.36190	
Adjusted R-squared	0.985193	S.D. dependent var	6.304401	
S.E. of regression	0.767155	F-statistic	424.1940	
Sum squared resid	10.00496	Prob(F-statistic)	0.000000	
Durbin-Watson stat	1.963416			

Table 8.8 Three-stage least squares estimation of the Klein model

Estimation method: three-stage least squares				
	Coefficient	Std. error	t-Statistic	Prob.
$C = C(1) + C(2) * \pi + C(3) * \pi(-1) + C(4) * (W_1 + W_2)$				
C(1)	16.44079	1.304549	12.60266	0.0000
C(2)	0.124890	0.108129	1.155013	0.2535
C(3)	0.163144	0.100438	1.624323	0.1105
C(4)	0.790081	0.037938	20.82563	0.0000
$I = C(5) + C(6) * \pi + C(7) * \pi(-1) + C(8) * K(-1)$				
C(5)	28.17785	6.793770	4.147601	0.0001
C(6)	−0.013079	0.161896	−0.080787	0.9359
C(7)	0.755724	0.152933	4.941532	0.0000
C(8)	−0.194848	0.032531	−5.989674	0.0000
$W_1 = C(9) + C(10) * Y + C(11) * Y(-1) + C(12) * TREND$				
C(9)	0.150802	1.014983	0.148576	0.8825
C(10)	0.400492	0.031813	12.58877	0.0000
C(11)	0.181291	0.034159	5.307304	0.0000
C(12)	0.149674	0.027935	5.357897	0.0000
Determinant residual covariance: 0.282997				

are always assigned the same signs, but the orders of magnitude vary slightly. However, even if the value taken by the coefficients is sometimes different, the weight of the variables is not modified in the sense that a variable that was not significant with the two-stage least squares method is also not significant with the three-stage least squares method. The same applies to significant variables.

Table 8.9 Estimation statistics, three-stage least squares method

Consumption equation			
R-squared	0.980108	Mean dependent var	53.99524
Adjusted R-squared	0.976598	S.D. dependent var	6.860866
S.E. of regression	1.049565	Sum squared resid	18.72696
Durbin-Watson stat	1.424939		
Investment equation			
R-squared	0.825805	Mean dependent var	1.266667
Adjusted R-squared	0.795065	S.D. dependent var	3.551948
S.E. of regression	1.607958	Sum squared resid	43.95398
Durbin-Watson stat	1.995884		
Labor-demand equation			
R-squared	0.986262	Mean dependent var	36.36190
Adjusted R-squared	0.983838	S.D. dependent var	6.304401
S.E. of regression	0.801490	Sum squared resid	10.92056
Durbin-Watson stat	2.155046		

Table 8.10 Full-information maximum likelihood estimation of the Klein model

Estimation method: full-information maximum likelihood (Marquardt)				
	Coefficient	Std. error	z-Statistic	Prob.
$C = C(1) + C(2) * \pi + C(3) * \pi(-1) + C(4) * (W_1 + W_2)$				
C(1)	15.83177	4.111036	3.851040	0.0001
C(2)	0.299937	0.412579	0.726980	0.4672
C(3)	0.042552	0.166547	0.255499	0.7983
C(4)	0.781083	0.078554	9.943317	0.0000
$I = C(5) + C(6) * \pi + C(7) * \pi(-1) + C(8) * K(-1)$				
C(5)	15.59875	14.40899	1.082571	0.2790
C(6)	0.382663	0.341197	1.121533	0.2621
C(7)	0.409364	0.248292	1.648721	0.0992
C(8)	−0.137156	0.071673	−1.913642	0.0557
$W_1 = C(9) + C(10) * Y + C(11) * Y(-1) + C(12) * TREND$				
C(9)	0.036159	4.171662	0.008668	0.9931
C(10)	0.370776	0.128994	2.874372	0.0040
C(11)	0.207497	0.090315	2.297480	0.0216
C(12)	0.184179	0.101391	1.816528	0.0693
Log likelihood: −69.25950				
Determinant residual covariance: 0.146976				

Full-Information Maximum Likelihood Estimation

Implementing the full-information maximum likelihood procedure involves assuming that the error terms are normally distributed. The results obtained are shown in Tables 8.10 and 8.11. In general, it can be seen that the t-statistics of the coefficients are significantly lower than those associated with the coefficients estimated by

Table 8.11 Estimation statistics, full-information maximum likelihood method

Consumption equation			
R-squared	0.979294	Mean dependent var	53.99524
Adjusted R-squared	0.975640	S.D. dependent var	6.860866
S.E. of regression	1.070813	Sum squared resid	19.49287
Durbin-Watson stat	1.260803		
Investment equation			
R-squared	0.926089	Mean dependent var	1.266667
Adjusted R-squared	0.913046	S.D. dependent var	3.551948
S.E. of regression	1.047396	Sum squared resid	18.64964
Durbin-Watson stat	1.895880		
Labor-demand equation			
R-squared	0.982884	Mean dependent var	36.36190
Adjusted R-squared	0.979864	S.D. dependent var	6.304401
S.E. of regression	0.894610	Sum squared resid	13.60556
Durbin-Watson stat	2.024727		

the other techniques (two-stage and three-stage least squares). In the consumption equation, the values taken by the coefficients of the two profit variables differ from those obtained by three-stage least squares, but remain insignificant. Conversely, in the investment equation, the variables that were significant with three-stage least squares are no longer significant with the maximum likelihood method. Finally, the results concerning the last equation of the Klein model remain similar to those obtained with the three-stage least squares technique.

Conclusion

This chapter has gone beyond the univariate framework by presenting multi-equation models, i.e., systems of equations. Simultaneous equations models, the subject of this chapter, are based on economic foundations and are therefore an alternative to VAR models (presented in the previous chapter), which are a-theoretical. We have seen that a prerequisite for estimating simultaneous equations models is identification: we need to check that the available data contain sufficient information for the parameters to be estimated. Once identification has been carried out, it is possible to proceed with estimation. Several procedures have been presented and/or applied, including indirect least squares, two-stage least squares, three-stage least squares, and full-information maximum likelihood.

The Gist of the Chapter

Simultaneous equations model	$\underset{(M,M)}{\boldsymbol{B}}\underset{(M,1)}{\boldsymbol{Y}} + \underset{(M,k)}{\boldsymbol{\Gamma}}\underset{(k,1)}{\boldsymbol{X}} = \underset{(M,1)}{\boldsymbol{\varepsilon}}$
	$\boldsymbol{Y}$: vector containing the M endogenous variables
	$\boldsymbol{X}$: vector containing the k exogenous variables
	$\boldsymbol{\varepsilon}$: vector of structural disturbances
Identification	Order condition (identification)
	Rank condition (uniqueness of the solution)
Estimation	
Limited information methods	Indirect least squares
	Two-stage least squares
Full-information methods	Three-stage least squares
	Full-information maximum likelihood
Specification	Hausman (1978) test

Further Reading

Developments on simultaneous equations models can be found in the textbooks by Gujarati et al. (2017) and Greene (2020). Theil (1978), Pindyck and Rubinfeld (1991), and Florens et al. (2007) will also prove useful.

Appendix: Statistical Tables

Standard Normal Distribution

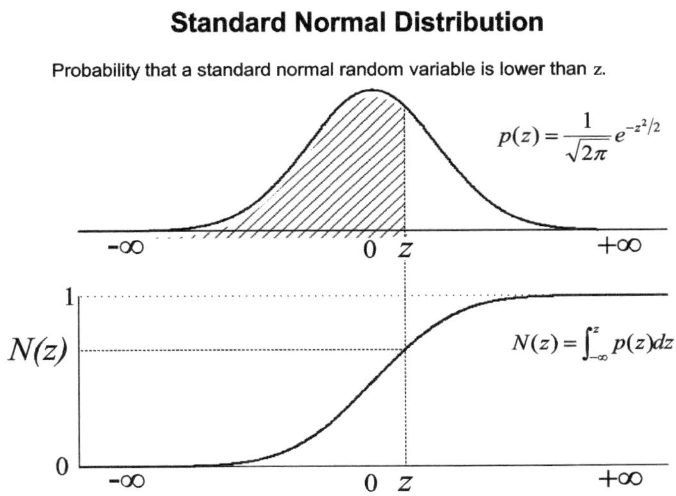

The table below shows the values for z positive, For z, negative, the value is $N(z) = I - N(-z)$.

V. Mignon, *Principles of Econometrics*, Classroom Companion: Economics,
https://doi.org/10.1007/978-3-031-52535-3

z	**0.00**	**0.01**	**0.02**	**0.03**	**0.04**	**0.05**	**0.06**	**0.07**	**0.08**	**0.09**
0.0	0.500000	0.503989	0.507978	0.511966	0.515953	0.519939	0.523922	0.527903	0.531881	0.535856
0.1	0.539828	0.543795	0.547758	0.551717	0.555670	0.559618	0.563559	0.567495	0.571424	0.575345
0.2	0.579260	0.583166	0.587064	0.590954	0.594835	0.598706	0.602568	0.606420	0.610261	0.614092
0.3	0.617911	0.621720	0.625516	0.629300	0.633072	0.636831	0.640576	0.644309	0.648027	0.651732
0.4	0.655422	0.659097	0.662757	0.666402	0.670031	0.673645	0.677242	0.680822	0.684386	0.687933
0.5	0.691462	0.694974	0.698468	0.701944	0.705401	0.708840	0.712260	0.715661	0.719043	0.722405
0.6	0.725747	0.729069	0.732371	0.735653	0.738914	0.742154	0.745373	0.748571	0.751748	0.754903
0.7	0.758036	0.761148	0.764238	0.767305	0.770350	0.773373	0.776373	0.779350	0.782305	0.785236
0.8	0.788145	0.791030	0.793892	0.796731	0.799546	0.802337	0.805105	0.807850	0.810570	0.813267
0.9	0.815940	0.818589	0.821214	0.823814	0.826391	0.828944	0.831472	0.833977	0.836457	0.838913
1.0	0.841345	0.843752	0.846136	0.848495	0.850830	0.853141	0.855428	0.857690	0.859929	0.862143
1.1	0.864334	0.866500	0.868643	0.870762	0.872857	0.874928	0.876976	0.879000	0.881000	0.882977
1.2	0.884930	0.886861	0.888768	0.890651	0.892512	0.894350	0.896165	0.897958	0.899727	0.901475
1.3	0.903200	0.904902	0.906582	0.908241	0.909877	0.911492	0.913085	0.914657	0.916207	0.917736
1.4	0.919243	0.920730	0.922196	0.923641	0.925066	0.926471	0.927855	0.929219	0.930563	0.931888
1.5	0.933193	0.934478	0.935745	0.936992	0.938220	0.939429	0.940620	0.941792	0.942947	0.944083
1.6	0.945201	0.946301	0.947384	0.948449	0.949497	0.950529	0.951543	0.952540	0.953521	0.954486
1.7	0.955435	0.956367	0.957284	0.958185	0.959070	0.959941	0.960796	0.961636	0.962462	0.963273
1.8	0.964070	0.964852	0.965620	0.966375	0.967116	0.967843	0.968557	0.969258	0.969946	0.970621
1.9	0.971283	0.971933	0.972571	0.973197	0.973810	0.974412	0.975002	0.975581	0.976148	0.976705
2.0	0.977250	0.977784	0.978308	0.978822	0.979325	0.979818	0.980301	0.980774	0.981237	0.981691

(continued)

2.1	0.982136	0.982571	0.982997	0.983414	0.983823	0.984222	0.984614	0.984997	0.985371	0.985738
2.2	0.986097	0.986447	0.986791	0.987126	0.987455	0.987776	0.988089	0.988396	0.988696	0.988989
2.3	0.989276	0.989556	0.989830	0.990097	0.990358	0.990613	0.990863	0.991106	0.991344	0.991576
2.4	0.991802	0.992024	0.992240	0.992451	0.992656	0.992857	0.993053	0.993244	0.993431	0.993613
2.5	0.993790	0.993963	0.994132	0.994297	0.994457	0.994614	0.994766	0.994915	0.995060	0.995201
2.6	0.995339	0.995473	0.995604	0.995731	0.995855	0.995975	0.996093	0.996207	0.996319	0.996427
2.7	0.996533	0.996636	0.996736	0.996833	0.996928	0.997020	0.997110	0.997197	0.997282	0.997365
2.8	0.997445	0.997523	0.997599	0.997673	0.997744	0.997814	0.997882	0.997948	0.998012	0.998074
2.9	0.998134	0.998193	0.998250	0.998305	0.998359	0.998411	0.998462	0.998511	0.998559	0.998605

(continued)

For values of z higher than 3:

z	3.0	3.1	3.2	3.3	3.4	3.5	3.6	3.8	4.0	4.5
$N(z)$	0.998650	0.999032	0.999313	0.999517	0.999663	0.999767	0.999841	0.999928	0.999968	0.999997

Student t Distribution: Critical Values of t

Student *t* distribution - Critical values of *t*

Values of t with probability P of being exceeded in absolute value.

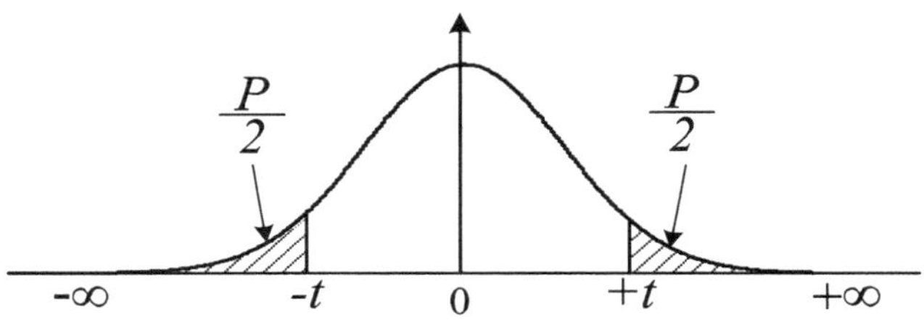

r	$P = 0.90$	$P = 0.80$	$P = 0.70$	$P = 0.60$	$P = 0.50$	$P = 0.40$	$P = 0.30$	$P = 0.20$	$P = 0.10$	$P = 0.05$	$P = 0.01$	$P = 0.005$
1	0.158	0.325	0.510	0.727	1.000	1.376	1.963	3.078	6.314	12.706	63.657	127.321
2	0.142	0.289	0.445	0.617	0.816	1.061	1.386	1.886	2.920	4.303	9.925	14.089
3	0.137	0.277	0.424	0.584	0.765	0.978	1.250	1.638	2.353	3.182	5.841	7.453
4	0.134	0.271	0.414	0.569	0.741	0.941	1.190	1.533	2.132	2.776	4.604	5.598
5	0.132	0.267	0.408	0.559	0.727	0.920	1.156	1.476	2.015	2.571	4.032	4.773
6	0.131	0.265	0.404	0.553	0.718	0.906	1.134	1.440	1.943	2.447	3.707	4.317
7	0.130	0.263	0.402	0.549	0.711	0.896	1.119	1.415	1.895	2.365	3.499	4.029
8	0.130	0.262	0.399	0.546	0.706	0.889	1.108	1.397	1.860	2.306	3.355	3.833
9	0.129	0.261	0.398	0.543	0.703	0.883	1.100	1.383	1.833	2.262	3.250	3.690
10	0.129	0.260	0.397	0.542	0.700	0.879	1.093	1.372	1.812	2.228	3.169	3.581
11	0.129	0.260	0.396	0.540	0.697	0.876	1.088	1.363	1.796	2.201	3.106	3.497
12	0.128	0.259	0.395	0.539	0.695	0.873	1.083	1.356	1.782	2.179	3.055	3.428
13	0.128	0.259	0.394	0.538	0.694	0.870	1.079	1.350	1.771	2.160	3.012	3.372
14	0.128	0.258	0.393	0.537	0.692	0.868	1.076	1.345	1.761	2.145	2.977	3.326
15	0.128	0.258	0.393	0.536	0.691	0.866	1.074	1.341	1.753	2.131	2.947	3.286
16	0.128	0.258	0.392	0.535	0.690	0.865	1.071	1.337	1.746	2.120	2.921	3.252
17	0.128	0.257	0.392	0.534	0.689	0.863	1.069	1.333	1.740	2.110	2.898	3.222
18	0.127	0.257	0.392	0.534	0.688	0.862	1.067	1.330	1.734	2.101	2.878	3.197
19	0.127	0.257	0.391	0.533	0.688	0.861	1.066	1.328	1.729	2.093	2.861	3.174
20	0.127	0.257	0.391	0.533	0.687	0.860	1.064	1.325	1.725	2.086	2.845	3.153
21	0.127	0.257	0.391	0.532	0.686	0.859	1.063	1.323	1.721	2.080	2.831	3.135
22	0.127	0.256	0.390	0.532	0.686	0.858	1.061	1.321	1.717	2.074	2.819	3.119

(continued)

r	$P=0.90$	$P=0.80$	$P=0.70$	$P=0.60$	$P=0.50$	$P=0.40$	$P=0.30$	$P=0.20$	$P=0.10$	$P=0.05$	$P=0.01$	$P=0.005$
23	0.127	0.256	0.390	0.532	0.685	0.858	1.060	1.319	1.714	2.069	2.807	3.104
24	0.127	0.256	0.390	0.531	0.685	0.857	1.059	1.318	1.711	2.064	2.797	3.091
25	0.127	0.256	0.390	0.531	0.684	0.856	1.058	1.316	1.708	2.060	2.787	3.078
26	0.127	0.256	0.390	0.531	0.684	0.856	1.058	1.315	1.706	2.056	2.779	3.067
27	0.127	0.256	0.389	0.531	0.684	0.855	1.057	1.314	1.703	2.052	2.771	3.057
28	0.127	0.256	0.389	0.530	0.683	0.855	1.056	1.313	1.701	2.048	2.763	3.047
29	0.127	0.256	0.389	0.530	0.683	0.854	1.055	1.311	1.699	2.045	2.756	3.038
30	0.127	0.256	0.389	0.530	0.683	0.854	1.055	1.310	1.697	2.042	2.750	3.030
40	0.126	0.255	0.388	0.529	0.681	0.851	1.050	1.303	1.684	2.021	2.704	2.971
80	0.126	0.254	0.387	0.526	0.678	0.846	1.043	1.292	1.664	1.990	2.639	2.887
120	0.126	0.254	0.386	0.526	0.677	0.845	1.041	1.289	1.658	1.980	2.617	2.860
∞	0.126	0.253	0.385	0.524	0.675	0.842	1.036	1.282	1.645	1.960	2.576	2.808

Chi-Squared Distribution: Critical Values of *c*

Chi-squared distribution - Critical values of *c*

Values of *c* with probability *P* of being exceeded in absolute value.

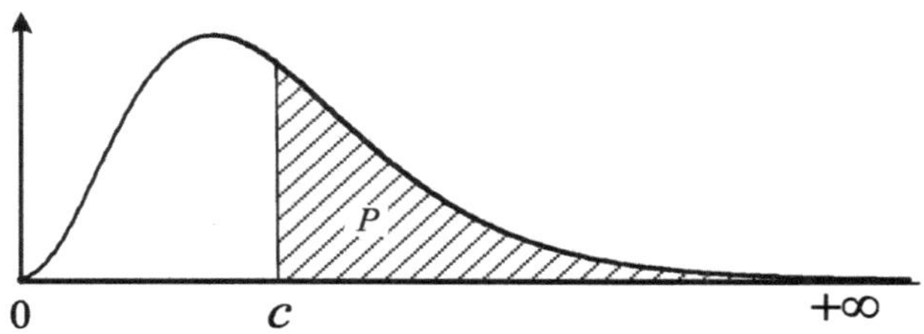

r	$P=0.990$	$P=0.975$	$P=0.950$	$P=0.900$	$P=0.800$	$P=0.700$	$P=0.500$	$P=0.300$	$P=0.200$	$P=0.100$	$P=0.010$	$P=0.005$	$P=0.001$
1	0.000	0.001	0.004	0.016	0.064	0.148	0.455	1.074	1.642	2.706	6.635	7.879	10.828
2	0.200	0.051	0.103	0.211	0.446	0.713	1.386	2.408	3.219	4.605	9.210	10.597	13.816
3	0.115	0.216	0.352	0.584	1.005	1.424	2.366	3.665	4.642	6.251	11.345	12.838	16.266
4	0.297	0.484	0.711	1.064	1.649	2.195	3.357	4.878	5.989	7.779	13.277	14.860	18.467
5	0.554	0.831	1.145	1.610	2.343	3.000	4.351	6.064	7.289	9.236	15.086	16.750	20.515
6	0.872	1.237	1.635	2.204	3.070	3.828	5.348	7.231	8.558	10.645	16.812	18.548	22.458
7	1.239	1.690	2.167	2.833	3.822	4.671	6.346	8.383	9.803	12.017	18.475	20.278	24.322
8	1.646	2.180	2.733	3.490	4.594	5.527	7.344	9.524	11.030	13.362	20.090	21.955	26.124
9	2.088	2.700	3.325	4.168	5.380	6.393	8.343	10.656	12.242	14.684	21.666	23.589	27.877
10	2.558	3.247	3.940	4.865	6.179	7.267	9.342	11.781	13.442	15.987	23.209	25.188	29.588
11	3.053	3.816	4.575	5.578	6.989	8.148	10.341	12.899	14.631	17.275	24.725	26.757	31.264
12	3.571	4.404	5.226	6.304	7.807	9.034	11.340	14.011	15.812	18.549	26.217	28.300	32.909
13	4.107	5.009	5.892	7.042	8.634	9.926	12.340	15.119	16.985	19.812	27.688	29.819	34.528
14	4.660	5.629	6.571	7.790	9.467	10.821	13.339	16.222	18.151	21.064	29.141	31.319	36.123
15	5.229	6.262	7.261	8.547	10.307	11.721	14.339	17.322	19.311	22.307	30.578	32.801	37.697
16	5.812	6.908	7.962	0.312	11.152	12.624	15.338	18.418	20.465	23.542	32.000	34.267	39.252
17	6.408	7.564	8.672	10.085	12.002	13.531	16.338	19.511	21.615	24.769	33.409	35.718	40.790
18	7.015	8.231	9.390	10.865	12.857	14.440	17.338	20.601	22.760	25.989	34.805	37.156	42.312
19	7.633	8.907	10.117	11.651	13.716	15.352	18.338	21.689	23.900	27.204	36.191	38.582	43.820
20	8.260	9.591	10.851	12.443	14.578	16.266	19.337	22.775	25.038	28.412	37.566	39.997	45.315

(continued)

21	8.897	10.283	11.591	13.240	15.445	17.182	20.337	23.858	26.171	29.615	38.932	41.401	46.797
22	9.542	10.982	12.338	14.041	16.314	18.101	21.337	24.939	27.301	30.813	40.289	42.796	48.268
23	10.196	11.689	13.091	14.848	17.187	19.021	22.337	26.018	28.429	32.007	41.638	44.181	49.728
24	10.856	12.401	13.848	15.659	18.062	19.943	23.337	27.096	29.553	33.196	42.980	45.559	51.179
25	11.524	13.120	14.611	16.473	18.940	20.867	24.337	28.172	30.675	34.382	44.314	46.928	52.620
26	12.198	13.844	15.379	17.292	19.820	21.792	25.336	29.246	31.795	35.563	45.642	48.290	54.052
27	12.879	14.573	16.151	18.114	20.703	22.719	26.336	30.319	32.912	36.741	46.963	49.645	55.476
28	13.565	15.308	16.928	18.939	21.588	23.647	27.336	31.391	34.027	37.916	48.278	50.993	56.892
29	14.256	16.047	17.708	19.768	22.475	24.577	28.336	32.461	35.139	39.087	49.588	52.336	58.301
30	14.953	16.791	18.493	20.599	23.364	25.508	29.336	33.530	36.250	40.256	50.892	53.672	59.703
40	22.164	24.433	26.509	29.051	32.345	34.872	39.335	44.165	47.269	51.805	63.691	66.766	73.402
80	53.540	57.153	60.391	64.278	69.207	72.915	79.334	86.120	90.405	96.578	112.329	116.321	124.839
120	86.923	91.573	95.705	100.624	106.806	111.419	119.334	127.616	132.806	140.233	158.950	163.648	173.617

Fisher–Snedecor Distribution: Critical Values of F

Fisher-Snedecor distribution - Critical values of *F*

Values of F with probability P of being exceeded in absolute value.

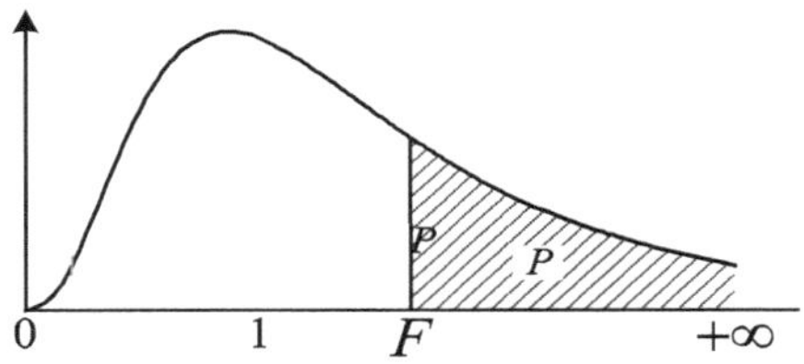

v_2	$v_1 = 1$		$v_1 = 2$		$v_1 = 3$		$v_1 = 4$		$v_1 = 5$		$v_1 = 6$	
	$P = 0.05$	**$P = 0.01$**	**$P = 0.05$**	**$P = 0.01$**	**$P = 0.05$**	**$P = 0.01$**	**$P = 0.05$**	**$P = 0.01$**	**$P = 0.05$**	**$P = 0.01$**	**$P = 0.05$**	**$P = 0.01$**
1	161.448	4052.181	199.500	4999.500	215.707	5403.352	224.583	5624.583	230.162	5763.650	233.986	5858.986
2	18.513	98.503	19.000	99.000	19.164	99.166	19.247	99.249	19.296	99.299	19.330	99.333
3	10.128	34.116	9.552	30.817	9.277	29.457	9.117	28.710	9.013	28.237	8.941	27.911
4	7.709	21.198	6.944	18.000	6.591	16.694	6.388	15.977	6.256	15.522	6.163	15.207
5	6.608	16.258	5.786	13.274	5.409	12.060	5.192	11.392	5.050	10.967	4.950	10.672
6	5.987	13.745	5.143	10.925	4.757	9.780	4.534	9.148	4.387	8.746	4.284	8.466
7	5.591	12.246	4.737	9.547	4.347	8.451	4.120	7.847	3.972	7.460	3.866	7.191
8	5.318	11.259	4.459	8.649	4.066	7.591	3.838	7.006	3.687	6.632	3.581	6.371
9	5.117	10.561	4.256	8.022	3.863	6.992	3.633	6.422	3.482	6.057	3.374	5.802
10	4.965	10.044	4.103	7.559	3.708	6.552	3.478	5.994	3.326	5.636	3.217	5.386
11	4.844	9.646	3.982	7.206	3.587	6.217	3.357	5.668	3.204	5.361	3.095	5.069
12	4.747	9.330	3.885	6.927	3.490	5.953	3.259	5.412	3.106	5.064	2.996	4.821
13	4.667	9.074	3.806	6.701	3.411	4.739	3.179	5.205	3.025	4.862	2.915	4.620
14	4.600	8.862	3.739	6.515	3.344	5.564	3.112	5.035	2.958	4.695	2.848	4.456
15	4.543	8.683	3.682	6.359	3.287	5.417	3.056	4.893	2.901	4.556	2.790	4.318
16	4.494	8.531	3.634	6.226	3.239	5.292	3.007	4.773	2.852	4.437	2.741	4.202
17	4.451	8.400	3.592	6.112	3.197	5.185	2.965	4.669	2.810	4.336	2.699	4.102
18	4.414	8.285	3.555	6.013	3.160	5.092	2.928	4.579	2.773	4.248	2.661	4.015
19	4.381	8.185	3.522	5.926	3.127	5.010	2.895	4.500	2.740	4.171	2.628	3.939
20	4.351	8.096	3.493	5.849	3.098	4.938	2.866	4.431	2.711	4.103	2.599	3.871
21	4.325	8.017	3.467	5.780	3.072	4.874	2.840	4.369	2.685	4.042	2.573	3.812
22	4.301	7.945	3.443	5.719	3.049	4.817	2.817	4.313	2.661	3.988	2.549	3.758

(continued)

	$v_1 = 1$		$v_1 = 2$		$v_1 = 3$		$v_1 = 4$		$v_1 = 5$		$v_1 = 6$	
v_2	**$P = 0.05$**	**$P = 0.01$**	**$P = 0.05$**	**$P = 0.01$**	**$P = 0.05$**	**$P = 0.01$**	**$P = 0.05$**	**$P = 0.01$**	**$P = 0.05$**	**$P = 0.01$**	**$P = 0.05$**	**$P = 0.01$**
23	4.279	7.881	3.422	5.664	3.028	4.765	2.796	4.264	2.640	3.939	2.528	3.710
24	4.260	7.823	3.403	5.614	3.009	4.718	2.776	4.218	2.621	3.895	2.508	3.667
25	4.242	7.770	3.385	5.568	2.991	4.675	2.759	4.177	2.603	3.855	2.490	3.627
26	4.225	7.721	3.369	5.526	2.975	4.637	2.743	4.140	2.587	3.818	2.474	3.591
27	4.210	7.677	3.354	5.488	2.960	4.601	2.728	4.106	2.572	3.785	2.459	3.558
28	4.196	7.636	3.340	5.453	2.947	4.568	2.714	4.074	2.558	3.754	2.445	3.528
29	4.183	7.598	3.328	5.420	2.934	4.538	2.701	4.045	2.545	3.725	2.432	3.499
30	4.171	7.562	3.316	5.390	2.922	4.510	2.690	4.018	2.534	3.699	2.421	3.473
40	4.085	7.314	3.232	5.179	2.839	4.131	2.606	3.828	2.449	3.514	2.336	3.291
80	3.960	6.963	3.111	4.881	2.719	4.036	2.486	3.563	2.329	3.255	2.214	3.036
120	3.920	6.851	3.072	4.787	2.680	3.949	2.447	3.480	2.290	3.174	2.175	2.956
∞	3.842	6.637	2.997	4.607	2.606	3.784	2.373	3.321	2.215	3.019	2.099	2.804

	$v_1 = 8$		$v_1 = 10$		$v_1 = 12$		$v_1 = 24$		$v_1 = 48$		$v_1 = \infty$	
v_2	**$P = 0.05$**	**$P = 0.01$**	**$P = 0.05$**	**$P = 0.01$**	**$P = 0.05$**	**$P = 0.01$**	**$P = 0.05$**	**$P = 0.01$**	**$P = 0.05$**	**$P = 0.01$**	**$P = 0.05$**	**$P = 0.01$**
1	238.883	5981.070	241.882	6055.847	243.906	6106.321	249.052	6234.631	251.669	6299.892	254.314	6365.861
2	19.371	99.374	19.396	99.399	19.413	99.416	19.454	99.458	19.475	99.478	19.496	99.499
3	8.845	27.489	8.786	27.229	8.745	27.052	8.639	26.598	8.583	26.364	8.526	26.125
4	6.041	14.799	5.964	14.546	5.912	14.374	5.774	13.929	5.702	13.699	5.628	13.463
5	4.818	10.289	4.735	10.051	4.678	9.888	4.527	9.466	4.448	9.247	4.365	9.020
6	4.147	8.102	4.060	7.874	4.000	7.718	3.841	7.313	3.757	7.100	3.669	6.880
7	3.726	6.840	3.637	6.620	3.575	6.469	3.410	6.074	3.322	5.866	3.230	5.650
8	3.438	6.029	3.347	5.814	3.284	5.667	3.115	5.279	3.024	5.074	2.928	4.859
9	3.230	5.467	3.137	5.257	3.073	5.111	2.900	4.729	2.807	4.525	2.707	4.311
10	3.072	5.057	2.978	4.849	2.913	4.706	2.737	4.327	2.641	4.124	2.538	3.909
11	2.948	4.744	2.854	4.539	2.788	4.397	2.609	4.021	2.511	3.818	2.404	3.602
12	2.849	4.499	2.753	4.296	2.687	4.155	2.505	3.780	2.405	3.578	2.296	3.361
13	2.767	4.302	2.671	4.100	2.604	3.960	2.420	3.587	2.318	3.384	2.206	3.165
14	2.699	4.140	2.602	3.939	2.534	3.800	2.349	3.427	2.245	3.224	2.131	3.004
15	2.641	4.004	2.544	3.805	2.475	3.666	2.288	3.294	2.182	3.090	2.066	2.868
16	2.591	3.890	2.494	3.691	2.425	3.553	2.235	3.181	2.128	2.976	2.010	2.753
17	2.548	3.791	2.450	3.593	2.381	3.455	2.190	3.084	2.081	2.878	1.960	2.653
18	2.510	3.705	2.412	3.508	2.342	3.371	2.150	2.999	2.040	2.793	1.917	2.566
19	2.477	3.631	2.378	3.434	2.308	3.297	2.114	2.925	2.003	2.718	1.878	2.489
20	2.447	3.564	2.348	3.368	2.278	3.231	2.082	2.859	1.970	2.652	1.843	2.421
21	2.420	3.506	2.321	3.310	2.250	3.173	2.054	2.801	1.941	2.593	1.812	2.360
22	2.397	3.453	2.297	3.258	2.226	3.121	2.028	2.749	1.914	2.540	1.783	2.305

(continued)

	$v_1 = 8$		$v_1 = 10$		$v_1 = 12$		$v_1 = 24$		$v_1 = 48$		$v_1 = \infty$	
v_2	**$P = 0.05$**	**$P = 0.01$**	**$P = 0.05$**	**$P = 0.01$**	**$P = 0.05$**	**$P = 0.01$**	**$P = 0.05$**	**$P = 0.01$**	**$P = 0.05$**	**$P = 0.01$**	**$P = 0.05$**	**$P = 0.01$**
23	2.375	3.406	2.275	3.211	2.204	3.074	2.005	2.702	1.890	2.492	1.757	2.256
24	2.355	3.363	2.255	3.168	2.183	3.032	1.984	2.659	1.868	2.448	1.733	2.211
25	2.337	3.324	2.236	3.129	2.165	2.993	1.964	2.620	1.847	2.409	1.711	2.169
26	2.321	3.288	2.220	3.094	2.148	2.958	1.946	2.585	1.828	2.373	1.691	2.131
27	2.305	3.256	2.204	3.062	2.132	2.926	1.930	2.552	1.811	2.339	1.672	2.097
28	2.291	3.226	2.190	3.032	2.118	2.896	1.915	2.522	1.795	2.309	1.654	2.064
29	2.278	3.198	2.177	3.005	2.104	2.868	1.901	2.495	1.780	2.280	1.638	2.034
30	2.266	3.173	2.165	2.979	2.092	2.843	1.887	2.469	1.766	2.254	1.622	2.006
40	2.180	2.993	2.077	2.801	2.003	2.665	1.793	2.288	1.666	2.068	1.509	1.805
80	2.056	2.742	1.951	2.551	1.875	2.415	1.654	2.032	1.514	1.799	1.325	1.494
120	2.016	2.663	1.910	2.472	1.834	2.336	1.608	1.950	1.463	1.711	1.254	1.381
∞	1.939	2.513	1.832	2.323	1.753	2.187	1.518	1.793	1.359	1.537	1.000	1.000

Durbin–Watson Critical Values

Significance level = 5%

k is the number of exogenous variables, and T is the sample size.

	$k=1$		$k=2$		$k=3$		$k=4$		$k=5$	
T	d_1	d_2	d_1	d_2	d_1	d_2	d_1	d_2	d_1	d_2
15	1.08	1.36	0.95	1.54	0.82	1.75	0.69	1.97	0.56	2.21
16	1.10	1.37	0.98	1.54	0.86	1.73	0.74	1.93	0.62	2.15
17	1.13	1.38	1.02	1.54	0.90	1.71	0.78	1.90	0.67	2.10
18	1.16	1.39	1.05	1.53	0.93	1.69	0.82	1.87	0.71	2.06
19	1.18	1.40	1.08	1.53	0.97	1.68	0.86	1.85	0.75	2.02
20	1.20	1.41	1.10	1.54	1.00	1.68	0.90	1.83	0.79	1.99
21	1.22	1.42	1.13	1.54	1.03	1.67	0.93	1.81	0.83	1.96
22	1.24	1.43	1.15	1.54	1.05	1.66	0.96	1.80	0.86	1.94
23	1.26	1.44	1.17	1.54	1.08	1.66	0.99	1.79	0.90	1.92
24	1.27	1.45	1.19	1.55	1.10	1.66	1.01	1.78	0.93	1.90
25	1.29	1.45	1.21	1.55	1.12	1.66	1.04	1.77	0.95	1.89
26	1.30	1.46	1.22	1.55	1.14	1.65	1.06	1.76	0.98	1.88
27	1.32	1.47	1.24	1.56	1.16	1.65	1.08	1.76	1.01	1.86
28	1.33	1.48	1.26	1.56	1.18	1.65	1.10	1.75	1.03	1.85
29	1.34	1.48	1.27	1.56	1.20	1.65	1.12	1.74	1.05	1.84
30	1.35	1.49	1.28	1.57	1.21	1.65	1.14	1.74	1.07	1.83
31	1.36	1.50	1.30	1.57	1.23	1.65	1.16	1.74	1.09	1.83
32	1.37	1.50	1.31	1.57	1.24	1.65	1.18	1.73	1.11	1.82
33	1.38	1.51	1.32	1.58	1.26	1.65	1.19	1.73	1.13	1.81
34	1.39	1.51	1.33	1.58	1.27	1.65	1.21	1.73	1.15	1.81
35	1.40	1.52	1.34	1.58	1.28	1.65	1.22	1.73	1.16	1.80
36	1.41	1.52	1.35	1.59	1.29	1.65	1.24	1.73	1.18	1.80
37	1.42	1.53	1.36	1.59	1.31	1.66	1.25	1.72	1.19	1.80
38	1.43	1.54	1.37	1.59	1.32	1.66	1.26	1.72	1.21	1.79
39	1.43	1.54	1.38	1.60	1.33	1.66	1.27	1.72	1.22	1.79
40	1.44	1.54	1.39	1.60	1.34	1.66	1.29	1.72	1.23	1.79

(continued)

T	$k=1$		$k=2$		$k=3$		$k=4$		$k=5$	
	d_1	d_2	d_1	d_2	d_1	d_2	d_1	d_2	d_1	d_2
45	1.48	1.57	1.43	1.62	1.38	1.67	1.34	1.72	1.29	1.78
50	1.50	1.59	1.46	1.63	1.42	1.67	1.38	1.72	1.34	1.77
55	1.53	1.60	1.49	1.64	1.45	1.68	1.41	1.72	1.38	1.77
60	1.55	1.62	1.51	1.65	1.48	1.69	1.44	1.73	1.41	1.77
65	1.57	1.63	1.54	1.66	1.50	1.70	1.47	1.73	1.44	1.77
70	1.58	1.64	1.55	1.67	1.52	1.70	1.49	1.74	1.46	1.77
75	1.60	1.65	1.57	1.68	1.54	1.71	1.51	1.74	1.74	1.77
80	1.61	1.66	1.59	1.69	1.56	1.72	1.53	1.74	1.51	1.77
85	1.62	1.67	1.60	1.70	1.57	1.72	1.55	1.75	1.52	1.77
90	1.63	1.68	1.61	1.70	1.59	1.73	1.57	1.75	1.54	1.78
95	1.64	1.69	1.62	1.71	1.60	1.73	1.58	1.75	1.56	1.78
100	1.650	1.69	1.63	1.72	1.61	1.74	1.59	1.76	1.57	1.78

References

Akaike, H. (1969), "Fitting Autoregressive Models for Prediction", *Annals of the Institute of Statistical Mathematics*, 21, pp. 243–247.

Akaike, H. (1973), "Information theory and an extension of maximum likelihood principle", *Second International Symposium on Information Theory*, pp. 261–281.

Akaike, H. (1974), "A new look at the statistical model identification", *IEEE Transactions on Automatic Control*, 19(6), pp. 716–723.

Almon, S. (1962), "The Distributed Lag between Capital Appropriations and Expenditures", *Econometrica*, 30, pp. 407–423.

Baltagi, B.H. (2021), *Econometric Analysis of Panel Data*, 6th edition, John Wiley & Sons.

Banerjee, A., Dolado, J., Galbraith, J.W. and D.F. Hendry (1993), *Cointegration, Error-Correction, and the Analysis of Nonstationary Data*, Oxford University Press.

Basmann, R.L. (1957), Generalized Classical Method of Linear Estimation of Coefficients in a Structural Equation", *Econometrica*, 25, pp. 77–83.

Bauwens, L., Hafner, C. and S. Laurent (2012), "Volatility models", in Bauwens, L., Hafner, C. and S. Laurent (eds), *Handbook of Volatility Models and their Applications*, John Wiley & Sons, Inc.

Beach, C.M. and J.G. MacKinnon (1978), "A Maximum Likelihood Procedure for Regression with Autocorrelated Errors", *Econometrica*, 46, pp. 51–58.

Belsley, D.A., Kuh, E. and R.E. Welsch (1980), *Regression Diagnostics: Identifying Influential Data and Sources of Collinearity*, John Wiley & Sons, New York.

Bénassy-Quéré, A. and V. Salins (2005), "Impact de l'ouverture financière sur les inégalités internes dans les pays émergents", *Working Paper CEPII*, 2005–11.

Beran, J. (1994), *Statistics for Long Memory Processes*, Chapman & Hall.

Blanchard, O. and S. Fischer (1989), *Lectures on Macroeconomics*, The MIT Press.

Bollerslev, T. (2008), "Glossary to ARCH (GARCH)", *CREATES Research Paper*, 2008–49.

Bollerslev, T., Chou, R.Y. and K.F. Kroner (1992), "ARCH modeling in finance: A review of the theory and empirical evidence", *Journal of Econometrics*, 52(1–2), pp. 5–59.

Bollerslev, T., Engle, R.F. and D.B. Nelson (1994), "ARCH Models", in Engle R.F. and D.L. McFadden (eds), *Handbook of Econometrics*, Vol. IV, pp. 2959–3038, Elsevier Science.

Box, G.E.P. and D.R. Cox (1964),17 "An Analysis of Transformations" , *Journal of the Royal Statistical Society*, Series B, 26, pp. 211–243.

Box, G.E.P. and G.M. Jenkins (1970), *Time Series Analysis: Forecasting and Control*, Holden Day, San Francisco.

Box, G.E.P. and D.A. Pierce (1970), "Distribution of Residual Autocorrelation in ARIMA Time Series Models", *Journal of the American Statistical Association*, 65, pp. 1509–1526.

Breusch, T.S. (1978), "Testing for Autocorrelation in Dynamic Linear Models", *Australian Economic Papers*, 17, pp. 334–335.

Breusch, T.S. and A.R. Pagan (1979), "A Simple Test for Heteroscedasticity and Random Coefficient Variation", *Econometrica*, 47, pp. 1287–1294.

V. Mignon, *Principles of Econometrics*, Classroom Companion: Economics,
https://doi.org/10.1007/978-3-031-52535-3

Brockwell, P.J. and R.A. Davis (1998), *Time Series. Theory and Methods*, 2nd edition, Springer Verlag.
Brown, R.L., Durbin, J. and J.M. Evans (1975), "Techniques for Testing the Constancy of Regression Relationship Over Time", *Journal of the Royal Statistical Society*, 37, pp. 149–192.
Campbell, J.Y. and P. Perron (1991), "Pitfalls and Opportunities: What Macroeconomists Should Know about Unit Roots", in Fisher, S. (ed.), *NBER Macroeconomic Annual*, MIT Press, pp. 141–201.
Chow, G.C. (1960), "Tests of Equality Between Sets of Coefficients in two Linear Regressions", *Econometrica*, 28, pp. 591–605.
Cochrane, D. and G.H. Orcutt (1949), "Application of Least Squares Regressions to Relationships Containing Autocorrelated Error Terms", *Journal of the American Statistical Association*, 44, pp. 32–61.
Davidson, R. and J.G. MacKinnon (1993), *Estimation and Inference in Econometrics*, Oxford University Press.
Dhrymes, P. (1973), "Restricted and Unrestricted Reduced Forms", *Econometrica*, 41, pp. 119–134.
Dhrymes, P. (1978), *Introductory Econometrics*, Springer Verlag.
Dickey, D.A. and W.A. Fuller (1979), "Distribution of the Estimators for Autoregressive Time Series With a Unit Root", *Journal of the American Statistical Association*, 74, pp. 427–431.
Dickey, D.A. and W.A. Fuller (1981), "Likelihood Ratio Statistics for Autoregressive Time Series With a Unit Root", *Econometrica*, 49, pp. 1057–1072.
Diebold, F.X. (2012), *Elements of Forecasting*, 4th edition, South Western Publishers.
Dowrick S., Pitchford R. and S.J. Turnovsky (2008), *Economic Growth and Macroeconomic Dynamics: Recent Developments in Economic Theory*, Cambridge University Press.
Duesenberry, J. (1949), *Income, Saving and the Theory of Consumer Behavior*, Harvard University Press.
Dufrénot, G. and V. Mignon (2002a), "La cointégration non linéaire : une note méthodologique", *Économie et Prévision*, n 155, pp. 117–137.
Dufrénot, G. and V. Mignon (2002b), *Recent Developments in Nonlinear Cointegration with Applications to Macroeconomics and Finance*, Kluwer Academic Publishers.
Durbin, J. (1960), "The Fitting of Time Series Models", *Review of the International Statistical Institute*, 28, pp. 233–244.
Durbin, J. (1970), "Testing for Serial Correlation in Least Squares Regression When some of the Regressors are Lagged Dependent Variables", *Econometrica*, 38, pp. 410–421.
Durbin, J. and G.S. Watson (1950), "Testing for Serial Correlation in Least Squares Regression I", *Biometrika*, 37, pp. 409–428.
Durbin, J. and G.S. Watson (1951), "Testing for Serial Correlation in Least Squares Regression II", *Biometrika*, 38, pp. 159–178.
Elhorst, J-P. (2014), *Spatial Econometrics*, Springer.
Engle, R.F. (1982), "Autoregressive Conditional Heteroscedasticity with Estimates of the Variance of United Kingdom Inflation", *Econometrica*, 50(4), pp. 987–1007.
Engle, R.F. and C.W.J. Granger (1987), "Cointegration and Error Correction: Representation, Estimation and Testing", *Econometrica*, 55, pp. 251–276.
Engle, R.F. and C.W.J. Granger (1991), *Long Run Economic Relationships. Readings in Cointegration*, Oxford University Press.
Engle, R.F. and S. Yoo (1987), "Forecasting and Testing in Cointegrated Systems", *Journal of Econometrics*, 35, pp. 143–159.
Farebrother, R.W. (1980), "The Durbin-Watson Test for Serial Correlation when There Is No Intercept in the Regression", *Econometrica*, 48, pp. 1553–1563.
Farrar, D.E. and R.R. Glauber (1967), "Multicollinearity in Regression Analysis: The Problem Revisited", *The Review of Economics and Statistics*, 49, pp. 92–107.
Farvaque, E., Jean , N. and B. Zuindeau (2007), "Inégalités écologiques et comportement électoral : le cas des élections municipales françaises de 2001", *Développement Durable et Territoires*, Dossier 9.

Feldstein, M. and C. Horioka (1980), "Domestic Saving and International Capital Flows", *Economic Journal*, 90, pp. 314–329.

Florens, J.P., Marimoutou, V. and A. Péguin-Feissolle (2007), *Econometric Modeling and Inference*, Cambridge University Press.

Fox, J. (1997), *Applied Regression Analysis, Linear Models, and Related Methods*, Sage Publications.

Friedman, M. (1957), *A Theory of the Consumption Function*, New York.

Frisch, R.A.K. (1933), Editorial, *Econometrica*, 1, pp. 1–4.

Gallant, A.R. (1987), *Nonlinear Statistical Models*, John Wiley & Sons.

Geary, R.C. (1970), "Relative Efficiency of Count Sign Changes for Assessing Residual Autoregression in Least Squares Regression", *Biometrika*, 57, pp. 123–127.

Giles, D.E.A. and M.L. King (1978), "Fourth Order Autocorrelation: Further Significance Points for the Wallis Test", *Journal of Econometrics*, 8, pp. 255–259.

Glejser, H. (1969), "A New Test for Heteroscedasticity", *Journal of the American Statistical Association*, 64, pp. 316–323.

Godfrey, L.G. (1978), "Testing Against Autoregressive and Moving Average Error Models when the Regressors Include Lagged Dependent Variables", *Econometrica*, 46, pp. 1293–1302.

Goldfeld, S.M. and R.E. Quandt (1965), "Some Tests for Homoskedasticity", *Journal of the American Statistical Association*, 60, pp. 539–547.

Goldfeld, S.M. and R.E. Quandt (1972), *Nonlinear Econometric Methods*, North-Holland, Amsterdam.

Gouriéroux, C. (1997), *ARCH Models and Financial Applications*, Springer Series in Statistics.

Gouriéroux, C. (2000), *Econometrics of Qualitative Dependent Variables*, Cambridge University Press.

Gouriéroux, C. and A. Monfort (1996), *Time Series and Dynamic Models*, Cambridge University Press.

Gouriéroux, C. and A. Monfort (2008), *Statistics and Econometric Models*, Cambridge University Press.

Granger, C.W.J. (1969), "Investigating Causal Relations by Econometric Models and Cross-Spectral Methods", *Econometrica*, 36, pp. 424–438.

Granger, C.W.J. (1981), "Some Properties of Time Series Data and their Use in Econometric Model Specification" , *Journal of Econometrics*, pp. 121–130.

Granger, C.W.J. and P. Newbold (1974), "Spurious Regressions in Econometrics", *Journal of Econometrics*, 26, pp. 1045–1066.

Granger, C.W.J. and T. Teräsvirta (1993), *Modelling Nonlinear Economic Relationships*, Oxford University Press.

Greene, W. (2020), *Econometric Analysis*, 8th edition, Pearson.

Griliches, Z. (1967), "Distributed Lags: A Survey", *Econometrica*, 36, pp. 16–49.

Griliches, Z. and M. Intriligator (1983), *Handbook of Econometrics*, Vol. 1, Elsevier.

Gujarati, D.N., Porter, D.C. and S. Gunasekar (2017), *Basic Econometrics*, McGraw Hill.

Hamilton, J.D. (1994), *Time Series Analysis*, Princeton University Press.

Hannan, E.J. and B.G. Quinn (1979), "The Determination of the Order of an Autoregression", *Journal of the Royal Statistical Society*, Series B, 41, pp. 190–195.

Harvey, A.C. (1990), *The Econometric Analysis of Time Series*, MIT Press.

Harvey, A.C. and G.D.A. Phillips (1973), "A Comparison of the Power of Some Tests for Heteroscedasticity in the General Linear Model", *Journal of Econometrics*, 2, pp. 307–316.

Hausman, J. (1975), "An Instrumental Variable Approach to Full-Information Estimators for Linear and Certain Nonlinear Models", *Econometrica*, 43, pp. 727–738.

Hausman, J. (1978), "Specification Tests in Econometrics", *Econometrica*, 46, pp. 1251–1271.

Hausman, J. (1983), "Specification and Estimation of Simultaneous Equation Models", in Griliches, Z. and M. Intriligator (eds), *Handbook of Econometrics*, North-Holland, Amsterdam.

Hendry, D.F. (1995), *Dynamic Econometrics*, Oxford University Press.

Hendry, D.F. and Morgan, M.S. (eds) (1995), *The Foundations of Econometric Analysis*, Cambridge University Press.

Hildreth, C. and J. Lu (1960), "Demand Relations with Autocorrelated Disturbances", Technical Bulletin no. 276, Michigan State University Agricultural Experiment Station.

Hoel, P.G. (1974), *Introduction to Mathematical Statistics*, John Wiley & Sons.

Hoerl, A.E. and R.W. Kennard (1970a), "Ridge Regression: Biased Estimation for Non-Orthogonal Problems", *Technometrics*, pp. 55–68.

Hoerl, A.E. and R.W. Kennard (1970b), "Ridge Regression: Applications to Non-Orthogonal Problems", *Technometrics*, pp. 69–82.

Hurlin, C. and V. Mignon (2005), "Une synthèse des tests de racine unitaire sur données de panel", *Économie et Prévision*, n 169–170-171, pp. 253–294.

Hurlin, C. and V. Mignon (2007), "Une synthèse des tests de cointégration sur données de panel", *Économie et Prévision*, n 180–181, pp. 241–265.

Hurlin, C. and V. Mignon (2022), *Statistique et probabilités en économie-gestion*, 2nd edition, Dunod.

Hurvich, C.M. and C.-L. Tsai (1989), "Regression and time series model selection in small samples", *Biometrika*, 76, pp. 297–307.

bib89 Intriligator, M.D. (1978), *Econometric Models, Techniques and Applications*, Prentice Hall.

Jarque, C.M. and A.K. Bera (1980), "Efficient Tests for Normality, Homoscedasticity and Serial Independence of Regression Residuals", *Economics Letters*, 6, pp. 255–259.

Johansen, S. (1988), "Statistical Analysis of Cointegration Vectors", *Journal of Economic Dynamics and Control*, 12, pp. 231–254.

Johansen, S. (1991), "Estimation and Hypothesis Testing of Cointegration Vectors in Gaussian Vector Autoregressive Models", *Econometrica*, 59, pp. 1551–1580.

Johansen, S. (1995), *Likelihood-based Inference in Cointegrated Vector Autoregression Models*, Oxford University Press.

Johansen, S. and K. Juselius (1990), "Maximum Likelihood Estimation and Inferences on Cointegration with Application to the Demand for Money", *Oxford Bulletin of Economics and Statistics*, 52, pp. 169–210.

Johnston, J. and J. Dinardo (1996), *Econometric Methods*, 4th edition, McGraw Hill.

Jorgenson, D. (1966), "Rational Distributed Lag Functions", *Econometrica*, 34, pp. 135–149.

Judge, G.G., Griffiths, W.E., Hill, R.C., Lutkepohl, H. and T.C. Lee (1985), *The Theory and Practice of Econometrics*, 2nd edition, John Wiley & Sons.

Judge, G.G., Griffiths, W.E., Hill, R.C., Lutkepohl, H. and T.C. Lee (1988), *Introduction to the Theory and Practice of Econometrics*, John Wiley & Sons.

Kaufmann, D., Kraay, A. and M. Mastruzzi (2006), "Governance Matters V: Aggregate and Individual Governance Indicators for 1996–2005", http://web.worldbank.org.

Kennedy, P. (2008), *A Guide to Econometrics*, 6th edition, MIT Press.

Keynes, J.M. (1936), *The General Theory of Employment, Interest, and Money*, Macmillan.

Klein, L.R. (1950), *Economic Fluctuations in the United States, 1921–1941*, John Wiley & Sons, New York.

Klein, L.R. (1962), *An Introduction to Econometrics*, Prentice-Hall, Englewood Cliffs.

Kmenta, J. (1971), *Elements of Econometrics*, Macmillan.

Koyck, L.M. (1954), *Distributed Lags and Investment Analysis*, North-Holland, Amsterdam.

Kullback, S. and A. Leibler (1951), "On information and sufficiency", *Annals of Mathematical Statistics* 22, pp. 79–86.

Lardic, S. and V. Mignon (1999), "La mémoire longue en économie : une revue de la littérature", *Journal de la Société Française de Statistique*, pp. 5–48.

Lardic, S. and V. Mignon (2002), *Économétrie des séries temporelles macroéconomiques et financières*, Economica.

Lardic, S., Mignon, V. and F. Murtin (2005), "Estimation des modèles à correction d'erreur fractionnaires : une note méthodologique", *Journal de la Société Française de Statistique*, pp. 55–68.

Leamer, E.E. (1983), "Model Choice and Specification Analysis", in Griliches, Z. and M.D. Intriligator (eds), *Handbook of Econometrics*, Vol. I, North Holland.

Lehnan, E.L. (1959), *Testing Statistical Hypothesis*, John Wiley & Sons.

LeSage, J. and R.K. Pace (2008), *Introduction to Spatial Econometrics*, Chapman & Hall.
Ljung, G.M. and G.E.P. Box (1978), "On a Measure of Lack of Fit in Time Series Models", *Biometrika*, 65, pp. 297–303.
MacKinnon, J.G. (1991), "Critical Values for Cointegration Tests", in Engle, R.F. and C.W.J. Granger (eds), *Long-Run Economic Relationships*, Oxford University Press, pp. 267–276.
Maddala, G.S. and I.-M. Kim (1998), *Unit Roots, Cointegration, and Structural Change*, Cambridge University Press.
Maddala, G.S. and A.S. Rao (1971), "Maximum Likelihood Estimation of Solow's and Jorgenson's Distributed Lag Models", *The Review of Economics and Statistics*, 53(1), pp. 80–89.
Matyas, L. and P. Sevestre (2008), *The Econometrics of Panel Data. Fundamentals and Recent Developments in Theory and Practice*, 3rd edition, Springer.
Mills, T.C. (1990), *Time Series Techniques for Economists*, Cambridge University Press.
Mittelhammer, R.C., Judge, G.G. and D.J. Miller (2000), *Econometric Foundations*, Cambridge University Press, New York.
Mood, A.M., Graybill, F.A. and D.C. Boes (1974), *Introduction to the Theory of Statistics*, McGraw-Hill.
Morgan, M.S. (1990), *The History of Econometric Ideas (Historical Perspectives on Modern Economics)*, Cambridge University Press.
Morgenstern, O. (1963), *The Accuracy of Economic Observations*, Princeton University Press.
Nelson, C.R. and C. Plosser (1982), "Trends and Random Walks in Macroeconomics Time Series: Some Evidence and Implications", *Journal of Monetary Economics*, 10, pp. 139–162.
Nerlove, M. (1958), *Distributed Lags and Demand Analysis for Agricultural and Other Commodities*, Agricultural Handbook 141, US Department of Agriculture.
Newbold, P. (1984), *Statistics for Business and Economics*, Prentice Hall.
Newey, W.K. and K.D. West (1987), "A Simple Positive Definite Heteroskedasticity and Autocorrelation Consistent Covariance Matrix", *Econometrica*, 55, pp. 703–708.
Palm, F.C. (1996), "GARCH Models of Volatility", in Maddala G.S. and C.R. Rao (eds), *Handbook of Statistics*, Vol. 14, pp. 209–240, Elsevier Science.
Phillips, A.W. (1958), "The Relationship between Unemployment and the Rate of Change of Money Wage Rates in the United Kingdom, 1861–1957", *Economica*, 25 (100), pp. 283–299.
Pindyck, R.S. and D.L. Rubinfeld (1991), *Econometric Models and Economic Forecasts*, McGraw-Hill.
Pirotte, A. (2004), *L'économétrie. Des origines aux développements récents*, CNRS Éditions.
Prais, S.J. and C.B. Winsten (1954), "Trend Estimators and Serial Correlation", *Cowles Commission Discussion Paper*, no. 383, Chicago.
Puech, F. (2005), *Analyse des déterminants de la criminalité dans les pays en développement*, Thèse pour le doctorat de Sciences Économiques, Université d'Auvergne-Clermont I.
Rao, C.R. (1965), *Linear Statistical Inference and Its Applications*, John Wiley & Sons.
Sargan, J.D. (1964), "Wages and Prices in the United Kingdom: A Study in Econometric Methodology", in Hart, P.E., Mills, G. and J.K. Whitaker (eds), *Econometric Analysis for National Economic Planning*, Butterworths, London.
Schmidt, P. (1976), *Econometrics*, Marcel Dekker, New York.
Schwarz, G. (1978), "Estimating the Dimension of a Model", *The Annals of Statistics*, 6, pp. 461–464.
Sims, C.A. (1980), "Macroeconomics and Reality", *Econometrica*, 48, pp. 1–48.
Solow, R.M. (1960), "On a Family of Lag Distributions", *Econometrica*, 28, pp. 393–406.
Spanos, A. (1999), *Probability Theory and Statistical Inference: Econometric Modeling with Observational Data*, Cambridge University Press.
Swamy, P.A.V.B. (1971), *Statistical Inference in Random Coefficient Regression Models*, Springer Verlag.
Teräsvirta, T., Tjøstheim, D. and C.W.J. Granger (2010), *Modelling Nonlinear Economic Time Series*, Oxford University Press.
Theil, H. (1953), "Repeated Least Squares Applied to Complete Equation Systems", Central Planning Bureau, The Hague, Netherlands.

Theil, H. (1971), *Principles of Econometrics*, John Wiley & Sons, New York.
Theil, H. (1978), *Introduction to Econometrics*, Prentice Hall.
Thuilliez, J. (2007), "Malaria and Primary Education: A Cross-Country Analysis on Primary Repetition and Completion Rates", *Working Paper Centre d'Économie de la Sorbonne*, 2007–13.
Tobin, J. (1950), "A Statistical Demand Function for Food in the USA", *Journal of the Royal Statistical Society*, Series A, pp. 113–141.
Wallis, K.F. (1972), "Testing for Fourth-Order Autocorrelation in Quarterly Regression Equations", *Econometrica*, 40, pp. 617–636.
White, H. (1980), "A Heteroscedasticity Consistent Covariance Matrix Estimator and a Direct Test of Heteroscedasticity", *Econometrica*, 48, pp. 817–838.
Wooldridge, J.M. (2010), *Econometric Analysis of Cross Section and Panel Data*, 2nd edition, MIT Press.
Wooldridge, J.M. (2012), *Introductory Econometrics: A Modern Approach*, 5th edition, South Western Publishing Co.
Zellner, A. (1962), "An Efficient Method of Estimating Seemingly Unrelated Regressions and Tests of Aggregation Bias", *Journal of the American Statistical Association*, 57, pp. 500–509.
Zellner, A. and H. Theil (1962), "Three Stage Least Squares: Simultaneous Estimation of Simultaneous Equations", *Econometrica*, 30, pp. 63–68.

Bringing together theory and practice, this book presents the basics of econometrics in a clear and pedagogical way. It focuses on the acquisition of the methods and skills that are essential for all students wishing to succeed in their studies and for all practitioners wishing to apply econometric techniques. The approach adopted in this textbook is resolutely applied. Through this book, the author aims to meet a pedagogical and operational need to quickly put into practice various concepts presented (statistics, tests, methods, etc.). This is why, after each theoretical presentation, numerous examples are given, as well as empirical applications carried out on the computer using existing econometric and statistical software.

This textbook is primarily intended for students of Bachelor's and Master's Degrees in Economics, Management, and Mathematics and Computer Sciences, as well as for students of Engineering and Business Schools. It will also be useful for professionals who will find practical solutions to the various problems they face.

Valérie MIGNON is Professor of Economics at the University of Paris Nanterre (France), Member of the EconomiX–CNRS research center, and Scientific Advisor to the leading French center for research and expertise on the world economy, CEPII (Paris, France). She teaches econometrics at undergraduate and graduate levels. Her econometric research focuses mainly on macroeconomics, finance, international macroeconomics and finance, and energy, fields in which she has published numerous articles and books.

Index

V. Mignon, *Principles of Econometrics*, Classroom Companion: Economics,
https://doi.org/10.1007/978-3-031-52535-3

N

www.ingramcontent.com/pod-product-compliance
Ingram Content Group UK Ltd.
Pitfield, Milton Keynes, MK11 3LW, UK
UKHW022225051026
15624UKWH00002B/365
* 9 7 8 3 0 3 1 5 2 5 3 4 6 *